Enzymology and Medicine

Enzymology and Medicine

D W Moss, MA, MSc, PhD
Reader in Enzymology at the
Royal Postgraduate Medical School,
University of London

and

P J Butterworth, MSc, PhD
Lecturer in Biochemistry,
Chelsea College of Science and Technology,
University of London

PITMAN MEDICAL

First published 1974

Pitman Medical Publishing Company Ltd
Sir Isaac Pitman and Sons Ltd
Pitman House, Parker Street, Kingsway, London WC2B 5PB
PO Box 46038, Banda Street, Nairobi, Kenya

Sir Isaac Pitman (Aust.) Pty Ltd
Pitman House, 158 Bouverie Street, Carlton, Victoria 3053, Australia

Pitman Publishing Corporation
6 East 43rd Street, New York, NY 10017, USA

Sir Isaac Pitman (Canada) Ltd
495 Wellington Street West, Toronto 135, Canada

The Copp Clark Publishing Company
517 Wellington Street West, Toronto 135, Canada

ISBN: 0 272 00094 9

Text set in 10pt IBM Univers Medium, printed by photolithography,
and bound in Great Britain at The Pitman Press, Bath.

21.5013:81

Preface

Medicine is becoming increasingly an exact and quantitative science, due to a growing understanding of the changes at the molecular level which underlie and give rise to human disease. This is recognized in the current reappraisals of the curriculum of medical studies, which are motivated by the need to ensure that the principles of basic biological science to which the student is introduced in his pre-clinical years continue to inform his thinking and rationalize his approach as a practising doctor. In the traditional pattern of medical studies there has in general existed a marked dissociation between the pre-clinical and the clinical years. Too often, the significance of the examinations at the end of the pre-clinical phase has appeared only to be that of an obstacle which must be surmounted for entry to the clinical phase and one which, once it is overcome, can be dismissed from the mind. To reshape this pattern of thought, therefore, it is desirable to integrate the pre-clinical and clinical parts of the curriculum, and to endeavour to bring out during the pre-clinical years the relevance to the practice of medicine of what is being taught. In clinical studies, the scientific principles on which modern diagnosis and treatment are founded are similarly emphasized.

The study of enzymes and its relevance to medicine seem to be a particularly apt field for such an integrated treatment. In his pre-clinical studies in biochemistry, the student learns something of the nature and properties of enzymes. Later, he learns to examine and manage patients whose diseases may be due to abnormalities in the production or functioning of some of the many enzymes of the human body; or he may use serum enzyme activities as an aid to diagnosis, often without interpreting these observations in the light of his formerly acquired (and perhaps subsequently forgotten) knowledge of enzyme biochemistry.

In these two phases of his career the student is faced with a choice of textbooks which by their nature tend to emphasize this division. The biochemistry of enzymes is dealt with fully in many textbooks which, however, usually make only minimal mention of the relevance of enzymology to human disease. Textbooks of medicine, on the other hand, are equally silent on the nature of the biochemical lesions which manifest themselves as specific diseases.

The aim of the present book is to describe the nature and properties of enzymes in such a way as to bring out the relevance of this branch of biochemistry to the practice of medicine. It is intended to be complementary to the standard textbooks of biochemistry and medicine on which the student will continue to rely for the main body of his organized knowledge, but it will, it is hoped, present a picture of a thread of knowledge running from the laboratory to the bedside.

Contents

1 The Biochemical Basis of Health and Disease

Life is a chemical process. Food of a definable range of chemical composition is taken in by living organisms, and waste products of a different chemical composition are excreted. Growth and reproduction entail the production by the organism of new tissue, again consisting of material of a definite chemical constitution: they are thus processes of chemical synthesis. Everyday experience shows the relationship between enhanced muscular activity and the increased expenditure of chemical energy (in the form of a greater intake of food). The elucidation of the sequences of chemical reactions which intervene between the input of raw materials into the living system (intake of food) and the appearance of the end products (new tissue, waste products, energy) lies within the ambit of biochemistry, and biochemical research has now succeeded in defining most of the reaction sequences, or metabolic pathways, which are followed in living matter. From these researches the realization has emerged that the main pathways of chemical change are essentially similar throughout the great diversity of living matter, and that while there are quantitative and qualitative differences in metabolism between different organisms and different cells, many reaction sequences by which energy is produced or components of living matter are synthesized are of universal occurrence. A further fundamental similarity is shared by the metabolic processes of all living things: almost every chemical reaction in living matter is a catalysed process, and all the catalysts concerned belong to a group of specific proteins possessing certain common properties and known as enzymes. From the biochemical point of view, therefore, life consists of integrated sequences of chemical reactions catalysed by enzymes.

The early history of enzymology is closely connected with the development of ideas on the nature of two processes, alcoholic fermentation by yeast cells, and digestion in higher animals. Digestion was recognized as a "chemical" event mediated by the action of juices such as saliva by the end of the 17th century, and its experimental investigation was begun by Réaumur (1683–1757), who induced a tame kite to swallow perforated metal tubes containing food of various kinds. When the tubes were regurgitated meat was found to be partly dissolved, though starchy substances were unaffected. Since enclosure in the metal tubes protected the food from any grinding action (at that time grinding was the principal non-chemical theory), it appeared that the active agent must be the liquid in which the food was seen to be soaked. Spallanzani (1729–99) confirmed this by achieving *in vitro* digestion of meat when it was warmed in samples of gastric juice obtained by placing sponges in the perforated metal tubes. Beaumont (1785–1853) was able to extend this type of observation to the digestive processes of the human stomach by taking advantage of an unusual shooting accident in which a wound in the body-wall healed leaving the victim with a permanent gastric fistula through which Beaumont could study the time taken to digest various kinds of food. Although the strongly acid nature of gastric juice was recognized, it was realized that this alone could not account for the digestive power of the juice, and in 1836 T. Schwann discovered the proteolytic enzyme pepsin. The principal proteases of the pancreatic juice, trypsin and chymotrypsin, were also identified during this early phase of enzymology, as were enzymes acting on starch (amylases) and fats (lipases). Digestion was thus the first vital process to be clearly seen to consist of a series of enzymic reactions.

The study of digestion was facilitated by the largely extracellular nature of this process. The presence of many of the enzymes in the various secretions of the gut made their characterization and isolation possible before techniques for the investigation of organized cells and tissues were developed. The proteases of the digestive tract have continued to figure in advances in enzymology and their ease of purification and comparatively small molecular weights placed them amongst the first enzymes to be crystallized in the late 1920s and early 1930s, while at the present time some of the first complete structural analyses of enzyme molecules have taken chymotrypsin as their subject.

The investigation of intracellular enzymic reactions may be regarded as beginning with Buchner's preparation in 1896 of a cell-free yeast extract which retained the power to ferment glucose to alcohol. Fractionation of the yeast juice into its component enzymes and metabolic intermediates occupied the next thirty years or so and was paralleled by work on minced and sliced avian and mammalian muscle which revealed the essentially similar reaction sequence by which glucose is fermented to lactic acid in this tissue. By the 1950s the main outlines and many of the details of metabolism in living organisms had been discovered, firmly establishing that the universal and unique secret of life is to be found in co-ordinated chemical reactions and in the nature and properties of the enzymes which catalyse them. The same period also included the first extensive enzyme purifications. The list of known enzymes has now grown to over 1000, of which more than 100 have been crystallized.

Since life is a chemical process dependent on the specific catalytic activities of enzymes, with a certain pattern of reactions and a particular balance between the rates at which they proceed corresponding to normal functioning, it follows that the fundamental nature of disease lies in derangements of the normal sequence of enzymic reactions. Pathological modification of the normal pattern of metabolism can arise in several ways. The metabolism of the host may be altered in response to invasion by an infective agent, e.g. a bacterium or a virus. Local injury may cause a temporary disturbance of the micro- and macro-architecture of tissues and organs resulting in a redistribution of enzymes and metabolites so that new reactions, or alterations in rates of reaction, can occur. Ingestion of poisons or drugs may directly influence the activities of particular enzymes. The normal balance of reactions may be distorted as a consequence of malignant changes taking place within the cells after exposure to ionizing radiations or chemical agents, or even apparently arising spontaneously. A congenital alteration in the properties of certain enzymes or even their complete absence may be compatible with life so that an abnormal pattern of metabolism is observed in the affected individuals.

The degree to which alterations in specific enzymic reactions can be identified as the causes or consequences of the various pathological states varies greatly. This relationship is most clearly identifiable in the case of certain inborn errors of metabolism, in which a genetically determined absence of one enzyme can be demonstrated as the single abnormality from which the observed pathological consequences are all derived. At the other extreme are the multiple changes which result from tissue damage, which are too complex to be analysed fully at present in terms of alterations in individual reactions. Malignant change, which appears to involve subtle quantitative shifts in the balance of reactions in the affected cells rather than readily identifiable qualitative differences from normal patterns, also falls into this category. The way in which the agents responsible for infectious diseases modify the host's metabolism is the subject of much research. Viruses impose new patterns of enzyme synthesis and metabolism on the cells which they infect, and the specific sites of action of certain bacterial toxins are currently being defined. Other poisons which gain entry to the body are known to have actions on individual enzymic reactions (an example is the inhibition by cyanide of the enzymes of cellular respiration), while the mode of action of other poisons and drugs at the molecular level is progressively becoming clear.

It is also relevant to ask how far an extension of understanding of the biochemical nature of disease is necessary for the management of individual patients, and furthermore, how far the underlying biochemical abnormalities accompanying disease can be diagnosed in a particular instance. It is clearly important to know the nature of the metabolic defect in sufferers from inborn errors of metabolism as a first step towards designing rational therapy, perhaps by avoiding the blocked reaction as far as possible, as in the dietary treatment of phenylketonuria or galactosaemia. Hopes of reversing or even preventing malignant change similarly rest on discovering those features which distinguish the biochemistry of a cancer cell from that of its normal counterpart. The treatment of infectious diseases is perhaps more usefully approached through a study of the biochemical characteristics of the pathogenic organism so that differences between its metabolism and that of the host can be exploited in the design of specific chemotherapeutic agents — Ehrlich's "magic bullets". The fruits of this approach, the sulphonamides and antibiotics, have transformed medicine within the space of a few decades. A knowledge of the biochemical consequences of mechanical injury to tissues is at the present time of only minor assistance in determining the pattern of treatment, although the known physiological action of substances released from damaged cells (e.g. histamine-like substances) accounts for some of the features of these cases. In addition, small biochemical changes in the blood are often of great value in detecting the occurrence of tissue damage in internal organs, such as may result from coronary artery disease.

The experimental methods that are available to the biochemist in elucidating the patterns of metabolism of normal living tissues or of abnormal forms (e.g. cancer cells) range from studies of the overall chemical exchanges of whole organs or thin slices of living tissue, to a dissection of the cells into their component enzymes and metabolites. Thus, in studying the metabolism of an animal liver (that of a rat, for instance), the ability of the organ as a whole to carry out certain chemical processes might be' assessed by making a preparation of the living, anaesthetized animal in which the normal circulation of blood to and from the liver is replaced by a pump perfusing the organ with either blood or mammalian Ringer solution. The experimenter can add various substances to the perfusing fluid and can analyse the resulting changes in composition of the fluid leaving the organ. A less physiological but experimentally simpler approach is to cut thin slices of liver taken from a freshly killed animal and incubate these in Ringer solution to which the substances being investigated can be added. If the medium is kept well oxygenated such slices will survive for some time. The anatomical relationships and access of substrates to the interior of the cells are altered in these slices and the results are thus not identical with those obtained from organ perfusion.

Proceeding a stage further, the experimenter can destroy the cellular structure of the tissue by homogenizing it and can then observe the metabolism of these homogenates. The homogenate itself can be fractionated with the object of preparing and studying either samples of the several types of subcellular particles — mito-

chondria, microsomes, etc. — which make up animal cells, or, in the final analysis, single, purified enzymes each of which catalyses a single reaction only or a small group of related reactions. All these procedures have their place in understanding the biochemistry of normal and abnormal cells and organs, and none is by itself completely adequate: the metabolism of a whole cell shows features which are additional to those deducible from the properties of its parts (i.e. its component enzymes), while the isolated, perfused organ lacks the hormonal and nervous control to which it is subject in the intact animal.

The biochemical investigations which the clinical biochemist and clinician can carry out in living human patients are of course limited in their scope, nor is it necessary in the majority of cases to demonstrate the biochemical lesion at the molecular level before the patient can be properly treated. The history, signs and symptoms elicited on examination, X-ray appearances, the isolation of a pathogenic organism in infectious disease, the characteristic histology of a section of a tumour removed at operation in suspected malignancy, are the foundations on which diagnosis and subsequent treatment are based. To these are added laboratory tests — haematological, dealing with the relative numbers and morphology of the blood cells, and chemical, directed towards revealing abnormalities in the composition of the blood and other body fluids and excreta. These laboratory tests, added to the clinician's own observations, confirm a diagnosis or decide between possible alternative diagnoses and assist in monitoring the course of an illness.

In carrying out analyses of the composition of the blood or other fluids, the clinical biochemist is attempting to infer from these extracellular measurements the course of events within the cells and tissues of the body. Thus, failure to metabolize a particular constituent of the normal diet may result in the appearance of that substance in abnormally large amounts in the blood or urine — the excess of galactose, for example, in the blood and urine of children who cannot metabolize that sugar. Biochemical disturbance may also result in the appearance of abnormal metabolites which are the products of enzymic reactions normally of minor importance or the substrates of enzymes, the activity of which is reduced or absent. In studying the composition of the blood, in particular, changes in the activities and properties of enzymes deriving from known intracellular locations are of great value in drawing conclusions about biochemical events within the tissues.

One technique by which it is possible to gain a more direct insight into the cells themselves is by the removal of small specimens of an organ by biopsy. At the present time the main value of biopsy specimens lies in their examination under the microscope, but as techniques for measuring and characterizing the enzymes in these small amounts of tissue are improved their biochemical examination will no doubt increase in importance. One group of cells which can readily be sampled in this way to give relatively large, homogeneous samples are the red blood corpuscles, and studies of the enzyme content of these cells or of the amounts of intermediate metabolites in them have proved of great value in the investigation of several congenital metabolic abnormalities.

The following chapters attempt to summarize the main facts about the nature and properties of enzymes as far as these are known at the present time and to relate them to the manifestations of human disease and its diagnosis and treatment.

Enzyme Nomenclature. The first systematic method of naming enzymes was a proposal, in 1898, by E. Duclaux that "—ase" should be added to a word, or part of a word, indicating the nature of the substance on which the enzyme acts. Apart from a few digestive enzymes which had already been given names (still in use today) ending in "—in", e.g. pepsin and trypsin, this suggestion remains the basis of enzyme terminology. The "—ase" system has had to be extended in some cases to include information about the type of reaction catalysed since more than one enzyme may act on a single substrate (the compound on which an enzyme acts is termed its "substrate"), catalysing different reactions. More recently a still more explicit system of naming and numbering all the known enzymes, with gaps to accommodate those which will undoubtedly be discovered in the future, has had to be devised so that unambiguous description of a particular enzyme is possible. For most purposes, however, the trivial name of an enzyme, based on the "—ase" terminology and with perhaps some additional information about the tissue and species from which the enzyme has been prepared, is sufficient and will be used in this book.

2 Experimental Methods in Enzymology

As in all branches of natural science, present knowledge and future advances in enzymology depend on experiments, particularly those of a quantitative nature. Experimental methods in enzymology are directed to the detection and measurement of enzyme activity and to the purification of enzymes so that their constitution and properties can be studied.

MEASUREMENT OF ENZYME ACTIVITY

Enzymes are catalysts, that is, they accelerate specific chemical reactions and are themselves unchanged in amount and chemical composition when the reaction is completed. It is this property which gives enzymes their unique biological importance and it also forms the basis of methods of estimating the amount of an enzyme which is present in a given system, e.g. a tissue extract, blood, etc. As with the inorganic catalysts known to chemists, enzymes exert their accelerating effect detectably on reaction rates when they are present in small traces only: the molar concentration of an enzyme *in vitro* is usually far lower than that of the substrate which it transforms. The presence of an enzyme is therefore revealed and its quantity is measured by its activity, and in most cases (except when dealing with highly purified enzymes) attempts to determine its absolute amount — for example, by making use of methods of estimating proteins, since enzymes are proteins — are doomed to failure because of their relative insensitivity. This ability sensitively to detect changes in the amount of enzyme present, e.g. in a sample of blood, in the presence of a large excess of other proteins is a characteristic which contributes greatly to the clinical value of enzyme estimations as indicators of tissue damage in a way that will be discussed more fully in a later chapter.

The basis of an enzyme activity determination is thus the comparison of the rate of a chemical reaction (i.e. the amount of product formed or substrate destroyed in a given time) in the presence of the active enzyme, with the rate of reaction in a control solution from which the enzyme is omitted or in which it has been inactivated (e.g. by heating or by addition of protein precipitants). Conditions such as the pH of the reaction medium and its temperature must be controlled within narrow limits by buffer solutions and thermostatic water-baths since variations in these factors have a marked influence on the rates of enzymic reactions, as do other components of the system, metal ions for example, which may accelerate or retard the reaction. Comparison of the relative amounts of a particular enzyme in different preparations or samples on the basis of the rate of the catalysed reaction under standard conditions depends on the assumption that, if a given number of enzyme molecules result in a certain reaction rate, twice or three times that number will increase the rate by a factor of two or three. Provided that the experimental conditions are correctly chosen and maintained so that other factors do not influence the rate, the proportionality between amount of enzyme and rate of reaction is almost always found to hold. The progress curve for an enzyme-catalysed reaction (chemical change expressed as a function of the duration of the reaction) typically consists of an initial linear portion merging into a curve as the rate falls off. The fall in rate is due to the combined effect of several factors — fall in substrate concentration, increasing reverse reaction, enzyme denaturation, etc. — and the shape of the curved portion of the progress curve thus cannot be represented by any fixed mathematical relationship applicable to all cases. When the progress curves are compared using different amounts of enzyme the slopes of the initial linear portions are proportional to the enzyme concentration, but this is usually not the case for the nonlinear portions of the curves (Fig. 2.1). It is therefore essential that all comparisons of reaction rates for estimation of enzyme activity must be based on linear progress curves. The reaction rate may be measured by allowing the reaction to proceed for a fixed length of time,

are possible in the experimental details of different methods. These variations lie principally in the techniques which are used to determine the amount of chemical change which has taken place. Gravimetric analysis is of little value in enzyme work since most of the transformations catalysed by enzymes involve organic compounds,

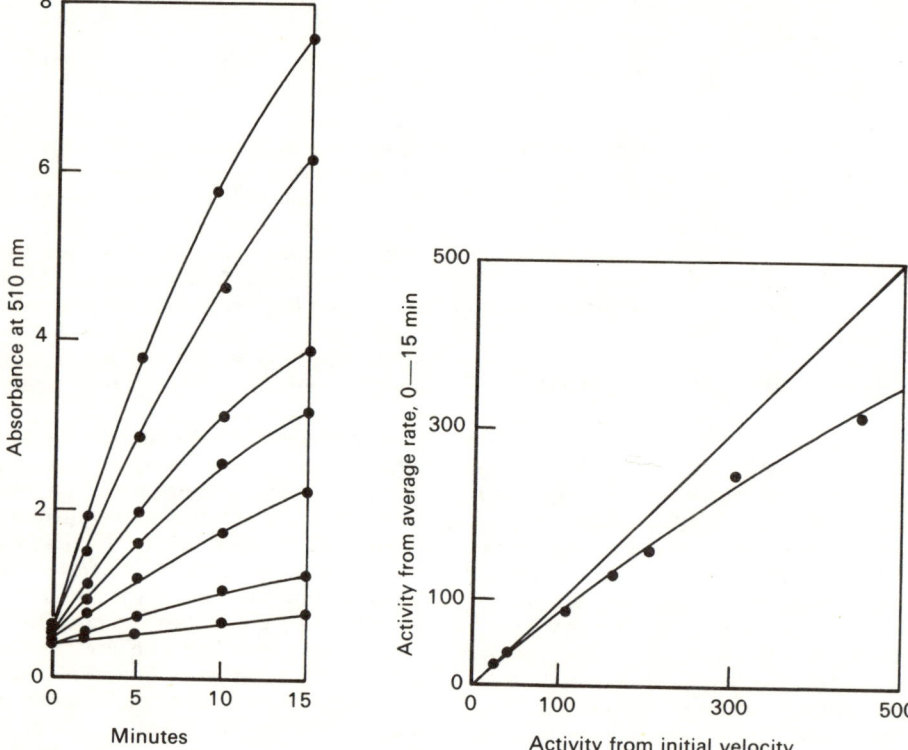

Fig. 2.1. Reaction progress—curves with different amounts of enzyme. Phenol released by hydrolysis of phenyl phosphate by alkaline phosphatase is estimated colorimetrically after different periods of reaction with increasing amounts of enzyme (left). Activities derived from the average rates over 15 min are lower than those derived from initial velocities, at higher enzyme concentrations (right).

stopping it (for example by adding a protein-precipitant) and estimating the amount of some product of the reaction which has been formed. However, it is difficult to be certain that the progress of the reaction has remained linear in this type of assay (termed a "two-point" or "fixed-time" assay), particularly since, in many reactions, the duration of linearity may be very short indeed. Consequently, methods are to be preferred in which the progress of reaction can be monitored continuously; these are often referred to as "kinetic" assays.

While all enzyme estimations are based on the same principles, a great many variations

and the limited sensitivity of this type of analysis is unsuitable for the low concentrations of reactants usual in enzymic reactions. On the other hand, the sensitivity and wide scope of colorimetric analysis has proved of great value in enzymology, and it is this form of analysis, together with its extension into the ultraviolet spectrum, which is applied in most of the enzyme estimations carried out from day to day for clinical purposes.

One of the products of the action of the enzyme may be made to undergo a second reaction to yield a coloured compound, the intensity of the colour (measured in a photoelectric colorimeter) being proportional to the amount of product formed. For example, many enzymes release phosphate from their substrates which can be detected and measured by the blue colour formed by reaction with molybdate and a reducing agent. The reagents involved in the colour-forming reaction often interfere with enzyme action, so that these methods can then only be applied on a fixed-time basis. More directly, the product of the enzyme's action may itself be coloured, obviating the need for a second reaction and allowing the reaction to be monitored continuously. When the enzyme concerned is one of wide specificity, i.e. one which will act on a variety of substrates and catalyse a similar reaction with each, the substrate can be chosen or adapted chemically so that a coloured product is formed. An example is the estimation of alkaline phosphatase, an enzyme of wide distribution in human tissues the activity of which in blood plasma varies significantly in liver and bone diseases (Chapter 9). This enzyme hydrolyses a wide range of phosphate esters, both naturally occurring and synthetic, in alkaline solution. One synthetic substrate is p-nitrophenyl phosphate which is split to p-nitrophenol and orthophosphate. p-Nitrophenyl phosphate is colourless, but p-nitrophenol is yellow at alkaline pH: the course of the enzymic reaction can therefore be followed by measuring the progressively deepening yellow colour of the solution:

$$O.PO_3^{2-} \xrightarrow[\text{pH 10}]{\substack{\text{Alkaline} \\ \text{phosphatase}}} O^- + PO_4^{3-}$$

p-nitrophenyl phosphate   p-nitrophenate ion (yellow)

In the case of enzymes of limited specificity the choice of substrates to make for easier chemical analysis is usually not possible.

Substances which form coloured solutions do so because they absorb light of a particular range of wavelengths in the visible region of the spectrum. When observations are extended into the ultraviolet range by the use of suitable sources of light and detectors other than the human eye (e.g. photoelectric cells), many more compounds are found to absorb light and the wavelengths at which they do so are characteristic of the presence in their molecules of particular groupings of atoms. Enzyme-catalysed reactions often involve rearrangements within the molecules which result in changes in the ultraviolet spectrum, and when these changes are sufficiently pronounced to be observed in a spectrophotometer they afford a convenient basis for the measurement of enzyme activity. An example is the action of uricase on uric acid. Uric acid has a strong absorption band in the ultraviolet spectrum centred at 293 nm. Oxidation by the enzyme uricase, which occurs in some animal tissues and yeasts, is accompanied by the progressive weakening and eventual disappearance of this band (Fig. 2.2). As well as providing a means of estimating uricase activity this property also constitutes a sensitive and specific method of measuring small amounts of uric acid, e.g. in blood samples, from the change in absorption in the presence of an excess of the enzyme.

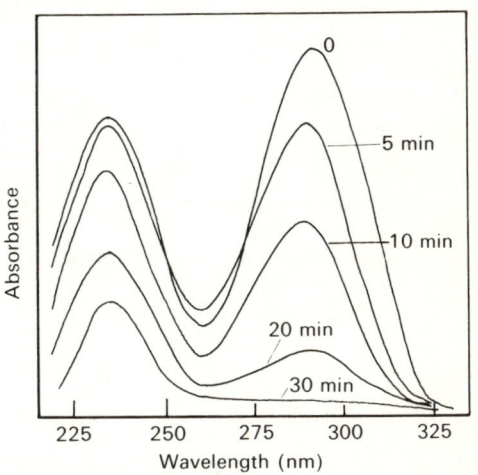

Fig 2.2. **Progressive changes in the ultraviolet absorption spectrum of uric acid during oxidation by uricase.**

The component of the reaction mixture that undergoes a change in its absorption spectrum need not be the substrate of the enzyme for the ultraviolet method to be applicable, provided that some participant in the reaction is changed at a rate proportional to the rate of change of the substrate. Many enzyme reactions involve parallel changes in components other than the substrate; these reactants are termed coenzymes and the same coenzyme is often involved in a wide range of reactions catalysed by many different enzymes. Two important coenzymes are nicotinamide adenine dinucleotide, abbreviated to NAD, and nicotinamide adenine dinucleotide phosphate (NADP), which take part in many reactions in which hydrogen atoms are removed from, or added to, the substrates under the influence of specific dehydrogenase enzymes. The coenzymes act as acceptors or donors of hydrogen atoms and are thus either reduced or oxidized during the exchanges. (For a fuller description, see Chapter 5.) Both coenzymes have similar ultraviolet spectra with a strong absorption maximum at 340 nm in the reduced form which disappears on oxidation. Thus, the action of the enzyme lactate dehydrogenase on lactic acid, which is accompanied by the transfer of hydrogen to NAD^+, can be followed by the increasing absorption due to NADH at 340 nm (Fig. 2.3). The molecular extinction coefficient of the reduced coenzymes is $6 \cdot 22 \times 10^6$. Thus, the appearance of one micromole of reduced coenzyme in a reaction volume of 1 ml results in a change in extinction at 340 nm of $6 \cdot 22$ in a 1 cm cuvette and changes of one hundredth part or less of this are readily measurable in a good spectrophotometer.

The measurement of the absorption changes of NAD and NADP is so sensitive and convenient that considerable ingenuity has been displayed in devising ways in which enzymic reactions which do not involve these coenzymes can be brought within the scope of the method. This can be done by coupling together two, three, or more reactions so that the product of the first forms the substrate of the next, and so on, until a reaction is reached which involves the oxidation or reduction of one of the coenzymes. Transaminases (now called aminotransferases) which are enzymes of amino acid metabolism of diagnostic value, are often estimated in this way:

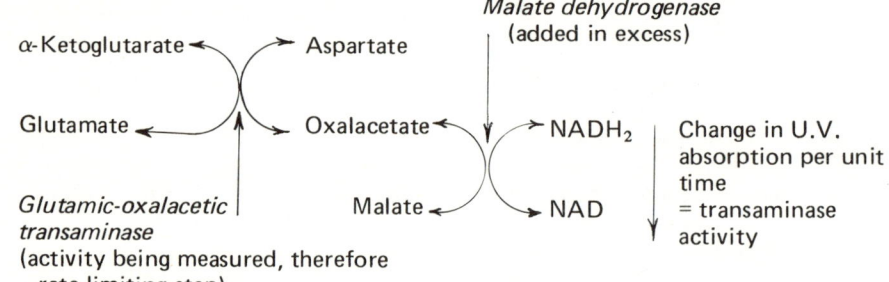

It is essential in setting up linked reactions that the rate-limiting step should be the one catalysed by the enzyme being assayed, and that the accessory enzymes, coenzymes and substrates should be present in sufficient amounts to ensure that this is so. However, the presence of additional substrates or coenzymes may modify the activity of certain enzymes and in these cases coupled assay systems may not be appropriate. The light which is absorbed by certain compounds is re-emitted, typically at lower energies and therefore longer wavelengths, as fluorescence. Measurements of fluorescence can usually be made with 10 or 100 times the sensitivity of absorbance measurements and when fluorescence changes accompany enzymic reactions they can be used to facilitate the assay of small amounts of activity. Thus, the reduced forms of NAD and NADP are fluorescent when irradiated with light of 340 nm wavelength whereas the oxidized coenzymes are not. Increased sensitivity becomes extremely valuable

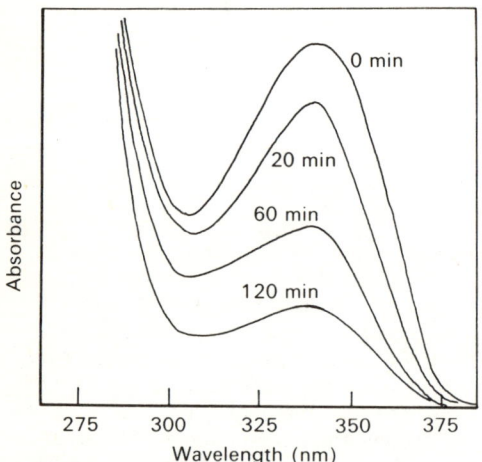

Fig 2.3. Progressive reduction in absorption of light at 340 nm during the oxidation of NADH to NAD^+, with transfer of hydrogen to pyruvate to form lactate, catalysed by lactate dehydrogenase.

when small tissue samples only (e.g. biopsy specimens) are available.

Changes in the ultraviolet or visible spectra of the reactants are not the only methods by which enzymic reactions can be monitored, although they are amongst the most sensitive and convenient. Many reactions catalysed by enzymes, particularly those concerned with tissue respiration, are accompanied by the uptake or evolution of gas, e.g. carbon dioxide. The change in gas volume (or pressure) in a closed reaction vessel can therefore be measured, suitable corrections being made for the effects of pressure and temperature on the gas volume. These manometric methods have an important place in the history of the elucidation of many of the metabolic pathways in animal tissues, but have now largely given way to optical techniques.

A number of enzymic reactions are accompanied by a change in the pH of the reaction mixture, which can be followed conveniently. However, the alteration of pH shifts the conditions from those that are optimal for enzyme activity, slowing the reaction down and eventually stopping it. A more satisfactory procedure, therefore, is to add alkali or acid at a controlled rate to the solution so that the pH is kept constant: the volume added in unit time thus becomes the index of the rate of reaction. An example of an enzyme of clinical interest which liberates acid by its action is pancreatic lipase and the acid can be titrated:

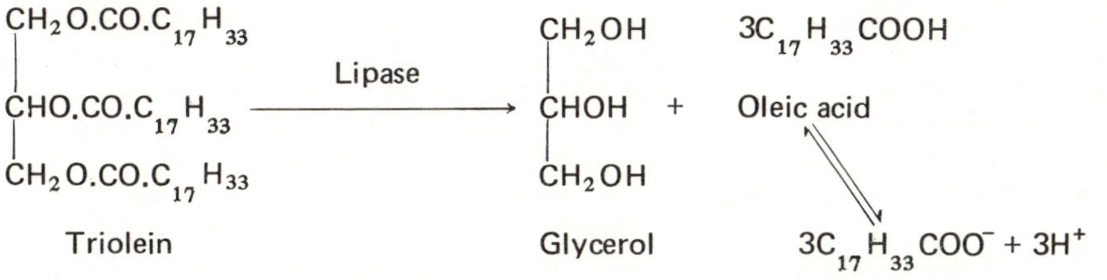

collection of urine. For enzymes which are of established clinical value, eponymous units named for the authors of particular methods are frequently encountered and with the continuing improvement and modification of methods of estimation the definitions of the different "units" in which enzyme activity may be reported can multiply at an alarming rate, since every change in reaction conditions almost inevitably alters the significance of the units — at least twenty different units for expressing serum alkaline phosphatase activity have been defined, for instance, of which three are in everyday use. The proliferation of units makes comparison of results obtained in different laboratories difficult or even impossible, a distinct disadvantage in long-term studies on individual patients or comparisons of groups of patients.

An attempt at rationalization has been made recently in which an "international enzyme unit" is defined as the number of micromoles of substrate transformed per minute, and enzyme concentration (as in serum, urine, etc.) as the number of units per ml or per litre. These units are not absolute, in that the conditions (nature of substrate, temperature, pH, etc.) under which the measurements were made must be specified, so that for a particular enzyme an "international unit" based on one method is not necessarily equivalent to an international unit based on a different method, but the

$$CH_2O.CO.C_{17}H_{33}$$
$$|$$
$$CHO.CO.C_{17}H_{33} \xrightarrow{\text{Lipase}} $$
$$|$$
$$CH_2O.CO.C_{17}H_{33}$$

Triolein

$$CH_2OH$$
$$|$$
$$CHOH \quad + $$
$$|$$
$$CH_2OH$$

Glycerol

$$3C_{17}H_{33}COOH$$

Oleic acid

$$3C_{17}H_{33}COO^- + 3H^+$$

These examples by no means exhaust the variety of ways in which enzyme activity can be measured but some methods are applicable to estimations of a few enzymes, or even of one only, and the methods outlined represent those most generally useful.

The results of enzyme estimations are expressed as the amount of chemical change in unit time — milligrammes of a product formed, or of a substrate destroyed, per minute, for example. A further term is added to convey the amount of material which contains this activity, e.g. per 100 ml of serum, or in a 24-hr

system of international units represents at least a step towards uniformity.

Some of the potential sources of error in estimating enzyme activity have already been mentioned; for instance, the marked dependence of enzyme activity on pH and temperature with the resulting need for careful control of these variables, and the requirement for adequate concentrations of intermediate reactants in linked reaction sequences. Other sources of error are not peculiar to enzyme estimations but are characteristic of the type of analysis, colorimetric, manometric, etc., which is being employed. Where more than one method of estimating a particular enzyme is available, the relative accuracy and reproducibility of the several procedures are important factors in deciding which shall be adopted. Also critical in the choice of an enzyme assay for clinical investigations is its sensitivity: many clinical applications involve the measurement of small changes in enzyme activity in samples which are available in limited volumes only, e.g. venous blood, and this can only be done successfully with techniques of adequate sensitivity. An example of the limitations that can be imposed by the lack of a suitable assay method is provided by the enzyme pancreatic lipase. The appearance in the circulation of enzymes produced by the cells of the pancreas is a valuable diagnostic sign in acute abdominal pain, when it strongly supports a diagnosis of acute pancreatitis. Lipase releases fatty acids from triglycerides and is usually estimated by titrating the acids (p. 9). However, this method is so insensitive (some procedures which have been suggested specify a 24-hr incubation period) as to be of little value for estimations on blood. Attempts have been made to adapt the substrate to permit a colori-metric assay, e.g. by synthesizing a fatty-acid ester of p-nitrophenol (cf. the estimation of alkaline phosphatase, p. 7); however, it is by no means certain that the enzymes which split these synthetic substrates are the same as the lipase which acts on triglycerides. For these reasons, estimations of the starch-splitting enzyme of the pancreas, amylase, for which a variety of more sensitive methods are available, are preferred in the investigation of acute pancreatitis although there is some reason to think that changes in plasma lipase would show a more consistent correlation with disease of the pancreas.

In enzyme analysis, as in other analytical procedures, the sum of the effects of the various sources of error makes up the total experimental error of the method, which is reflected in its accuracy and precision. Accuracy is the degree to which a particular result approaches the true or most probable value for the constituent being estimated (in this case the enzyme), while precision refers to the extent to which repeated estimations on the same sample give results which are in agreement. The accuracy and precision of an analytical method can be assessed by repeated analysis of specimens of known composition, but when the assay is used to analyse large numbers of samples over a long

period of time, as is the case when enzyme estimations are carried out on serum samples for diagnostic purposes, quality-control programmes are needed to ensure that the degrees of accuracy and precision of which the method is capable are being maintained. This is achieved for analyses involving substances which can be obtained in pure and stable form (e.g. sodium, urea) by preparing solutions of known consti-tution by weighing, then analysing a sample of these solutions along with each batch of specimens of unknown composition. However, when enzymes are the subject of analysis their nature and properties complicate this simple approach. Few if any enzymes of clinical interest have been obtained in a sufficiently pure form for solutions of known enzyme activity to be pre-pared by weighing known amounts of the puri-fied material. Accuracy in an enzyme assay method is therefore difficult to assess, but the degree to which results obtained by the method under assessment agree with those given by a reference method is often used for this purpose. Monitoring the reproducibility of enzyme assays is rendered difficult by the instability of enzyme preparations. Enzymes are proteins, and are consequently susceptible to denaturation with accompanying loss of activity (Chapter 3). When relatively concentrated solutions of a pure enzyme are available the occurrence of denatur-ation can be detected by changes in certain properties of the solution such as its effect on the rotation of plane-polarized light, but for the dilute enzyme solutions used for quality-control purposes in clinical enzymology, with their admixture of other proteins, there is no way in which a low result due to errors in analysis can be distinguished from one due to deterioration of the control preparation. Future research may produce enzyme solutions of assured stability; meanwhile, it is desirable that the activity of enzyme solutions used to monitor the precision of enzyme assays should be checked regularly by the reference method.

Human operators who have to carry out large numbers of repetitive analyses are liable to fatigue, with a consequent increase in experi-mental error. The advantages of automated methods in eliminating this source of variation and improving the consistency and rate of production of results have been well demon-strated in analysis for non-enzymic constituents, and a number of systems for the mechanization of colorimetric analysis have been devised, notably the Technicon Auto Analyzer system invented by L. T. Skeggs in which reactions take place in flowing streams of liquid. Automatic analysis of this type can readily be applied to enzyme assays of a colorimetric, fixed-time type. However, the advantages of continuous monitoring of progress curves mentioned earlier, which are particularly noticeable when samples of widely varying enzyme activity are to be analysed, make it desirable to investigate the possibilities of mechanizing this form of assay.

A spectrophotometric assay for a coenzyme linked enzymic reaction, e.g. lactate dehydrogenase, may be broken down into several stages. First, the enzyme-containing sample (usually serum) is measured accurately into a known volume of buffer solution containing all but one of the essential components of the reaction (e.g. pyruvate). After a period of incubation to allow any changes in extinction due to the presence of endogenous substrates in the sample to take place and a constant extinction value to be reached, the reaction proper is initiated by adding a measured volume of a solution of pyruvate and mixing. The rate of change of extinction at 340 nm is then recorded and the gradient of the initial linear portion of the recorder trace is computed to give the rate of reaction. Temperature must of course be constant from the moment recording starts, so that preheating prior to this stage is necessary. Instruments are available in which various stages in the process are mechanized, from the successive transfer of samples to the spectrophotometer in simpler instruments to the complete sequence in the more complex ones. The fully automatic machines require the operator merely to load into them unmeasured samples in a known sequence, results being printed out at the completion of analysis. This degree of complexity is necessarily expensive, but where large numbers of enzymic analysis are carried out savings in operators' time, as well as improvements in precision of results, justify the expenditure.

PURIFICATION OF ENZYMES

One of the first requirements in starting to extract and purify an enzyme from the tissues in which it occurs is the availability of a suitable method of assaying it, so that the precipitates, fractions of solutions, etc., containing the enzyme can be identified during the various stages of the purification process. Next, the material from which the extraction of the enzyme is to be attempted must be chosen. If the intention is to prepare a quantity of enzyme having a particular activity, perhaps for therapeutic purposes or to serve as quality-control materials in analysis, a source material is selected which is known to be rich in the enzyme, e.g. a bacterial strain which can be cultured in quantity. More often, however, the object of the investigation is to study the properties of an enzyme occurring in a particular organ or tissue, perhaps in order to compare it with analogous enzymes from other tissues or to reveal abnormalities which may be associated with a genetically determined disease. In cases such as these, limited amounts only of the starting material may be available, particularly when human tissues are the object of study, and purification procedures then have to be adapted to use on a small scale.

The majority of enzymes act intracellularly and, moreover, are in many cases firmly attached to the various structural elements of the cells in which they occur. The first step in enzyme purification is therefore to release the enzyme into solution freed from the organized structures of the cells. When the enzyme is one which occurs in the cytoplasmic fluid, breaking the outer membrane and filtering off the cellular debris achieves the desired result. The cell membranes can be broken by grinding the minced tissue with sand in a mortar, or by homogenizing it either in a rotating-blade blender or in the Potter-Elvehjem type in which the cells are subjected to shearing forces in the liquid layer between a motor-driven pestle and the wall of a closely fitting glass tube. Ultrasonic disintegration is also useful. The cell walls of bacteria are often very tough and vigorous homogenization may be needed to break them: alternatively, recourse may be had to treatment with organic solvents or even to the addition of enzymes that specifically attack chemical linkages in the cell-wall material.

Release of those enzymes which occur in subcellular particles is more difficult, as in some of them, e.g. mitochondria, the enzymes seem to be part of the structure of the particles themselves. The construction of the particles often involves lipid materials and treatment with fat solvents or detergents is usually effective in bringing the enzymes into solution. Samples of a particular type of sub-cellular particle relatively free from other structures can be obtained by differential ultracentrifugation, in which the tissue homogenate is subjected to increasing gravitational fields in an ultracentrifuge so that the particles sediment in descending order of their particle weights, the heaviest (i.e. the nuclei) at the lower centrifuge speeds and progressively down to small bodies such as microsomes which require high-speed centrifugation to sediment them (Chapter 8).

When a true solution of the contents of the cells of a tissue or of the components of a particular class of sub-cellular organelles has been obtained, it will contain not only the enzyme under investigation but also a complex mixture of other substances both of small molecular weight (inorganic ions, simple sugars and lipids, amino acids, etc.) and macromolecules (mainly proteins, with some polysaccharides and also nucleic acids when the solution has been made from nuclei and microsomes). Small molecules can be removed by dialysis and nucleoprotein by adjusting to pH 5 and centrifuging. The problem of enzyme purification thus resolves itself into the isolation of a single type of protein molecule from a mixture of many different types.

Protein molecules contain many groups of atoms which are capable of ionization under appropriate pH conditions to acquire positive or negative charges. These groups consist of the basic and acidic radicals (e.g. amino and carboxyl) in the side chains of the amino acids from which the protein molecules are built up and those at the ends of the protein chains: the nature, relative proportions and arrangement of the several amino acids in a given protein molecule are responsible for its characteristic properties (Chapter 3). Each of the ionizable groups in a particular protein molecule displays a characteristic variation of ionic charge with pH, the precise form of which depends on the chemical nature of the group and the nature and arrangement of the neighbouring groups in the protein molecule. The tendency of an acidic or basic group to ionize is expressed as its pK value, that is, the negative logarithm of its ionization constant. The pK value has the dimensions of pH, and it is that pH at which half of all the groups of that type present in solution are ionized, since

$$pH = pK + \log \frac{[\text{ionized form}]}{[\text{un-ionized form}]}$$

and when the concentration of the ionized form equals that of the un-ionized form, pH = pK.

The different pK values of the many ionizing groups in a protein molecule have the effect that, at a particular pH, some of these radicals will carry positive charges, some will have negative charges and others will be undissociated. As the pH changes, so the degree of ionization of the groups varies accordingly. Under any given conditions of pH, therefore, the protein molecule itself has a net charge that is the algebraic sum of the individual positive and negative charges; when the sum is zero, i.e. when the number of positive charges just balances the number of negative charges, the protein is at its isoelectric pH. The isoelectric point is a characteristic of the type of protein molecule concerned, and differs from one kind of protein to another. At the isoelectric pH protein molecules do not migrate in an electric field, their solubility is at a minimum and the viscosity, osmotic pressure and conductivity of their solutions also show minimum values. Many of the methods used to fractionate mixtures of proteins exploit the differences in degree of ionization and net charge which exists between different types of protein molecules under particular conditions.

The solubility of protein molecules depends on the interaction of their charged groups with water molecules. Addition of organic solvents such as ethyl alcohol or acetone reduces this interaction by competition for the water molecules and so lowers the solubility of the protein, resulting in precipitation if enough alcohol or acetone is added. The point at which precipitation occurs depends on the nature and degree of ionization of the protein; thus, fractionation of a mixture of proteins can be effected by stepwise increases in the amount of organic solvent in the solution, collecting the protein precipitated after each addition. Removal of water of hydration by organic solvents has a powerful denaturing effect on proteins, and if this is to be avoided it is essential to keep the temperature below 0°C during precipitation. A similar precipitation effect is obtained by the addition of inorganic salts in high concentrations to protein solutions. Ammonium sulphate is often used for this purpose since its high solubility allows the production of concentrated solutions, while this salt is generally without deleterious effects on protein molecules. The amount of ammonium sulphate required to precipitate a particular protein is usually expressed as a percentage of the amount required to produce a saturated solution; serum globulins, for example, are precipitated at 50 per cent saturation.

Precipitation with organic solvents or with ammonium sulphate is a convenient means of obtaining a broad separation of protein mixtures and it forms the first stage of most enzyme purification procedures. However, for fine resolution, methods are needed which are able to exploit quite small differences in net charge between various types of protein molecules; one such method is ion-exchange chromatography.

Ion-exchange materials consist of an insoluble and chemically inert base to which ionizable chemical groups are attached. The insoluble base is usually cellulose in the materials used for chromatography of proteins because this polymer has a low degree of cross-linking between the chains, giving an open structure of a hydrophilic nature to which the protein molecules can gain access. The active groups are chosen so that they ionize either in acid solution, with the acquisition of a positive charge (an example is diethyl-

aminoethyl, $—O.CH_2 CH_2 N{\displaystyle <}^{C_2 H_5}_{C_2 H_5}$, which

becomes ionized to $O.CH_2 CH_2 \overset{+}{N}H{\displaystyle <}^{C_2 H_5}_{C_2 H_5}$), or

of positive or negative charge can be obtained. Protein molecules, or other ions, which have a net charge of opposite sign to that of the exchange material will attach themselves to it by electrostatic attraction and will be released from this attachment either by alteration of the pH so as to reduce the ionization of the exchange groups or by addition of an excess of an ion of similar charge which competes for attachment to the exchanger. An ion-exchanger with positively charged groups, i.e. one which attracts negative ions, is called an anion-exchanger, while one with negative groups is a cation-exchanger.

It is usual to carry out the separation of protein mixtures by packing the ion-exchange material, suspended in buffer solution, into a long, vertical glass tube. The protein mixture is introduced at the top of the column and the protein molecules become attached to the exchanger. The pH of the buffer flowing down the column is then progressively changed, or

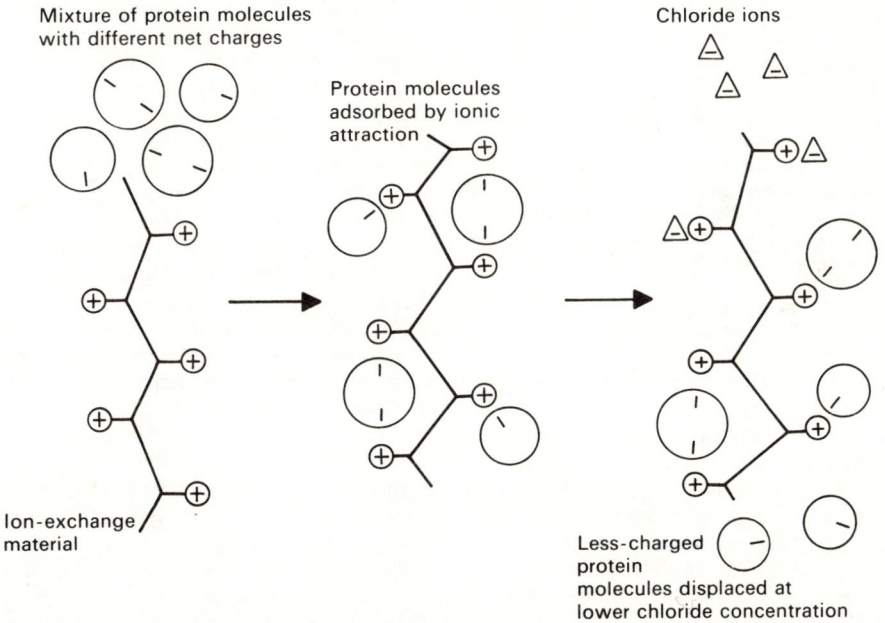

Fig 2.4. Schematic representation of the reactions taking place during ion-exchange chromatography, showing displacement of protein molecules by a competing ion.

in alkaline solution to gain a negative charge (e.g. carboxyl, $—COOH$, which dissociates to $—COO^-$). Thus, by choosing a particular ion-exchange material and soaking it in a buffer solution at the appropriate pH, an insoluble stationary phase

the concentration of a competing ion is gradually increased. The reactions taking place in anion-exchange chromatography are represented in Fig. 2.4. The protein molecules with the smallest net charge are displaced most easily, followed by proteins of progressively greater charge. The effluent from the column is collected in a fraction collector, and by estimation of the protein content of each fraction a profile can be plotted (Fig. 2.5) and the components of the mixture can be identified.

Another separation method which depends on the differences in net charge carried by different types of protein molecules is electrophoresis. This technique has been applied to

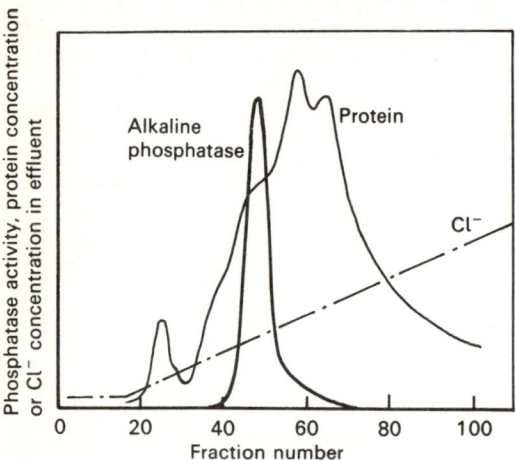

Fig 2.5. An ion-exchange chromatogram obtained during purification of human liver alkaline phosphatase on a column of diethylaminoethyl (DEAE) cellulose, showing separation of the enzyme-activity peak from inactive protein peaks.

problems of clinical importance ever since its introduction in 1933 by Tiselius, who showed that the globulins of blood serum could be separated in this way into several fractions. The principle of electrophoresis is that if positive and negative electrodes are dipped into a solution containing ions (e.g. protein molecules), the ions with a net negative charge will move towards the anode and those with a net positive charge towards the cathode, at rates which are proportional to the magnitude of their charges and the strength of the electrical field.

At first electrophoresis was carried out in free solution, the presence of migrating fronts of the different protein components being detected by optical methods such as changes in refractive index of the solution. Thermal diffusion of the protein molecules prevents the different components from being completely separated, so that electrophoresis in free solution is not applicable to preparative work, but it remains a

valuable analytical tool for testing the homogeneity of protein samples. Diffusion can be restricted by causing the protein molecules to migrate through a buffer solution which has soaked into a porous supporting medium such as a strip of filter paper or cellulose acetate, or in which a paste of starch grains has been prepared. If, at the beginning of the separation process, the protein mixture is applied as a narrow zone it becomes separated into zones of proteins of different mobilities. Zone electrophoresis, as this is called, is a very simple technique, particularly when the support is filter paper; the protein zones can be readily made visible by staining the strip with suitable dyes, and the relative intensities of the coloured bands are approximately proportional to the amounts of protein they contain. Enzyme activity can be visualized directly on the supporting medium in many cases by colour reactions of the type used in histochemistry (Chapter 8). For preparative work, as in the purification of an enzyme, the proteins are recovered from the supporting material by cutting it into segments from which the protein is eluted with buffer solution. Generally, the supporting materials used in zone electrophoresis have little effect on the nature of the separation taking place in them. In recent developments in zone electrophoresis, however, the restriction on diffusion and migration in the supporting medium is increased so as to bring into play another factor in determining the nature of the separation, namely, the size and shape of the protein molecules. In gels, the size of the pores in the medium approaches the dimensions of protein molecules, so that large proteins can move through media of this type less freely than smaller ones. Gels can be prepared by heating partially hydrolysed starch in buffer solution and then allowing the mixture to cool: when electrophoresis is carried out in the starch gel, a different separation (e.g. of serum proteins) is obtained in which many more protein zones are seen than with paper electrophoresis, the additional bands resulting from the retardation of the larger molecules in the gel, even though these may have the same net charge as other, smaller proteins present in the mixture. Besides starch gels, gels made by chemical polymerization of acrylamide are used for electrophoresis. Polyacrylamide gels have the advantage that the degree of cross-linking of the polymer chains, and hence the pore size, can be adjusted chemically when the gel is made so that the extent to which proteins of different sizes are retarded can be controlled. There are experimental difficulties associated with recovering protein zones from gel media so that this very highly resolving procedure is valuable only for small scale preparations at present. Zone electrophoresis is a very useful technique in the separation of the multiple molecular forms, or isoenzymes, in which many enzymes are now known to exist (Chapter 3), and determination of the distribution of isoenzymes in serum samples in this way is a valuable diagnostic aid (Chapter 9).

Differences in the sizes of protein molecules have been used in separating mixtures of them since the introduction of the ultracentrifuge by Svedberg in 1925. With the application of gravitational fields of the order of 100 000 x g or more made possible by the ultracentrifuge, protein molecules in solution sediment at rates proportional to their molecular weights; these rates are to some extent also affected by the shape of the molecules. As in electrophoresis in free solution, the protein molecules are free to diffuse and at the start of the centrifuge run molecules of all sizes are uniformly distributed throughout the solution. During centrifugation the boundaries of the zones of sedimenting proteins can be detected by optical methods, but the zones remain overlapping. For this reason the use of the ultracentrifuge in the purification of enzymes and other proteins has mainly been in determining the number of proteins present in a mixture and in confirming the purity of a final preparation. A recent development, however, analogous to the introduction of zone electrophoresis, allows protein zones to be separated. In zonal centrifugation the protein mixture is introduced as a narrow band near the axis of a spinning rotor shaped like a hollow disc. As the protein zones move outwards towards the periphery they are drawn off while the rotor is still in motion.

Protein molecules of different sizes can be separated more simply, however, by passing them through a "molecular sieve". The use of materials with pores approaching molecular dimensions was mentioned in connection with gel electrophoresis. Various naturally occurring linear polysaccharides, e.g. dextrans, which consist of glucose monomers, or agarose, which is made up mainly of galactose units, can be modified chemically so that cross-linkages are formed between adjacent chains. Alternatively,

entirely synthetic polymers of acrylamide can be produced and cross-linked. When placed in water these materials swell to form gels, the size of the pores in the gel being determined by the degree of cross-linking. If such a gel is packed into a glass column and the protein mixture is washed through it, the larger proteins emerge first from the bottom of the column since they are excluded from the pores of the gel, while smaller and smaller molecules, which have a greater solution space available to them, are progressively later in reaching the bottom. If a gel is chosen which has a suitable pore-size (i.e. one with an appropriate degree of cross-linkage), the process of gel-filtration or "exclusion chromatography" can offer considerable resolution of protein mixtures on the basis of molecular size with little denaturation. The technique can also be used as an alternative to dialysis in separation of proteins from small molecules or ions.

Purification of an enzyme involves the application of these methods one after another, beginning usually with salt or alcohol-fractionation then proceeding through gel-filtration, ion-exchange and electrophoresis stages. Enzymes are readily denatured and lose their activity as a result of even slight rises in temperature, and this sensitivity generally increases as they become more pure. The various purification procedures are therefore carried out at low temperatures in refrigerated rooms. After each stage the activity of the preparation is checked and the increase in purity is assessed by calculating the specific activity; that is, the enzyme activity per milligramme of protein. It is difficult to decide when a protein is completely pure. Its chemical composition cannot be compared with an "expected" formula, as is the practice in the preparation or purification of smaller molecules, since the absolute composition of only a very few proteins is known with certainty, while the formation of crystals is not a reliable guide to protein purity as proteins can crystallize while still quite impure. In enzyme purification it is usual to continue the process until further stages fail to raise the specific activity any more, and, if possible, until only one protein component is seen during electrophoresis or ultracentrifugation.

Scores of enzymes have now been extensively purified and many have been crystallized. Most of the present knowledge of the nature and function of enzymes derives from studies on these purified preparations.

3 Structure and Biosynthesis of Enzyme Molecules

The chemical composition of enzymes has been a subject of lively discussion in the past but now the protein nature of enzymes has become firmly established and universally accepted. Some early workers believed that enzymes were carried on high molecular weight colloids, not necessarily protein in structure, and that these large molecules were non-specific and acted as interchangeable "carriers" of smaller "enzyme" molecules of unknown chemical composition. These views were apparently supported by observations that enzyme activity could be detected in dilute solutions which did not give positive chemical colour tests for protein. In cases where positive colour tests, such as the biuret test or the xanthoproteic reaction, were obtained the protagonists of the "carrier" theory argued that the protein was either the non-specific carrier or an impurity but was not the true enzyme. Other workers were equally convinced that enzymes were protein in nature but some felt that it was unlikely that this view could ever be proved absolutely by purification since most enzymes seemed to become less stable with increasing purity. The break-through came in 1926 with the purification and crystal-lization of urease from jackbean meal by J. B. Sumner and his demonstration that the crystals consisted of protein. Though convincing in retrospect to the present generation of bio-chemists, this discovery was less readily accepted by some scientists of the late 1920s. However, many other enzymes have been isolated in crys-talline form and in each case have been shown to be protein. Particularly important pioneer studies were those carried out by J. H. Northrop and his colleagues on digestive proteases and much of our present background information on the characteristics of enzymes as proteins is due to these early researches.

These and other classical studies showed that all enzymes are large molecules made up of carbon (55 per cent), hydrogen (7 per cent), oxygen (20 per cent), nitrogen (16 per cent) and usually some sulphur (2 per cent). They are colloidal, non-dialysable ampholytes with vari-able solubility in water and buffer solutions. Characteristically, like other proteins, enzymes are usually least soluble at or about their iso-electric point. They are precipitated from solution by protein precipitants such as trichloro-acetic acid or ammonium sulphate and they are subject to denaturation, which may be reversible, on exposure to extremes of pH or temperature. Enzymes are also more stable in concentrated than in dilute solution. Another important property demonstrated in early work was the observation that pure preparations of enzymes were like other proteins and acted as antigens when injected into other animal species. Studies on the interaction of enzyme and antibody have been of particular value since then, not only in enzyme chemistry but also in immunology since the possibility of monitoring catalytic activity adds an extra parameter not available for study with non-enzyme proteins. A non-protein prosthetic group may sometimes be involved in catalysis (Chapter 5), but many enzymes are entirely protein in composition and no case of enzyme action has been sub-stantiated in which protein molecules are not involved.

The outlines of the chemical structure of proteins have been known for many years, but it is only within the last two decades that methods have been developed which have made it possible to examine in detail the structure and shape of individual protein molecules. The results of these studies are beginning to reveal why particular protein molecules, including enzymes, have their specific biological functions and why even slight modifications to the structure of a protein molecule introduced by genetic mutation can so profoundly affect its activity.

TABLE 3.1. FORMULAE OF AMINO ACIDS OCCURRING IN ANIMAL PROTEINS

Glycine (Gly)	$CH_2(NH_2).COOH$	Cystine (Cy SSCy)	$S—CH_2.CH(NH_2).COOH$ $S—CH_2.CH(NH_2).COOH$
Alanine (Ala)	$CH_3.CH(NH_2).COOH$	Methionine (Met)	$CH_2(S.CH_3).CH_2.CH(NH_2).COOH$
Valine (Val)	$CH_3.CH(CH_3).CH(NH_2).COOH$		
Leucine (Leu)	$CH(CH_3)_2.CH_2.CH(NH_2).COOH$	Phenylalanine (Phe)	
Isoleucine (Ileu)	$CH_2(CH_3).CH(CH_3).CH(NH_2).COOH$	Tyrosine (Tyr)	
Serine (Ser)	$CH_2(OH).CH(NH_2).COOH$		
Threonine (Thr)	$CH_3.CH(OH).CH(NH_2).COOH$	Tryptophan (Try)	
Aspartic acid (Asp)	$CH_2(COOH).CH(NH_2).COOH$		
Asparagine (Asn)	$CH_2(CONH_2).CH(NH_2).COOH$	Histidine (His)	
Glutamic acid (Glu)	$CH_2(COOH).CH_2.CH(NH_2).COOH$		
Glutamine (Glu)	$CH_2(CONH_2).CH_2.CH(NH_2).COOH$	Proline (Pro)	
Lysine (Lys)	$CH_2(NH_2)CH_2.CH_2CH_2.CH(NH_2)COOH$		
Hydroxylysine (Hyl)	$CH_2(NH_2).CH(OH).CH_2.CH_2.CH(NH_2).COOH$		
Arginine (Arg)	$HN:C(NH_2).NH.CH_2.CH_2.CH_2.CH(NH_2).COOH$	Hydroxyproline (Hyp)	
Cysteine (Cys)	$CH_2(SH).CH(NH_2).COOH$		

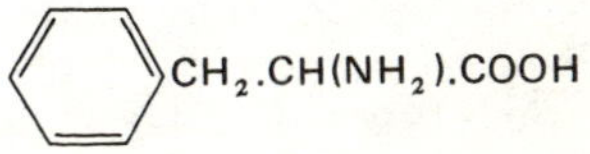

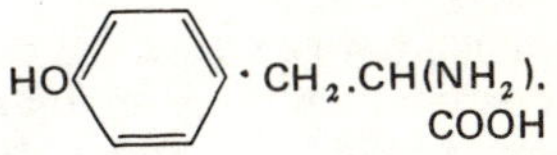

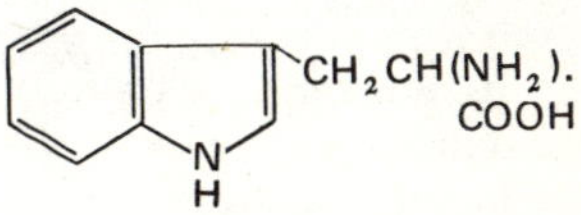

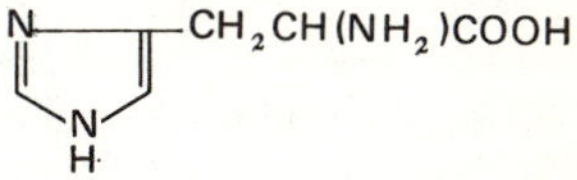

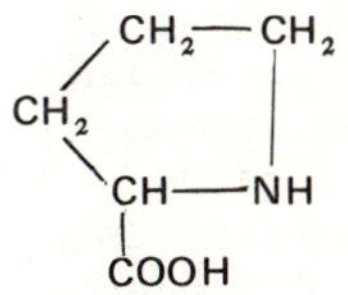

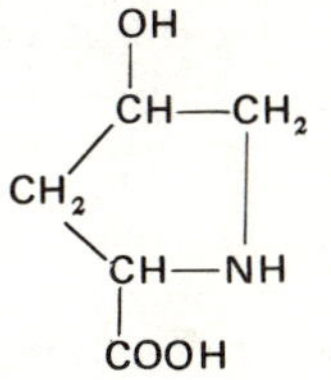

Protein molecules are composed of long-chains of α-amino acids. Approximately twenty different α-amino acids occur in animal proteins and the formulae of these are given in Table 3.1. The amino acids are linked together covalently by peptide bonds formed between the α-carboxyl group of one amino acid and the α-amino group of its neighbour (right):

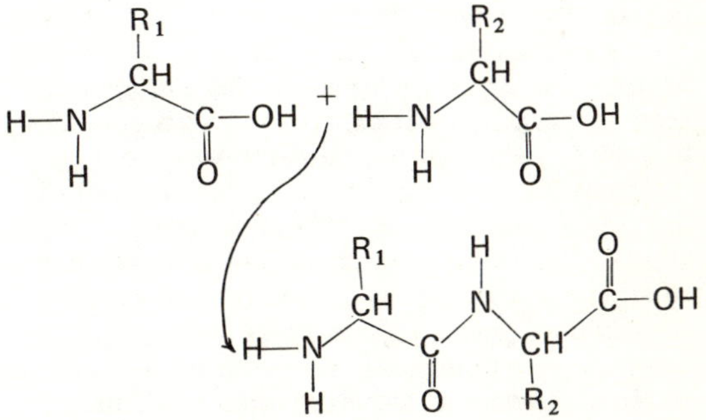

The polypeptide chains of protein molecules may be several hundred amino acids in length, and the alternating carbon and nitrogen atoms form a backbone from which the amino acid sidechains project sideways (right):

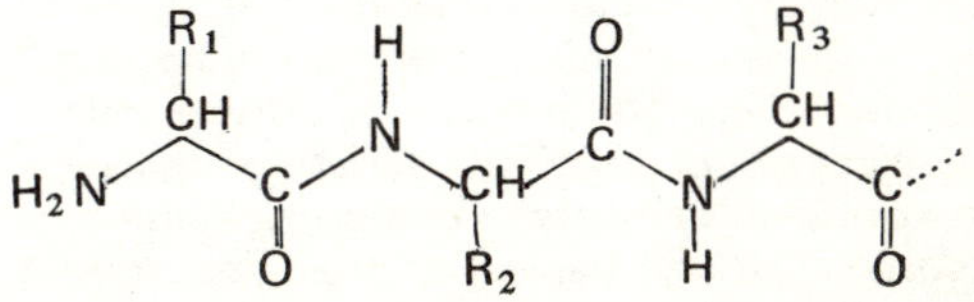

At one end, the N-terminal end, is a free α-amino group and at the other, the C-terminal end, a free carboxyl group. The projecting side-chains are responsible for the physicochemical properties of protein molecules, such as electrophoretic mobility, colour reactions, etc.

All proteins have this structure of chains of peptide-bonded amino acids, and it is clear therefore that the different properties of the various types of proteins must result from differences in the relative proportions of the several amino acids which they contain and the order in which these amino acids succeed one another along the chains. Much information has been gathered about the amino acid composition of different proteins by hydrolysis of the peptide bonds and release of the individual amino acid molecules, but differences in amino acid composition cannot by themselves explain the great range of properties and functions of protein molecules, including enzyme catalysis. Determination of the amino acid sequence (also called the *primary structure*) of protein molecules is a much more difficult problem of which the solution was not forthcoming until, in 1953, F. Sanger was able to announce the primary structure of insulin. Insulin is a small protein (mol. wt 5800) with a molecule consisting of two different polypeptide chains made up of 30 and 21 amino acids, respectively, while enzymes and other proteins may have molecular weights reaching into hundreds of thousands or even millions. Nevertheless, sequence determinations have been extended to include many more proteins consisting of as many as 150 amino acids linked in a single chain. These studies have indicated the high degree of specificity that is characteristic of the primary structure of a particular type of protein; in other words, all molecules of human insulin seem to have an identical amino acid sequence, but the sequence of the human hormone is different in some respects from that from other animal species, although there are inter-species similarities. There is no pattern or arrangement of amino acids which is common to all proteins (as had been thought possible at one time), but there are resemblances between short sections of the sequences of closely related proteins.

The primary structure constitutes only one aspect of the organization of protein molecules. Long polymeric molecules such as polypeptide chains can exist in an infinite variety of shapes, from extended filaments to compact globules, and observations on the loss of biological activity when proteins are subjected to heat gave rise many years ago to the idea that this process of denaturation involves a distortion of the shape of a protein molecule, and consequently biological activity in proteins is associated in each case with a particular molecular shape, or conformation. Methods such as measurements of the viscosity of protein solutions or the rate of sedimentation of protein molecules in the ultracentrifuge allowed broad inferences to be drawn about the shapes of protein molecules; whether they are long and thin, or spherical, for example, and these results suggested that most enzyme molecules are globular. A definitive picture of the shapes of protein molecules was not possible, however, until the method of X-ray crystallography was applied to the study of protein fibres and crystals.

The diffraction of X-rays by the regular arrangement of atoms in crystals was discovered in 1912 by M. von Laue and interpretation of the diffraction patterns to deduce the structure of crystals was later developed by W. L. Bragg and others. By the 1930s the first X-ray diffraction photographs of protein crystals had been taken but, while these proved that protein crystals like those of smaller molecules consist of regularly arranged atoms, it was impossible at the time to translate the diffraction data into a picture of the three-dimensional arrangement of the atoms in the crystals. In 1953, however, M. F. Perutz discovered that by attaching heavy metal atoms to definite positions in the protein molecule the additional information provided by the modified X-ray diffraction pattern so obtained allows the structure of the molecule to be calculated. The first successes of the new technique were the three-dimensional structures of myoglobin, deduced by J. C. Kendrew, and of haemoglobin, elucidated by Perutz himself, and the shapes of a number of small enzyme molecules have now been revealed. The first of these was of the enzyme lysozyme, as a result of the investigations of D. C. Phillips and his colleagues. Lysozyme is an enzyme obtainable from tears, nasal mucus and egg white, as well as from other sources, which breaks down a complex polysaccharide found in the cell walls of certain bacteria. The bacteria are thus dissolved, or "lysed", and Sir Alexander Fleming, later the discoverer of penicillin, who in 1922 found lysozyme in nasal mucus, was for a time hopeful that the enzyme would prove a valuable bactericidal agent: the bacteria which are attacked by lysozyme are not, however, among those which cause important diseases in man. Lysozyme is a protein of molecular weight 14 400 with 129 amino acids arranged in a single chain. The sequence of the amino acids was established by work in Paris and New York. The important property of lysozyme from the X-ray crystallographer's point of view is that suitable crystals can be prepared, both of the native enzyme and of derivatives of it containing atoms of heavy metals.

The picture of the lysozyme molecule that emerges from the X-ray diffraction data is one of a compact, globular molecule formed by a very complex folding of the polypeptide chain. (Fig. 3.1). Adjacent loops of the chain are held together at four places by disulphide (—S—S—) bridges, formed between the sulphydryl groups

Fig 3.1. Model of the lysozyme molecule obtained by X-ray analysis at 6Å resolution. A molecule of the inhibitor tri-N-acetylchitotriose (dark) is positioned in the cleft in the molecule which contains the active centre. The wire follows the course of the polypeptide chain. (From C. C. F. Blake *et al*. (1967) *Proc. R. Soc.B.* 167, 378, by courtesy of Prof. D. C. Phillips, FRS, and the Royal Society.)

of molecules of cysteine (Table 3.2). The existence of these links was known from chemical determinations of the amino acid sequence and they thus assist in matching the sequence with the crystallographic picture. Some sections of the chain are coiled into the particular helical configuration (the "α-helix") which was pre-

dicted in 1951 by L. Pauling and R. B. Corey as a probable stable configuration for protein chains on the basis of their studies on simple derivatives of amino acids. Not much of the lysozyme chain is coiled into α-helices, however, compared with the large proportions of the chains of myoglobin and haemoglobin which occur in this form, and some other enzymes (notably chymotrypsin) appear to have even less α-helix than lysozyme. Another feature of the structure of lysozyme, which it shares with all other globular proteins whose three-dimensional structures are known, is that amino acids with non-ionizable, hydrophobic side-chains (e.g. leucine, valine, phenylalanine) tend to have their side-chains directed towards the interior of the molecule and away from the surrounding medium while the ionizable side-chains of hydrophilic amino acids (e.g. aspartic and glutamic acids, serine and histidine) in general are on the outside of the molecule. But the most interesting aspect of the lysozyme molecule from the point of view of its enzymic function is the cleft, or pocket, which runs approximately through the middle of it.

Phillips and his colleagues have shown that the enzyme molecule will react with the amino sugars that are the monomers from which the cell-wall polysaccharide, the enzyme's natural substrate, is built up. These amino sugar molecules become bound to the enzyme in the cleft in the molecule, which is long enough to accommodate a section of the polysaccharide substrate six sugar residues long. Substrate binding appears to be accompanied by a very slight closing of the cleft. The polysaccharide chain is then broken between the fourth and fifth sugar molecules counting from the top of the cleft. From the three-dimensional picture of the enzyme-substrate complex which X-ray crystallography reveals, those amino acid residues in the lysozyme chain which probably take part in breaking bonds in the substrate molecule can be identified.

The structures of some seven or eight enzyme molecules have been solved by X-ray crystallography at the present time, the largest, carboxypeptidase A from bovine pancreas, having 307 amino acid residues in its polypeptide chain with an atom of zinc as part of the active centre, while α-chymotrypsin has three intertwined polypeptide chains totalling 241 residues. Few generalizations about enzyme structure can yet be drawn from comparisons of the enzyme molecules whose structures have been defined.

TABLE 3.2 STABILIZATION OF PROTEIN SECONDARY AND TERTIARY STRUCTURE

Type of Bond

Comments

(1) *Disulphide* (covalent)
Disulphide link of cystine residues

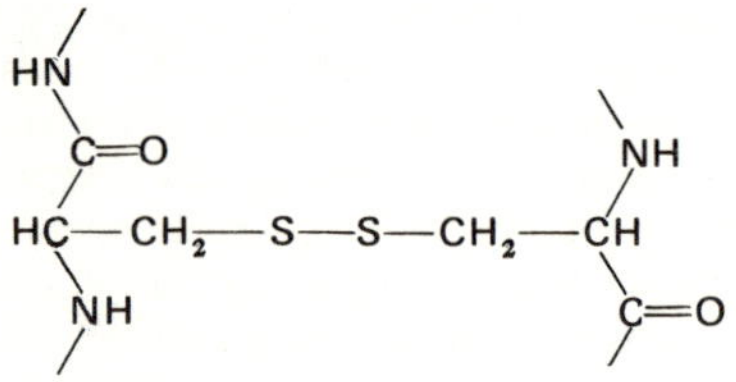

Forms cross links between parts of same polypeptide chain or bridges different chains. The bond limits the flexibility of the polypeptide.

(2) *Hydrogen* (non-covalent)

Hydrogen bond between N atom of one peptide bond and O atom of another is very important in stabilizing secondary structure, e.g. helices and pleated sheets. Hydroxyl groups of serine residues can also partake in hydrogen bonding.

(3) *Salt Bridge* (non-covalent)

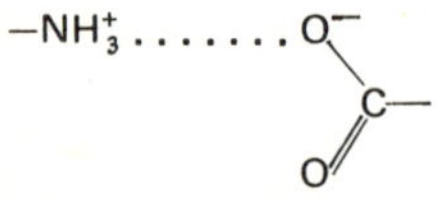

Formed by electrostatic interaction between charged groups in proteins, e.g. N-terminal amino group or ϵ-amino of lysine with carboxyl group of C-terminal amino acid or side-chain carboxyl of aspartic and glutamic acids.

(4) *Hydrophobic interactions* (non-covalent)

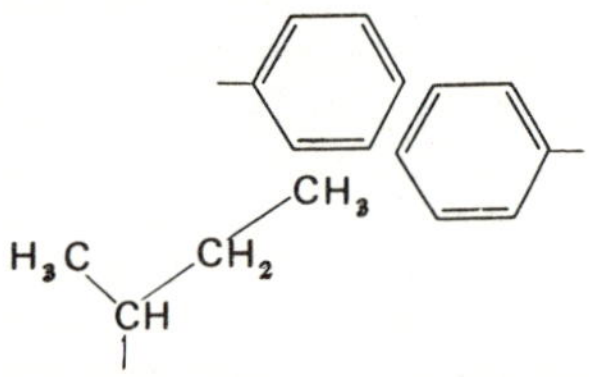

Attraction between non-polar side chains of amino acids, with expulsion of water, contributes to the maintenance of tertiary structure. Non-polar side chains seem to be concentrated in the core of all proteins making this region impenetrable to solvent. Important residues are those of leucine, isoleucine, valine, phenylalanine and trytophan.

All are irregular and of great complexity, and even enzymes which act on chemically similar substrates do not necessarily have similar structures: chymotrypsin hydrolyses the synthetic dipeptide, tyrosylglycine (i.e. tyrosine and glycine combined so that the amino group of tyrosine is free), but this enzyme has an entirely different structure from carboxypeptidase which splits the same pair of amino acids combined in the reverse order, as glycyltyrosine. However, all follow the rule already mentioned that non-polar residues tend to predominate in the interior of the molecule with polar residues in contact with the surrounding medium. In most of the molecules a cleft or depression can be identified into which the substrate fits when it combines with the enzyme, and amino acid side-chains concerned with the attack on the substrate (in other words those forming the active centre) are located in the cleft. The non-polar environment of the interior of the enzyme molecule enhances the reactivity of the side-chains within the cleft and thus facilitates their interaction with the substrate.

The results of these studies on lysozyme and other proteins show that to the primary structure, or amino acid sequence, must be added other, higher levels of organization: *secondary*, referring to the way in which short lengths of polypeptide chain are arranged (e.g. as α-helices);

tertiary, describing the folding and interaction of adjacent sections of such a chain; and *quaternary*, signifying the association of different polypeptide chains (or sub-units) which is a feature of many, if not all, large protein molecules. It seems that the structure of a protein molecule has to be closely defined at all these levels for it to possess and retain its specific biological activity. Since the method of X-ray analysis requires the enzyme or other protein to be in crystalline form it is important to know whether the structures existing in the crystals are retained when the protein passes into solution, which is probably closer to its normal functional state in the living cell. Fortunately, there is some evidence that this is the case; for example, the amino acid side-chains which are most reactive in solutions of the protein and which therefore might be expected to be on the outside of the molecule can be shown by X-ray data to occupy this position, while the amount of α-helix in a protein molecule in solution (which can be independently estimated) often agrees with the amount present in the crystal.

The existence of higher orders of protein structure raises problems of how these three-dimensional arrangements become established when proteins are synthesized in the living cell and how they are stabilized in the completed molecule. The present view is that the sequence of amino acids — the primary structure — largely determines the final shape of the molecule. If a protein molecule is denatured under controlled conditions by the action of strong solutions of urea or by a change in pH or by other chemical methods, it becomes distorted, assuming a shape different from its native form and losing its biological activity. When the original conditions are restored, however, provided that the polypeptide chain is still intact, it may regain its original shape and properties. The importance of the amino acid sequence in maintaining the structural integrity of a protein molecule can also be appreciated from studies on the genetic variants of proteins. For example, in the human haemoglobins, the deletion of a portion of the polypeptide sequence in the variant haemoglobin Gun Hill, or even the replacement of just one single amino acid by a different residue in the region where the haem prosthetic group is bound (e.g. the variant Hb Hammersmith), can lead to considerable changes within the protein molecule and associated changes in its biological properties.

The complex foldings of the protein chains are held in place by interactions between the projecting acid side-chains, and there are three types of bonding which are particularly important in this respect (Table 3.2). Covalent chemical bonding can occur between the sulphydryl (—SH) groups of cysteine residues in neighbouring sections of polypeptide chains with formation of disulphide bridges (—S—S—), as occur in the lysozyme molecule. A second type of bond arises from the tendency of a hydrogen atom which is attached to a nitrogen or an oxygen atom to be attracted by the electron cloud surrounding another oxygen atom nearby. The hydrogen atom has only one orbiting electron and when this electron is engaged in bond formation with another atom the hydrogen atom is left with an excess of positive charge. Although the hydrogen bond is weaker than covalent chemical bonds, many opportunities for its formation occur in the close-packed structure of protein molecules and these bonds are particularly important in stabilizing the α-helix. Probably even more significant than hydrogen bonds, however, in stabilizing the higher orders of protein structure are the interactions which take place between the non-ionizable amino acid side-chains which are brought into close proximity in the interior of the molecule. These non-polar or hydrophobic bonds arise from the mutual attractions of the groups of carbon and hydrogen atoms which make up these particular side chains. Other types of bonding could occur such as peptide-bond formation between the second carboxyl groups of dicarboxylic amino acids (e.g. aspartic or glutamic acids) and a second amino group (e.g. of lysine) but this seems rarely to happen; also possible are salt-like attractions between ionized side-chains, but as these groups tend to be on the outside of the molecule their influence on the stability of the structure is probably also small.

As the size of the protein molecules under study by X-ray diffraction methods increases (provided that suitable crystals of the native protein and derivatives of it can be prepared), the tasks of measuring the patterns and computing the structures reach formidable proportions even with the aid of electronic computers. The successful applications of this definitive method are therefore likely to increase in number rather slowly. By contrast, the methods for primary structure determination undergo progressive refinement so that this approach, while still representing a considerable chemical feat, yields information at an increasing rate. It is therefore attractive to speculate how far a knowledge of the primary structure can be used to predict the higher structural levels of the molecule. A long polypeptide chain with a given amino acid sequence can exist in an enormous number of configurations: the fact that one form is preferred to all the others suggests that the chosen form is the most stable. Theoretically, therefore, it should be possible to calculate the relative stabilities of the possible conformations and select the most stable as indicating the three-dimensional structure of the molecule. In practice, however, the amount of computation is too great because of the great number of possible forms, but as more rules of protein structure are discovered (such as the tendency for ionizable side-chains to be on the outside of the molecule and non-ionizable ones inside) they can be used to restrict the number of alternative forms and so reduce the computation required.

While the X-ray diffraction analysis of protein structures has yielded information of great fundamental importance which goes a long way towards explaining why particular enzymes should have their specific catalytic functions, the enzymes that have so far been studied in this way are not ones in which alterations or deficiencies result in important diseases in man. A good deal of information about the structure of protein molecules can be obtained, however, by methods which stop short of a full structural analysis, or even of a complete determination of primary structure, and these methods have in some cases revealed the underlying molecular abnormality which shows itself as human disease. The first demonstration of the applicability of these methods in understanding disease came, not from the study of an enzyme, but from investigations into variations in the haemoglobin molecule.

Sickle-cell anaemia is a genetically determined condition in which characteristic crescent-shaped red blood corpuscles are seen (sickling), as a result of a change in the solubility of the haemoglobin of the affected individuals. Sickle-cell haemoglobin is also different from the normal pigment in its electrophoretic mobility (i.e. in its net electric charge at a given pH); hence, a change from the normal amino acid composition was suspected. This was confirmed and the nature of the change identified by V. M. Ingram in 1958 by application of the simple but powerful technique of peptide mapping, or "fingerprinting" as it has come to be called. When a protein molecule is attacked by a proteolytic

enzyme (e.g. trypsin) a number of peptide fragments of various sizes are produced. These fragments can be separated by chromatography or electrophoresis and, since the specificity of proteolytic enzymes ensures that their action is reproducible (whereas the effects of partial acid or alkaline hydrolysis are not), a given protein gives rise under these conditions to a characteristic peptide pattern or map. The peptide map of normal haemoglobin produced by two-dimensional paper chromatography and high-voltage electrophoresis of a tryptic digest was compared by Ingram with the corresponding map derived from sickle-cell haemoglobin and it was seen that the difference resided in a single peptide. Subsequent analysis of this peptide showed that a substitution of one amino acid for another, valine for glutamic acid at position 6 of the β-chain, out of nearly 300 amino acids in the two chains, is the cause of the abnormal properties of sickle-cell haemoglobin.

Some 200 variants of human haemoglobin resulting from specific alterations in the amino acid sequences of the molecule have now been described. Some of these have no apparent pathological effect but others carry with them a serious physiological handicap, in the form of reduced stability of the haemoglobin molecule, shortened red-cell survival and haemolytic anaemia. The disadvantageous mutations illustrate the fundamental nature of the principle that protein molecules consist of a hydrophobic interior and hydrophilic exterior. Substitution of a charged amino acid residue for a non-charged one in the interior regions would probably produce a completely non-functional molecule, but even the replacement of one non-polar side-chain by another of different dimensions can so distort the molecule that it becomes unstable and has an altered affinity for oxygen.

An example of the potentialities of the fingerprint method in the analysis of genetically determined enzyme variants has been provided in the case of glucose-6-phosphate dehydrogenase. A change of a single amino acid (asparagine to aspartic acid) has been shown to account for the differences between the normal enzyme and a variant of it (G6PD A⁻) which occurs in negro populations. Peptide mapping is a fairly microscale technique but nevertheless one which requires a pure sample of the protein under study; it is this requirement that limits its rapid extension to human material, which may be obtainable in only small amounts, e.g. in the form of biopsy specimens.

Significant variations in the structure and properties of an enzyme can also arise from genetic mutations and in other ways without resulting in obvious disease. Before considering the nature and origins of multiple molecular forms of enzymes, however, the normal process of protein biosynthesis will be outlined. The dramatic growth of understanding of this process which has taken place in the last decade has been widely recorded, so that only a brief recapitulation is needed here.

BIOSYNTHESIS OF PROTEINS

Early views on this process included the possibility that a reverse of proteolysis might take place, with peptide fragments being linked by proteases. However, the limited specificity of proteases and the energy-barriers which oppose their action in reverse rendered these hypotheses untenable. The modern phase of study of protein

so that the process of protein biosynthesis could be seen as one which involves the assembly of amino acids in a specific linear sequence. This in turn suggested the concepts of a template for each protein and a code for each amino acid, and the reality of these concepts has now been repeatedly demonstrated in many kinds of cells. The templates are now identified with several forms of nucleic acid present in the cell while the sequence of purine and pyrimidine bases in the nucleic acids codes for individual amino acids. Although some details of the process have still to be worked out (notably the complex events which take place in the ribosome during translation of the RNA message, and the way in which the whole process is regulated), the main events in mammalian cells can be summarized as in Fig. 3.2. The specific linear sequence of bases in nuclear DNA (the structural gene for that protein) is copied as a complementary strand of messenger RNA, to which the ribosome

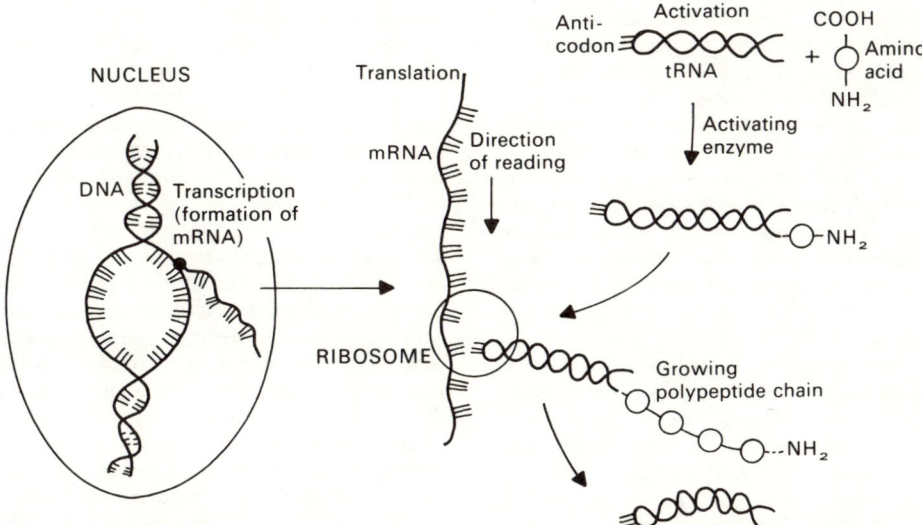

Fig 3.2. Summary of the events in protein biosynthesis in animal cells. The scheme is much simplified, particularly with respect to the events taking place in the ribosome during translation and chain-elongation. (mRNA, messenger RNA; tRNA, amino acid-specific transfer RNA.)

biosynthesis dates from the use of isotopically-labelled amino acids shortly after the Second World War; these gave proof that individual amino acids are the precursors of protein molecules and demonstrated the involvement of sub-cellular fractions (microsomes) and energy-rich compounds in protein biosynthesis.

During the same period the nature of the primary structure of proteins was revealed,

becomes attached during the process of translation. Several ribosomes are engaged in translating the message at any moment, each with an attached polypeptide chain in process of completion. Individual amino acid molecules enter the process after being "activated" by attachment to specific transfer-RNA molecules.

Armed with a knowledge of the structure of enzyme molecules and of the way in which they are made in the living cell, it is possible to discuss the nature and origin of diversity in enzyme molecules.

MULTIPLE MOLECULAR FORMS OF ENZYMES: ISOENZYMES

It has been recognized for many years that enzymes exist which have similar catalytic functions, but which can nevertheless be distinguished from each other by other criteria such as solubility, molecular weight or response to inhibitors or activators. Classically the different enzymes have been prepared from distinct sources, e.g. from yeast cells on the one hand and animal cells on the other. Differences between yeast and liver alcohol dehydrogenases are well established, for example. Similarly, variations between functionally similar enzymes prepared from different tissues or organs of a single species have been shown to exist, such as those which differentiate the acid phosphatases of human prostate and red blood cells. However, recent improvements in techniques of separating mixtures of similar proteins, particularly in ion-exchange chromatography and electrophoresis, have shown that a given type of catalytic activity may be associated with more than one form of protein molecule, even within a single cell, and the term "isoenzymes" has been introduced to describe this phenomenon. Opinion is not unanimous at the present time as to where the line should be drawn between multiple forms of enzymes which can be classed as isoenzymes on the one side, and distinct enzymes of similar or overlapping substrate-specificities on the other. Most authorities confine the use of the term "isoenzymes" to catalytically similar forms of an enzyme occurring within a single species, however, while some prefer an even more restricted use, applying it only to the separate forms of a single enzyme within a single tissue. The former, more general use is employed here.

There are many different ways in which isoenzymes may arise but these can be broadly classified into two categories, genetic and non-genetic, though each of these categories is itself heterogeneous.

GENETIC CAUSES OF ISOENZYMES

At the genetic level (Table 3.3) isoenzymes may arise due to the occurrence of multiple gene loci and also due to the occurrence of multiple alleles at a particular gene locus. A noteworthy difference between the two causes is that the isoenzymes attributable to multiple loci are usually common to the species as a whole, whereas isoenzymes due to multiple alleles have a more restricted distribution within the species and can be regarded as truly individual characteristics. The situation may be more complicated when an enzyme consists of two or more polypeptide chains since hybrid isoenzymes may occur, consisting of unlike polypeptides determined by different gene loci or by different alleles.

These possibilities have been very thoroughly explored with regard to many different enzymes in man and in other species and some examples will serve to illustrate the general picture of isoenzyme formation.

(i) Lactate dehydrogenase (LDH)

About 15 years ago it was discovered in several laboratories that LDH activity was not the property of one type only of enzyme protein, but that up to five different LDH molecules existed in human and other animal tissues and these isoenzymes could be separated by a variety of means, such as electrophoresis. Furthermore, it was found that the relative proportions of the five isoenzymes were to some extent characteristic of the tissue from which they were extracted so that three main groups of tissues could be recognized: those with predominantly isoenzymes 1 and 2 (e.g. heart muscle, red blood cells), those with predominantly isoenzymes 4 and 5 (e.g. liver, skeletal muscle), and a group in which there is an excess of the intermediate isoenzyme forms. The existence of some degree of tissue-specificity in the distribution of LDH isoenzymes has proved useful in determining the origin of LDH activity leaking into the blood in various diseases (Chapter 9).

The isoenzymes of LDH differ in many respects, apart from the differences in net charge which form the basis of their separation by electrophoresis or chromatography (Table 3.4), and structural studies on the separated isoenzymes have shown that they originate by combination of two different polypeptide subunits in groups of four to give active LDH molecules (Fig. 3.3). The H-type sub-unit (prepared from pig heart) contains a greater proportion of the amino acids glutamic, aspartic, threonine, serine and valine than the M-type polypeptide

TABLE 3.3 POTENTIAL SOURCES OF MULTIPLE MOLECULAR FORMS OF ENZYMES ARISING AT THE GENETIC LEVEL

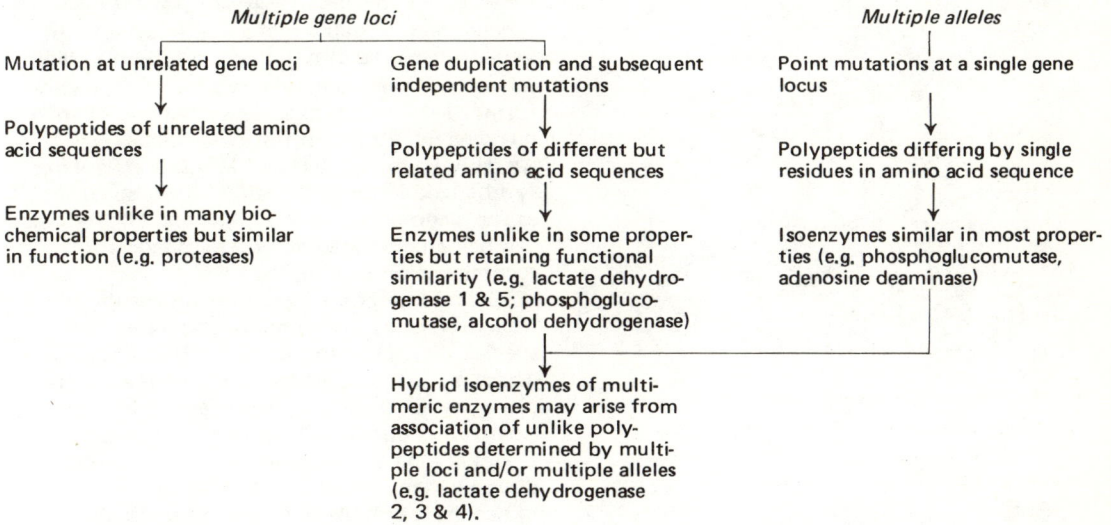

TABLE 3.4 PROPERTIES OF MAJOR ISOENZYMES OF LACTATE DEHYDROGENASE

	LD1	LD2	LD3	LD4	LD5
Separation (depends on net charge differences)					
(a) electrophoresis at pH 8·6	Most anodal	Decreasing order of anodal mobility			At origin or slightly cathodal
(b) anion-exchange chromatography	Most strongly retained	Increasing ease of elution			Most readily eluted
Stability	Most resistant to denaturation by heat or urea	Increasing resistance to denaturation			Most readily denatured by heat or urea
Catalytic properties	Km (lactate) 1×10^{-5} M	Intermediate values			Km (lactate) $2 \cdot 5 \times 10^{-5}$ M
	Ratio (pyruvate: oxobutyrate) reduction about 1	Intermediate values			Ratio (pyruvate: oxobutyrate) reduction > 3
	Most strongly inhibited by oxalate	Intermediate values			Least strongly inhibited by oxalate
Immunochemical properties	Not inhibited by "anti-5" antiserum	Increasing inhibition			Complete inhibition by "anti-5" antiserum
	Complete inhibition by "anti-1" antiserum	Increasing inhibition			Not inhibited by "anti-1" antiserum

Isoenzyme

(prepared from skeletal muscle or liver), which has more basic amino acids than the H-type. The two polypeptides can hybridize randomly with each other to form active tetramers, so that the proportions of the five isoenzymes in

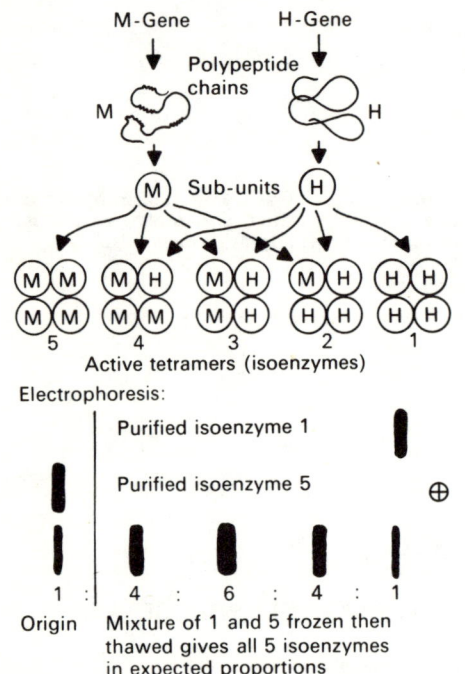

Fig 3.3. Schematic representation of structure of lactate dehydrogenase isoenzymes. Two distinct polypeptide chains are associated in different combinations of four to produce active enzyme tetramers (top). Dissociation and random reaggregation can be reproduced experimentally (bottom).

a particular cell or tissue depend on the relative abundance of the two types of sub-unit. This in turn appears to depend *in vivo* both on the rates of synthesis of the two forms and also on their rates of degradation. Thus, LD_5 is synthesized four times faster in rat liver than in heart, but is degraded only one-tenth as fast, so that the half-life of LD_5 in liver is ten times that in heart. The distribution of LDH and other iso-enzymes within particular tissues is often found to undergo change during pre- or post-natal development (Chapter 8).

The random association of H- and M-type sub-units to form heteropolymers can be demonstrated *in vitro* under suitable conditions, and hybridization between LDH sub-units from

different species can also be brought about, suggesting a considerable conservation of LDH structure during evolution. A further isoenzyme of LDH of intermediate electrophoretic mobility, isoenzyme x, occurs in mature testes of many species, and probably represents a tetramer of yet another type of polypeptide sub-unit, determined by a third genetic locus for LDH. Hybrid forms of this x polypeptide with H and M are not seen in extracts of testicular tissue, presumably because the H and M sub-units are not synthesized in the same cell or at the same time as the x sub-unit. However, MX and HX hybrids can be demonstrated *in vitro* by dissociation and recombination techniques.

Genetic variants of LDH M and H sub-units have been described in humans and in other species, giving rise to further LDH isoenzymes. These fall into the second group of the genetic classification of isoenzymes (Table 3.3), namely isoenzymes due to the occurrence of multiple alleles at a particular gene locus. Isoenzymes due to variant alleles at the locus which determines the x-polypeptide have been described in some species.

(ii) Phosphoglucomutase (PGM)

Another example of an enzyme determined by multiple gene loci is phosphoglucomutase. Three different sets of PGM isoenzymes can be separated from human tissues by electrophoresis and can be attributed to the existence of three separate structural gene loci. As well as being electrophoretically distinct, the three sets of isoenzymes differ in molecular size, substrate specificity and stability to heat.

Allelic variation occurs at each of the three loci and gives rise to yet further electrophoretic heterogeneity in this enzyme (Fig. 3.4). However, unlike the previous example, phosphoglucomutase appears to be a monomer and hybrid isoenzymes of the type seen with LDH do not seem to occur.

NON-GENETIC CAUSES OF ISOENZYMES

Certain multiple molecular forms of enzymes do not depend apparently on genetically determined alterations in primary structure. These secondary, or derived, isoenzymes are in general not as significant or as interesting as isoenzymes of genetic origin, but recognition of their existence can be important in the interpretation of isoenzyme patterns.

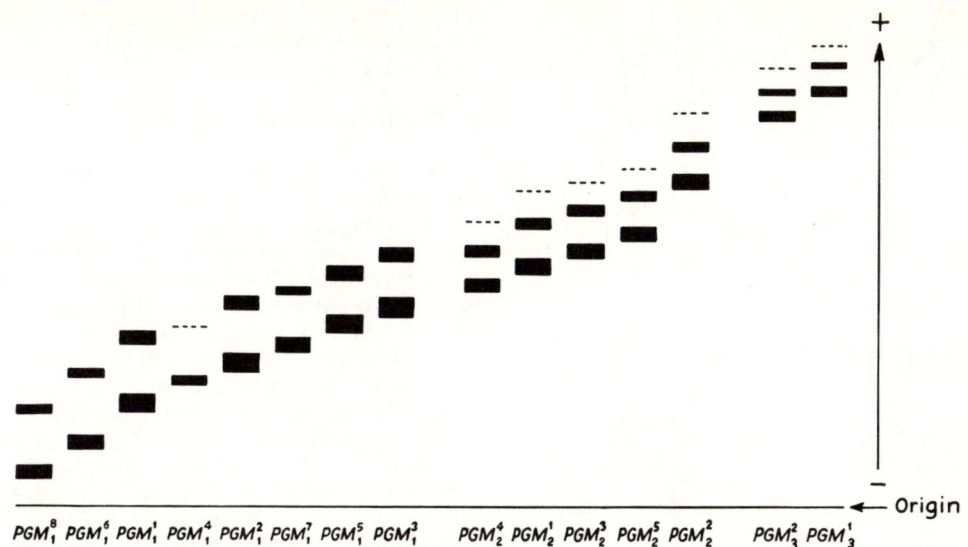

Fig 3.4. Diagram of a starch-gel electrophoresis pattern showing the isoenzymes of phosphoglucomutase (PGM) determined by eight different alleles at the PGM_1 locus, five different alleles at the PGM_2 locus and two different alleles at the PGM_3 locus. (From D. A. Hopkinson and H. Harris (1968) *Ann. Hum. Genet.* **31**, 359 by courtesy of the authors and Cambridge University Press.)

Some secondary isoenzymes may result from alterations in the covalent structure of the parent enzyme, with consequent changes in net charge or other properties. Thus, deamination is thought to account for some of the isoenzymes of carbonic anhydrase, and oxidation of sulphydryl groups has been shown to be responsible for the changes in isoenzyme patterns of red cell acid phosphatase and adenosine deaminase when haemolysates are stored *in vitro*. Some forms of yeast hexokinase may be due to the action of proteolytic enzymes on a single enzyme form. In man, proteolysis in the plasma alters the normal haemoglobin A to the variant haemoglobin Koellicker in intravascular haemolysis. Combination with different amounts of non-protein material may account for some isoenzymic forms: alkaline phosphatases are glycoproteins, and the presence of various amounts of sialic acid in the molecule contributes wholly or partly to the range of electrophoretic forms of this enzyme in tissues such as kidney.

Non-covalent modification of enzymes may include aggregation of enzyme molecules with each other to form complexes — this can happen with alkaline phosphatase, forming slowly-migrating zones on starch-gel electrophoresis.

Binding of small molecules may also alter molecular charge.

Conformational isomerism has been suggested to account for the existence of some sets of isoenzymes. Although a specific amino acid sequence (primary structure) is assumed to be associated typically with a single, most stable three-dimensional conformation into which the polypeptide chain becomes folded, it is possible that several distinct conformations of a single chain of nearly equal stabilities may exist. This explanation has been invoked in the case of the isoenzymes of mitochondrial malate dehydrogenase. For this explanation to hold, reversible denaturation of a single, isolated isoenzyme should generate the complete set, since the different configurations would be assumed randomly (Fig. 3.5). The respective properties of the isoenzymes would be due to the different exposed amino acid side-chains. While the conformer hypothesis is an attractive one, it has not proved easy to demonstrate reproducibly the expected isoenzyme interconversions and no fully established example of this phenomenon has yet been described.

Non-genetic isoenzymes could also arise from errors in copying the DNA code into the form of messenger RNA and in the translation of the

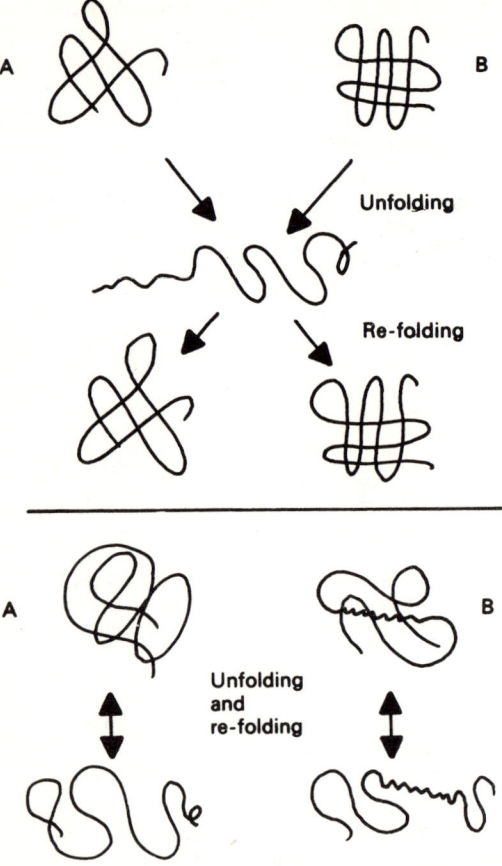

Fig 3.5. Possible formation of conformational isoenzymes ("conformers") when two stable forms A and B are separated by an energy barrier (top). Structurally different enzyme variants regain their original conformations on re-folding (bottom).

message into protein by the ribosomes. However, protein biosynthesis appears to be a remarkably accurate process and an error-rate of not more than one amino acid residue in 3000 has been estimated.

Secondary isoenzymes appear in red blood cells *in vivo* as a consequence of ageing. These cells lose their nuclei and the power to synthesize protein early in their relatively long life of about 120 days. As the cells age, alterations in the electrophoretic mobilities of some red-cell enzymes, e.g. pyruvate kinase can be observed. Changes in pH dependence and kinetic properties also occur with this enzyme. Although the mole-

cular changes underlying these modifications are unknown, deamidation or partial proteolysis may take place.

FUNCTION OF ISOENZYMES

In many cases it is not possible to decide whether the existence of multiple enzyme forms confers any advantage on the organism, or whether these isoenzymes are merely evolutionary accidents perpetuated from one generation to the next with no pressure of natural selection favouring or opposing their existence. Many of the products of allelic modification appear to come into this latter category. On the other hand, certain modified enzymes have disastrous consequences for their host. For some isoenzymes, e.g. those of LDH or aldolase, their tissue-specific localization and changes in relative proportions during development indicate an adaptation to changing metabolic patterns. The slow isoenzyme of LDH, LD_5, made up of four M-type sub-units, shows catalytic properties which seem to be more suitable than those of the H-isoenzyme for catalysis in an environment in which substrate (pyruvate) may accumulate. This is more likely to occur under anaerobic conditions, and those tissues which have an anaerobic pattern of metabolism (e.g. some types of skeletal muscle) do appear to be rich in LD_5. Similarly, aldolase B is active towards fructose-1-phosphate and therefore is adapted to gluconeogenesis, an important process in the liver in which this isoenzyme is abundant, whereas the predominant muscle isoenzyme, A, is more active in splitting fructose-1, 6-diphosphate, i.e. its action is appropriate to glycolysis (Chapter 8).

A regulatory role for LDH isoenzymes has also been suggested, since the presence of certain tricarboxylic acid cycle intermediates (e.g. citrate, isocitrate or α-oxoglutarate) changes the shape of the Michaelis curve of LD_5 from sigmoidal to hyperbolic, while not affecting the shape of the curve given by LD_1 (Chapter 7).

Differences in properties between cytoplasmic and mitochondrial isoenzymes also appear to represent adaptations to the different functional needs of the several subcellular compartments. The possible functions of the mitochondrial and extra-mitochondrial isoenzymes of malate dehydrogenase and aspartate transaminase are discussed in Chapter 7.

Enzyme function is now seen to be intrinsic

to the structure of the protein molecules concerned, which itself is of several levels of complexity. Some variation in the structure of a particular type of enzyme molecule is possible while still retaining broadly the same type of catalytic activity, and this limited variation gives rise to isoenzymic forms; for a given enzyme these may have a tissue-specific distribution which may be exploited for diagnostic purposes. The structure of enzyme molecules, like that of all other proteins, is determined genetically. Mutation produces specific alterations in enzyme molecules which may modify certain of their properties without impairing their physiological effectiveness, thus producing isoenzymic variants, but mutation may also result in the appearance of ineffective enzyme molecules giving rise to diseases of a wide range of severity, depending on the metabolic role of the enzymes concerned.

4 Factors Affecting Enzyme Activity

The speed at which a chemical reaction proceeds is markedly influenced by the prevailing conditions. The concentrations of the reactants, the nature, pH and temperature of the reaction medium, and whether or not a catalyst is present are all important factors in determining the reaction rate. The rate may range from a value that is so slow that virtually no reaction is detectable to one that is too fast to measure by the usual experimental techniques, i.e. the reaction is practically instantaneous. The study of rates of chemical reactions and of the effects of changes in conditions on them constitutes the subject of reaction kinetics.

A large volume of literature has been built up describing the kinetics of enzyme-catalysed reactions. Much of the work has resulted from the desire of enzymologists to understand as fully as possible the ways in which enzymes act as catalysts, but some knowledge is required of the kinetic parameters of a particular enzyme before its activity can be measured with confidence and reliability. Certain features of the kinetics of enzymic reactions will be discussed in this chapter together with their applications in enzyme assay procedures and their implications for the functioning of enzymes in living cells.

THE EFFECT OF SUBSTRATE CONCENTRATION ON REACTION VELOCITY

For a reversible chemical reaction represented by the equation

$$A \rightleftharpoons B$$

the rates of both the forward and reverse reactions are, by the Law of Mass Action, proportional to the concentration of the reacting species. If the proportionality constants for the forward and reverse reactions are k_{+1} and k_{-1} respectively, then:

the rate of the forward reaction = $k_{+1}a$
and the rate of the reverse reaction = $k_{-1}b$

where a and b represent the concentrations of A and B respectively. When chemical equilibrium is attained the rates in the forward and reverse directions will be equal, so that:

$$k_{+1}a = k_{-1}b$$
$$\text{or } \frac{a}{b} = \frac{k_{-1}}{k_{+1}} = K$$

where K is now the equilibrium constant of the reversible reaction between A and B.

Considering the rate in the forward direction only, this will decrease with time because the concentration of the reactant A decreases continuously as a result of its conversion to the product B (Fig. 4.1). The reaction rate can be expressed mathematically by a differential equation of the form:

$$\frac{-da}{dt} = k_{+1}a \qquad \text{(i)}$$

The differential term $\frac{-da}{dt}$ denotes the velocity of the reaction at any time t and includes a minus sign to indicate that the concentration a is decreasing. The proportionality constant, k_{+1} is usually known as the "rate constant" and has the dimensions of $(\text{time})^{-1}$. Since in this example only a single reacting species is involved, substance A, the reaction is said to be "first-order" with respect to reactants. It is a characteristic of first-order reactions that the velocity is proportional to the concentration of a single reactant and the expression of this statement by equation (i) gives the form of the rate equation for such a reaction. Reactions of higher orders, i.e. where two or more substances react together, or where several molecules of a single type combine are not uncommon and it is possible for the order to approach zero, i.e. for the rate of the reaction to become independent of the concentration of the reactants.

As has been mentioned already, the velocity of the reaction decreases with time as the reactant is converted to products. A progress curve obtained by plotting the concentrations

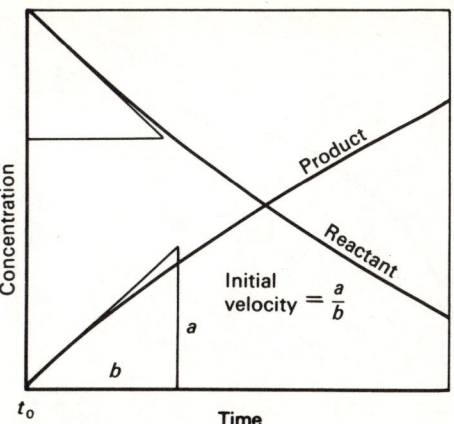

Fig 4.2. Change in concentrations of reactant and product with time. The tangents to the curves at t_0 give the "initial reaction velocity". Until a significant amount of reactant has been used up the fall in concentration of the reactant and rise in concentration of the product are linear with time.

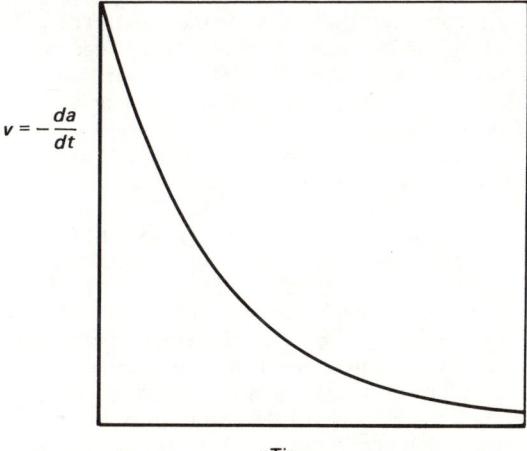

$$v = -\frac{da}{dt}$$

Fig 4.1. Variation of reaction rate with time. The reaction velocity V at any time t is given by $\frac{da}{dt}$, but since a is decreasing continuously as it is converted to products the reaction velocity also decreases.

of the reactant and product of a first-order reaction as a function of time is shown in Fig. 4.2. The velocity at any particular time t can be obtained by drawing tangents to the curve. The slope of the tangent at time $t=0$ gives the initial reaction velocity. This is the reaction rate corresponding to the concentration of reactant put in at the start of the reaction. Until approximately 20 per cent of the reactant has been converted to product there is very little deviation from a linear relationship between the amount of substrate broken down and time. Initial reaction velocities, i.e. those determined during the period of linearity are the most useful in enzyme work because the simple mathematical relationship makes interpretation of the effects of factors which modify the velocity relatively easy. Enzyme

activity is very susceptible to changes occurring in the medium so that progress curves constructed from observations over a lengthy time course can seldom be represented by a simple and reproducible mathematical equation. In the following discussions of the factors which affect the velocity of enzyme catalysed reactions, initial reaction velocities are implied unless otherwise stated.

If the concentration of an enzyme is held constant and measurements of initial reaction velocities are made over a wide range of substrate concentrations, the plot of the results for most enzymes will lie on a curve of the type shown in Fig. 4.3. At very low substrate concentrations the plot is almost linear, i.e. the velocity is proportional to the substrate concentration and the reaction is first-order with respect to substrate. At higher substrate concentrations, a limiting velocity is reached and the reaction rate is then independent of substrate concentration, i.e. the reaction has now become zero-order. In 1902, V. Henri obtained a curve of this type as a result of a study of the action of invertase on sucrose. To account for the shape of the curve he assumed that, prior to the breakdown of substrate, enzyme and substrate combined to form an intermediate complex that decomposed in a subsequent step to give rise to the products of the reaction. At very high concentrations of substrate

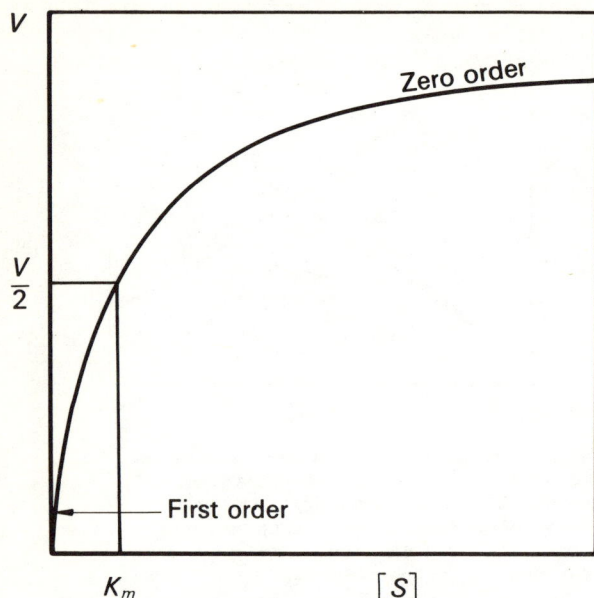

Fig 4.3. Dependence of reaction velocity v on substrate concentration $[S]$. The curve is a rectangular hyperbola and reaches a limiting velocity V at infinite substrate concentration. As the velocity approaches V the reaction becomes zero order with respect to $[S]$. At very low values of $[S]$, the velocity is first order with respect to $[S]$. When $v = \dfrac{V}{2}$, $[S] = K_m$

all the enzyme molecules would be present in the form of intermediate complexes, the concentration of which could not be increased further without the addition of more enzyme. Thus Henri considered that high substrate levels led to a saturated condition whereas at much lower concentrations enzyme saturation would be incomplete, and the concentration of the important intermediate complex would increase with the addition of more substrate. Although the existence of an enzyme-substrate intermediate was not proved for any enzymic reaction until 1943, when Chance demonstrated spectroscopically the formation of a complex in reactions catalysed by catalase and peroxidase, the concept was soon taken up by enzymologists after Henri's suggestion because it was difficult to see how a catalyst could increase the velocity of a reaction unless it, or some part of it, took part in the reaction itself. Subsequent experimental work has firmly established the soundness of the assumption of the formation of an enzyme-substrate complex as the basis of catalysis by enzymes.

An enzymic reaction may be represented as follows:

$$E + S \; \underset{k_{-1}}{\overset{k_{+1}}{\rightleftharpoons}} \; (ES) \; \xrightarrow{k_{+2}} \; \text{Products} + E$$

Only one molecule of substrate is bound at each region on the enzyme where catalysis occurs. If there are several of these regions or "active sites" then it is assumed that the occupation of one site by a molecule of substrate has no effect on the binding of substrate to the remaining sites. This assumption is not valid for every multi-site enzyme but discussion of the kinetics of reactions where interaction occurs between sites is left to Chapter 7. E, S, (ES) represent free enzyme (or total active sites) substrate and enzyme-substrate complex respectively, k_{+1} is the rate constant for the formation of (ES) and k_{-1} is the rate constant for the decomposition of (ES) to form the original reactants, i.e. E and S. The rate constant k_{+2} describes the decomposition of (ES) to form the products of the reaction. Michaelis and Menten (1913) assumed that equilibrium is rapidly attained between E, S and (ES), i.e. the effect of product formation on the concentration of (ES) is assumed to be negligible, and derived an equation relating the velocity of the reaction to substrate concentration.

Let the total concentration of enzyme active sites in the system be e_0 and total concentration of substrate be s. Then, if the concentration of (ES) *at equilibrium* is x, the amount of *free* enzyme in the system will equal $(e_0 - x)$. Provided that s is very much greater than e_0, the concentration of *free* substrate can be equated with s since the amount contained in (ES) will be a negligible proportion of the total.

The dissociation constant of (ES) K_s, can then be written:

$$K_s = \frac{(e_0 - x)s}{x}$$

$$\text{therefore } x = \frac{e_0 s}{K_s + s}$$

The measured velocity v of the overall reaction is determined by the rate of decomposition of the intermediate complex into products.

$$v = k_{+2}x$$

$$\text{so that } v = \frac{k_{+2}e_0 s}{K_s + s} \qquad (1)$$

The highest concentration of (ES) is reached when the enzyme is saturated with substrate, i.e.

all the enzyme is present in the form of the intermediate. When this condition applies:

$$v = k_{+2}e_0$$

and v will have its maximum value which can be denoted by the symbol V. Substitution into equation (1) gives

$$v = \frac{Vs}{K_s + s} \qquad (2)$$

For a particular set of conditions, K_s and V are constants and the equation is that of a rectangular hyperbola, the shape of the curve illustrated in Fig. 4.3.

The constant K_s is frequently referred to as the Michaelis constant and, since it is the dissociation constant of (ES), it is a measure of the binding strength or "affinity" of the enzyme for its substrate. The smaller is K_s the greater is the affinity of the enzyme for its substrate, since a small value for the constant implies that the equilibrium between E, S and (ES) lies far in the direction of (ES) formation.

If in equation (2) v is made equal to one half of the possible maximum velocity at a fixed enzyme concentration, i.e.

$$v = \frac{V}{2}$$

then rearrangement of equation (2) leads to the relationship:

$$K_s = s \qquad (3)$$

This is how the Michaelis constant is defined operationally. It is the substrate concentration that is required for the reaction to proceed at half its maximum velocity. As equation (3) and the definition imply, K_s has the dimensions of concentration and its approximate value may be deduced from velocity – substrate curves (Fig. 4.3.). Better methods of obtaining values for the Michaelis constant from experimental data will be discussed later in the chapter.

Two points of interest arising from the general form of the Michaelis-Menten equation are illustrated by first making s very large compared to K_s and second, making s very much smaller than K_s. In the first instance equation (1) simplifies to:

$$v = k_{+2}e_0 \qquad (4)$$
$$\text{or } v = V$$

This is the condition of saturation of the enzyme considered by Henri. The velocity of the reaction is directly proportional to the concentration of the enzyme and is of zero order with respect to substrate. Equation (4) is an important relationship for it is the basis of all methods of enzyme assay. The catalytic actions of enzymes are usually demonstrable at concentrations of enzyme that are very small in terms of molarity and almost invariably it is a much easier task to measure the velocity of an enzyme catalysed reaction than to determine the absolute concentration of enzyme. Thus enzyme "concentration" is usually expressed in terms of the reaction rate at saturating levels of substrate. When K_s is large compared to s equation (1) becomes:

$$v = \frac{k_{+2}e_0 s}{K_s}$$

and the reaction velocity is proportional to the concentration of substrate. The reaction is first-order with respect to substrate and is represented by the part of the curve close to the origin in Fig. 4.3.

STEADY STATE THEORY

Although the Michaelis-Menten equation provided an excellent fit for data obtained from a number of different enzymes, Briggs and Haldane in 1925 considered that certain of the assumptions made in the derivation of the equation were unjustified in the light of what was known about enzymic catalysis. Because many enzymes have very high catalytic activities, it seems unlikely that (ES) will necessarily remain in equilibrium with E and S for an appreciable interval during the reaction. Instead, the complex may perhaps be transformed to products so rapidly that equilibrium is never established. Briggs and Haldane did not assume an obligatory equilibrium step but derived a rate equation on the basis of an assumption of a "steady-state" condition. The "steady state" refers to the concentration of (ES) which is assumed to remain constant during the period of time in which initial velocity measurements are taken. The steady state implies that the rates of formation and decomposition of (ES) are exactly equal although E, S and (ES) need not be in equilibrium. Before considering the steady-state hypothesis it is necessary to consider all the rate equations of the reaction system:

$$E + S \underset{k_{-1}}{\overset{k_{+1}}{\rightleftharpoons}} (ES) \xrightarrow{k_{+2}} \text{Products} + E$$

Let e_0, s and x have the same meaning as before. The rate of disappearance of substrate, $\dfrac{-ds}{dt}$, includes two terms; $k_{+1}(e_0 - x)s$ arising from the formation of the (ES) complex and $k_{-1}x$ from the breakdown of (ES). Therefore:

$$\frac{-ds}{dt} = k_{+1}(e_0 - x) - k_{-1}x \qquad (5)$$

The change in concentration of (ES) with time can be written:

$$\frac{dx}{dt} = k_{+1}(e_0 - x)\,s - (k_{-1} + k_{+2})x \qquad (6)$$

The first term in equation (6) derives from considering the formation of (ES) from E and S while the second term arises from the breakdown of (ES) into E plus S together with decomposition of (ES) to form products. The rate of formation of product is given by $\dfrac{dp}{dt}$ where p is the concentration of product. This is the velocity v of the overall reaction.

$$\frac{dp}{dt} = v = k_{+2}x \qquad (7)$$

The equations (5), (6) and (7) should, in principle, be solvable for three unknowns e_0, s and x. Where there is no restriction in the relative magnitudes of k_{+1}, k_{-1} and k_{+2}, the relative concentration s and the time t when the velocity of the reaction is to be established however, the solution is impossible. The equations can be integrated numerically with the aid of a computer by assigning particular values to each of the unknowns and the rate constants. The change in concentration of the reacting species with time can then be calculated and plotted as shown in Fig. 4.4. When k_{+1}, k_{-1} and k_{+2} are approximately equal and s is greater than e_0, Fig 4.4 shows that over a finite period of time, indicated approximately by the vertical lines, $\dfrac{dx}{dt} = o$ and hence from equation (6);

$$k_{+1}(e_0 - x)s - (k_{-1} + k_{+2})x \qquad (8)$$

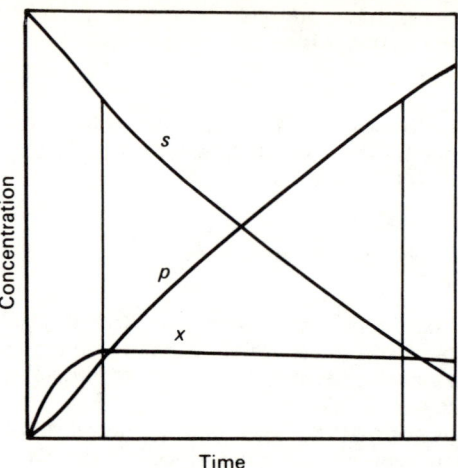

Fig. 4.4. **Progress curves for the reaction** $e + s \underset{k-1}{\overset{k+1}{\rightleftharpoons}} x \xrightarrow{k_{+2}} e + p$, **in which** $k_{+1} \sim k_{-1} \sim k_{+2}$. **For a considerable portion of the reaction the concentration of** x **is almost constant, i.e.** $\dfrac{dx}{dt} = 0$ **and a "steady state" level of** x **is maintained.**

i.e. the rates of formation and breakdown of (ES) are equal. This is an algebraic instead of a differential equation, and it allows a solution to be found for the rate of the overall reaction.

The period during which the rates of formation and breakdown of the (ES) complex are exactly balanced corresponds to a steady state, the existence of which was assumed by Briggs and Haldane in their method for the derivation of the rate equation. The time period required to allow steady-state conditions to be established is very short and measurements within this period, although they can be very useful for enzymologists, require detection devices sensitive to changes occurring within milli- or even nanoseconds. In practice, therefore, using ordinary methods of enzyme assay, pre-steady-state changes lie outside the means of detection and steady-state conditions can be assumed to prevail immediately after the mixing of the enzyme and substrate.

From equation (8):

$$x = \frac{k_{+1}\,e_0 s}{k_{+1}s + k_{-1} + k_{+2}} \qquad (9)$$

By making use of equation (7)

$$v = \frac{k_{+2} \cdot k_{+1} \cdot e_0 \cdot s}{k_{+1}s + k_{-1} + k_{+2}}$$

Dividing through by k_{+1}

$$v = \frac{k_{+2}e_0 s}{s + \dfrac{k_{-1} + k_{+2}}{k_{+1}}}$$

Substituting K_m for $\dfrac{k_{-1} + k_{+2}}{k_{+1}}$ and V for $k_{+2}e_0$,

$$v = \frac{Vs}{K_m + s} \tag{10}$$

Comparison of this equation with the Michaelis equation (equation 2) shows them to be of identical form. The physical significance of K_s and K_m differs, however. K_m is not a dissociation constant and is thus not a measure of substrate affinity unless k_{+2} is very small compared to k_{-1}. This is an important point to remember because in the large majority of cases where K_m values are deduced from the relatively simple experiment of determining the dependence of v on the substrate concentration, it is impossible to decide exactly whether (ES) is in equilibrium with E and S. Further data are usually needed before K_m can be equated with the dissociation constant for (ES). The change of symbol from K_s to K_m has been deliberately. In any future discussion, when K_m is used no restriction is implied whereas K_s is retained for the special case where equilibrium conditions are assumed to apply.

Many, if not all, enzyme reactions proceed via the formation of more than one intermediate complex. For example, a reaction may take the following form.

$$E + S \underset{k_{-1}}{\overset{k_{+1}}{\rightleftharpoons}} (ES)_1 \underset{k_{-2}}{\overset{k_{+2}}{\rightleftharpoons}} (ES)_2 \overset{k_{+3}}{\longrightarrow} \text{Products} + E$$

Each step is characterized by a forward and backward rate constant and a rate equation for the overall reaction can be derived assuming steady-state levels of $(ES)_1$ and $(ES)_2$ are maintained throughout. The rate law for the example just quoted becomes:

$$v = \frac{\dfrac{k_{+2}k_{+3}e_0 s}{k_{+2} + k_{-2} + k_{+3}}}{s + \dfrac{k_{-1}(k_{-2} + k_{+3}) + k_{+2}k_{+3}}{k_{+1}(k_{+2} + k_{-2} + k_{+3})}} \tag{11}$$

$$= \frac{k_0 e_0 s}{K_m + s}$$

where k_0 and K_m represent the combinations of rate constants in the numerator and denominator of equation (11). The example illustrates how a rate law conforming to the Michaelis-Menten pattern is generated no matter how many intermediate steps are interpolated in the reaction mechanism. The value of K_m can be obtained from velocity-substrate concentration data but it should be noted, however, that K_m is far removed from a dissociation constant and from a measure of binding affinity. In certain cases, however, the relative values of the individual rate constants may be such that K_m approximates to $\dfrac{k_{-1}}{k_{+1}}$, i.e. the binding constant for substrate, but to test whether such a condition applies for a particular enzyme is not usually a simple matter.

The maximum velocity V is equal to $k_0 e_0$. The constant k_0 is sometimes referred to as the "catalytic rate constant" since it is associated with the rate of the overall catalytic process. If the molarity of the enzyme in the reaction mixture is known, then k_0 can be calculated from $\dfrac{V}{e_0}$.

In the model reaction schemes discussed so far, the reaction leading to the formation of products has been assumed to be irreversible. All catalysts, enzymes included, bring about a rate enhancement of both the forward and reverse reactions of potentially reversible processes. Thus in the presence of a finite amount of product, the rate law for a reversible reaction will be modified because of the reverse reaction. Rate measurements are usually made, however, in the absence of products and the reactions terminated before a significant proportion of the substrate has been converted. In these circumstances the simple Michaelis equation is adequate to describe the rate law.

REACTIONS INVOLVING MORE THAN ONE SUBSTRATE

Many enzyme-catalysed reactions involve the participation of more than one substrate and a bisubstrate reaction, for example, may be represented in a general form by

$$A + B \rightleftharpoons C + D$$

In many instances one of the substrates is water, i.e. the reaction is a hydrolytic process. If the hydrolysis occurs in an aqueous medium, the number of molecules of water that participate in the reaction itself is infinitely small compared to the total number present, and the molarity of water remains constant during the reaction. As a consequence, the part played by water in the reaction is kinetically unimportant and the reaction can be considered as being *pseudo—first-order* with respect to the other substrate. A large number of enzymes, dehydrogenases for example, require a coenzyme before they can function (see Chapter 5). Such reactions are true representatives of bisubstrate processes.

Consider a bisubstrate reaction in which the enzyme combines with the first substrate A to form an enzyme-substrate complex (EA) that in turn combines with the second substrate B to yield a ternary complex (EAB) leading to the release of the products C and D. The reaction sequence can be represented by

$$A + E \rightleftharpoons (EA) + B \rightleftharpoons (EAB) \rightleftharpoons C + D$$

Applying the restrictions that the intermediates reach a steady-state concentration and that the initial concentration of the products is zero, a rate law may be derived for the forward reaction. For this example it takes the form of:

$$v = \frac{Vab}{ab + bK_m^A + aK_m^B + K_s^A K_m^B} \quad (12)$$

$$\text{or } v = \frac{V}{1 + \dfrac{K_m^A}{a} + \dfrac{K_m^B}{b} + \dfrac{K_s^A K_m^B}{ab}} \quad (13)$$

where a and b are the concentrations of A and B, K_m^A and K_m^B are the "Michaelis constants" for each substrate and K_s^A is the equilibrium constant for the reaction in which (EA) is formed from E and A. The Michaelis constants and V include a group of rate constants descriptive of various steps in the reaction sequence. If the concentration of one of the reactants, B, say, is made infinite then equation (13) simplifies to

$$v = \frac{V}{a + K_m^A} \quad (14)$$

Similarly for high levels of A,

$$v = \frac{V}{b + K_m^B} \quad (15)$$

Thus bisubstrate reactions can usually be described by a Michaelis-Menten rate law if the concentration of one of the substrates is maintained at a saturating level. K_m^A is the concentration of A required, when B is present in excess, for the reaction to proceed at half the maximum velocity and K_m^B is given by the concentration of B at half-maximum velocity when the system is saturated with A.

GRAPHICAL METHODS OF DETERMINING K_m AND V

To ensure that the enzyme is saturated with substrate during routine assays a knowledge of K_m is essential. A concentration of substrate is chosen that is approximately ten times the K_m value so that the velocity is as close to the theoretical maximum as possible: in practice theoretical values of V are rarely attained, even at very high substrate concentrations. It might be thought that any large excess of substrate could suffice to saturate the enzyme, but many enzymes are inhibited by excess amounts of substrate especially those that are concerned with the regulation of metabolic pathways *in vivo* (Chapter 7).

Experimental data may be plotted in the form of v against s to allow K_m and V to be estimated, but for most purposes plotting data according to linear transformations of the Michaelis-Menten equation is preferable because it is difficult to make an exact measurement of V from a curve such as the one illustrated in Fig 4.3. A popular method, usually attributed to Lineweaver and Burk (1934) consists of plotting $\frac{1}{v}$ against $\frac{1}{s}$. Rearrangement of the Michaelis equation in reciprocal form gives the equation:

$$\frac{1}{v} = \frac{K_m}{V} \cdot \frac{1}{s} + \frac{1}{V} \quad (16)$$

Thus a graph of $\frac{1}{v}$ against $\frac{1}{s}$ yields a straight line whose slope equals $\frac{K_m}{V}$. The intercept on the $\frac{1}{v}$ axis corresponds to $\frac{1}{V}$ and the negative intercept on the $\frac{1}{s}$ axis equals $\frac{-1}{K_m}$ (Fig 4.5.2). Multiplication of both sides of equation (16) by s gives an equation that is the basis of a second graphical procedure for K_m and V that was suggested by Hanes (1932):

$$\frac{s}{v} = \frac{K_m}{V} + \frac{s}{V} \qquad (17)$$

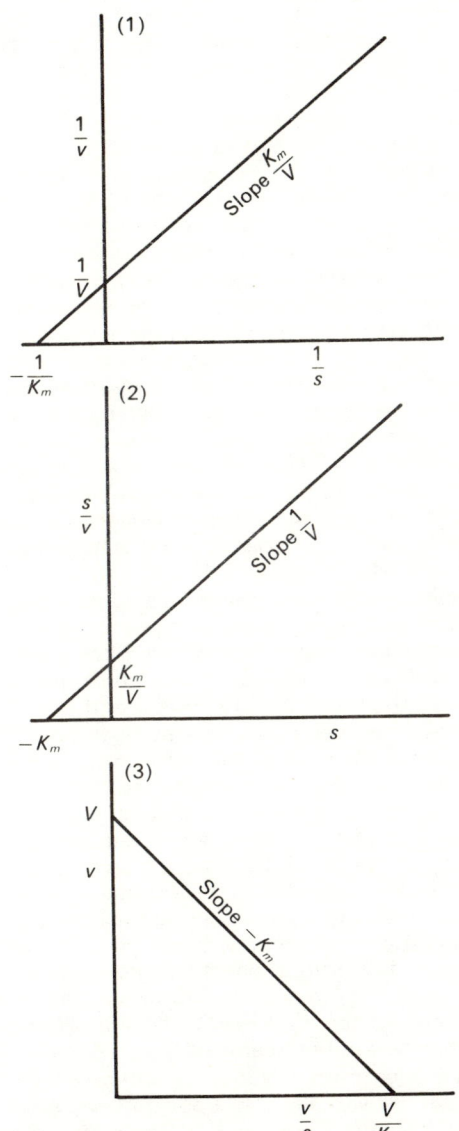

Fig 4.5. Plots of linear transformations of the Michaelis-Menten equation used to facilitate the determination of K_m and V from experimental data.

A straight line results from plotting $\frac{s}{v}$ against s; the slope is $\frac{1}{V}$ and the intercept on the vertical axis gives $\frac{K_m}{V}$. The intercept on the s axis equals $-K_m$ (Fig 4.5.2).

A third method of plotting was recommended by Hofstee. The Michaelis-Menten equation is rearranged to the form:

$$v = V - K_m \cdot \frac{v}{s} \qquad (18)$$

When v is plotted against $\frac{v}{s}$, the resulting straight line has a slope of $-K_m$, an intercept on the vertical axis at V, and one on the horizontal axis at $\frac{V}{K_m}$ Fig. 4.5.3).

The Lineweaver-Burk method has been used most widely but has been criticized on the grounds that too much reliance is placed on data obtained at the lowest substrate concentrations. Measurements of the relatively slow velocities at the lower substrate concentrations will be more subject to error than those taken at higher substrate concentrations. Because reciprocals are taken, the more reliable data are crowded close to the origin and have a smaller effect on the position of the best straight line drawn to fit the experimental points. The $\frac{s}{v}$ against s plot is free from this particular objection but does tend to overemphasize the higher-substrate data. For reliable estimates of K_m, a range of substrate concentrations is chosen such that K_m lies somewhere in the middle of the range. The third method gives equal weight to all data but seems to have remained unpopular because the plot does not make immediately obvious the relationship between v and s. Computer programs are now available for weighting individual observations according to their reliability and when this is done all three methods are equally valid and thus there seems little point in abandoning the Lineweaver-Burk and Hanes plots, particularly as their appearance is familiar to most biochemists. Also, for the interpretation of data collected in the presence of enzyme inhibitors they tend to be more useful than the Hofstee plot.

For a particular set of experimental conditions K_m values provide a means of characterizing enzymes, e.g. enzymes of similar catalytic function may be present in several different organs and tissues in the body and comparison of K_m values can help to decide whether the different tissue enzymes are identical. The several forms of lactate dehydrogenase which occur in human and other mammal tissues and which all catalyse the inter-conversion of lactate and pyruvate have been shown to have different Michaelis constants for lactate, and also for pyruvate when the enzymes are catalysing the reverse reaction. Similarly, alkaline phosphatases from different organs do not have identical K_m values. These differences in Michaelis constants may sometimes be very small and require careful measurement to establish them. When such is the case, K_m measurements made on the enzyme present in serum, for example, in order to identify the tissue of origin, reach beyond the bounds of routine chemical pathology.

THE EFFECT OF pH ON THE RATE OF ENZYMIC REACTIONS

If the velocity of an enzyme-catalysed reaction is measured at a number of different pH values, a plot of the results is usually bell-shaped (Fig. 4.6). The curve may be symmetrical with equal

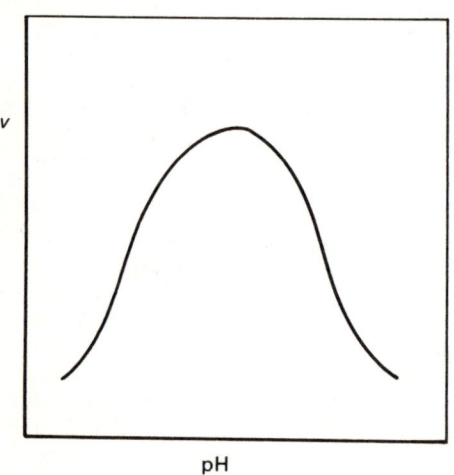

Fig 4.6. Effect of pH on enzyme activity.

rise and fall on both acid and alkaline sides, or, in some cases, may decline rapidly on one side. The pH at which the activity is greatest is called the pH optimum. With some enzymes the optimum is very sharp but in others may extend over a range of two or more pH units (e.g. erythrocyte acid phosphatase, which has a very flat activity curve between pH 4—6). The optimum is not always close to neutrality although many enzymes do have optima at pH 6·5—7·5. The proteolytic enzyme pepsin which is normally present in the stomach of mammals is most active at pH 1—2, and there is a sharp fall in activity with increasing pH. The enzyme seems to be particularly well suited to act in the acid conditions that normally prevail in the stomach.

The experimentally determined pH optimum does not necessarily reflect the pH of the environment in which the enzyme is likely to act *in vivo*. In some instances where optima have been demonstrated at extreme pH values it seems unlikely that the cell contents could reach such values. For example, alkaline phosphatases from liver, bone, intestine, kidney and placenta are optimally active at pH 10, when saturating levels of substrate are present. Little is known about the effects of the micro-environment on the pH of a particular region of the cell or organelle so that it could be argued that certain regions of high and low pH may exist. From the general properties of biological materials, however, it is evident that wide deviations from neutrality are usually harmful and would thus seem unlikely to occur.

Bell-shaped curves arise from a number of different factors, all of which, however, are connected with the protein nature of enzymes. Extremes of pH are known to denature proteins. The secondary and tertiary structure of enzymes circumscribes their catalytic activity and disturbance of the three-dimensional shape of enzyme molecules nearly always results in loss of catalytic activity. Hence a decline in activity at high and low pH is not unexpected. Loss of activity as a result of denaturation can usually be tested for by returning the enzyme to its optimum pH and then measuring the activity. Denaturation is often irreversible, and even when reversible full restoration of activity usually takes some time to achieve. Thus if denaturation has occurred, the residual activity is usually much less than before the treatment at an extreme pH.

A change in activity over a pH range where the enzyme is totally stable obviously does not derive from denaturation. Proteins contain many ionizable groups and changing the pH of the environment alters the degree of ionization of the groups. An enzyme can exist in a number of different ionic species but the existence of a pH optimum implies that one particular ionic form of the enzyme is most active. Interactions between the substrate and charged groups on the enzyme are believed to be critical in the binding of substrate and in the catalytic process itself. Changes in the charge of groups participating directly in catalysis will be expected to have a greater effect on activity than changes in groups located at sites on the enzyme molecule distant from the point of substrate binding. Thus a study of the effects of pH on enzyme activity can allow the probable nature of the groups involved in catalysis to be inferred.

If an enzyme is represented by EH in its active state and it contains two groups important in catalysis one of which is protonated and the other unprotonated, changing the pH brings about the following ionizations:

$$
\begin{array}{lll}
\text{Acid} & K_1 \quad K_2 & \text{Alkaline side} \\
\text{side of} & HEH \rightleftharpoons EH \rightleftharpoons E & \text{of optimum} \\
\text{optimum} & &
\end{array}
$$

(both groups (both groups
protonated) unprotonated)

Let e_0, be the total enzyme concentration and K_1, K_2 the acid dissociation constants for the important catalytic groups, then

$$e_0 = [HEH] + [EH] + [E]$$

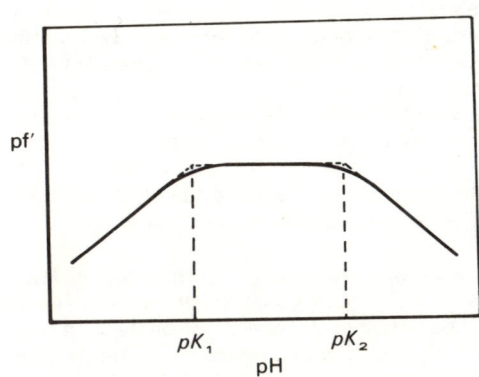

Fig 4.7. Plot of pf' against pH, where f' represents the Michaelis pH function $1 + \dfrac{[H^{+1}]}{K_1} + \dfrac{K_2}{[H^+]}$. $pf' = -\log f'$

$= \log \dfrac{1}{f'}$ The plot consists of three linear portions which intersect at pK_1 and pK_2.

and it can be shown that the fraction of the total enzyme present as the active form, EH, is given by the equation

$$[EH] = \frac{e_0}{1 + \dfrac{[H^+]}{K_1} + \dfrac{K_2}{[H^+]}} \qquad (19)$$

The term in the denominator of equation (19) is a pH function and was first derived by Michaelis. The relative importance of each term in the function depends on the H^+ ion concentration At high H^+ ion concentrations (low pH), $\dfrac{[H^+]}{K_1}$ predominates but at high pH, $\dfrac{K_2}{[H^+]}$ is dominant. When $\dfrac{[H^+]}{K_1} = \dfrac{K_2}{[H^+]}$ both terms are small compared to 1. If the negative logarithm of the whole function is plotted against pH a plot of 3 linear portions results (Fig. 4.7). Each portion corresponds to a region where only one term is important. Intersections of the straight lines occur at pK_1 and pK_2.

Assuming that the substrate S binds to EH, a reaction scheme that includes ionizations of reactive groups, substrate binding and the release of products can be represented as follows:

$$
\begin{array}{lcl}
& E & ES \\
K_2 \updownarrow \pm H^+ & & \pm H^+ \updownarrow K_2' \\
& EH + S \rightleftharpoons EHS \xrightarrow{\ k\ } \text{Products} + EH \\
K_1 \updownarrow \pm H^+ & & \pm H^+ \updownarrow K_1' \\
& HEH & HEHS
\end{array}
$$

The acid dissociation constants associated with the enzyme-substrate complex (K_1', K_2') are not necessarily identical to those of the free enzyme (K_1 and K_2). Involvement in substrate binding may change the proton accepting and donating tendencies of a particular group. At saturating levels of substrate the maximum velocity for the enzymic mechanism above is proportional to the fraction of the enzyme present as the most active form EHS. As shown for EH in equation (19) the fraction in the form of EHS is governed by a Michaelis pH function. Thus a plot of log V against pH consists of linear portions that intersect at pK_2' and pK_1' (Fig. 4.8). Similar arguments can be used to show that plots of $-\log K_m$ (i.e. pK_m) against pH have inflexions

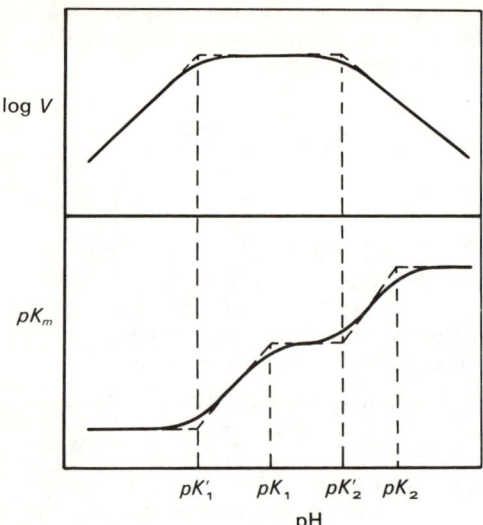

Fig 4.8. Effect of pH on K_m and V. The logarithmic method of plotting helps in the identification of ionizing groups in the enzyme and enzyme — substrate complex. Ionizations occurring in the free substrate will also show as downward bends in the pK_m plot. These can usually be identified from the known ionization characteristics of the substrate.

corresponding to pK_2 and pK_1 as well as pK_2' and pK_1' (Fig. 4.8). Treatment of data obtained for V and K_m at different pH values in this manner was suggested by Dixon and is widely used by enzymologists.

It may prove possible to identify the groups involved in substrate binding by comparison of the experimentally determined values for pK_1' and pK_2' with known values for reactive groups present in the side chains of some amino acids. The method suffers from the disadvantage that the pK of a group on an amino acid in free solution may be an unreliable guide to the value for the same group when present in a protein and subject to the influence of a number of neighbouring groups. Involvement in hydrogen bonding or location in a region of the protein relatively shielded against the penetration of solvent by the particular arrangement of secondary and tertiary structure are but two examples of factors that may result in a shift of pK.

The identification of reactive groups from pH studies alone is unreliable and confirmation of the results using a different method is usually

sought. The pH studies often provide the first suggestive evidence however, and provide the basis for more direct techniques, some of which will be considered in Chapter 6.

THE EFFECT OF TEMPERATURE

A rise in temperature nearly always increases the rate of chemical reactions, including those catalysed by enzymes. As a rough guide it is usually stated that a rise of 10°C doubles the rate of a reaction, but the actual change varies with the nature of the reaction. Enzymes are denatured at high temperatures so that the results obtained from a study of the rate of an enzymic reaction as a function of temperature reflect a balance between the reversible increase in the rate of the catalysed reaction and a decrease in activity associated with the onset of irreversible denaturation. This balance appears to confer on enzymes the property of a temperature optimum, i.e. a certain temperature can be found at which a particular enzyme works best. The optimum temperature is often, quite erroneously, assumed to be 37°C for human and most other mammalian enzymes. The whole concept of temperature optimum can be misleading. The value obtained for the optimum will depend on the time period over which measurements are taken. An enzyme can tolerate a high temperature for a short time and during this time the catalysed reaction will be greatly accelerated. Keeping the enzyme at a high temperature for a longer time, however, will bring about denaturation so that the *average* rate of release of products over the whole time period may well be less than at a much lower temperature at which, although the catalysis is much slower, denaturation is not occurring. Fig. 4.9 shows how different optima can arise depending on the period over which measurements are taken.

In order to function in human tissues and those of other warm-blooded animals, enzymes must be active at 37°C and not be denatured too rapidly at this temperature. Consequently, most determinations of the activities of human enzymes have been made at 37°C and many enzymes appear to be reasonably stable at this temperature. There are wide differences in stability, however, from enzyme to enzyme, particularly when isolated and purified preparations are compared. For example, alkaline phosphatase

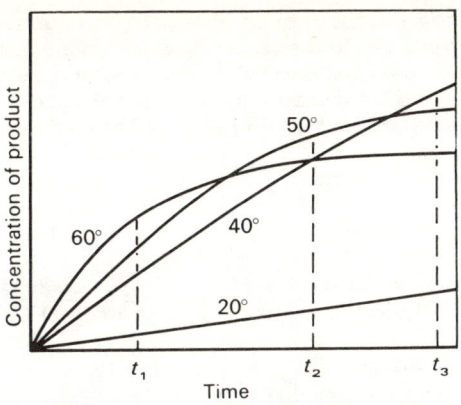

Fig 4.9. **Progress curves of an enzymic reaction at different temperatures. The apparent optimum temperature depends on the time at which measurements are taken. At t_1, t_2 and t_3 the optimum temperatures would be 60°, 50° and 40° respectively.**

from liver is fairly stable up to 50° whereas liver glucose-6-phosphatase is almost totally inactivated after 5 minutes treatment at 50° and pH 5. The local environment within the cell almost certainly aids the stability of many enzymes. Isolation of an enzyme free from other contaminating protein is usually accompanied by an increased sensitivity to heat inactivation.

THE EFFECT OF TEMPERATURE ON THE RATE OF THE CATALYSED REACTION

Arrhenius put forward an empirical equation linking the rate of a chemical reaction with the absolute temperature. The equation is of the form:

$$2 \cdot 303 \frac{d}{dt} (\log k) = \frac{E}{RT^2} \qquad (20)$$

k is the rate constant for the reaction, R is the gas constant (1·987 cal per degree per mole), T the absolute temperature and E is a constant called the Energy of Activation. Eyring in 1935 developed an "activated-complex theory of reaction rates" that gave physical significance to E. Eyring assumed that in a reaction involving one or more molecules a definite amount of energy has to be acquired by the reactants to bring them to an intermediate activated state that transiently has sufficient energy to react further and break down to form the products. Without promotion to an activated state a mole-

cule cannot react. Figure 4.10 shows diagrammatically the energy changes associated with a reaction represented by

$$A + B \rightleftharpoons AB \overset{*}{\rightleftharpoons} C$$
activated
complex

The reaction can be speeded up either by increasing the number of molecules that exist in the activated state or by lowering the activation energy so that a much lower energy barrier is presented to the reactants on the route to product formation. Molecules obtain the energy required for activation by colliding with their neighbours. Raising the temperature will increase the kinetic energy of molecules and so make intermolecular collisions more numerous. Catalysts increase the rate of a reaction because they decrease the activation energy required. Provided the enzyme is stable, a rise in temperature will accelerate a catalysed reaction because

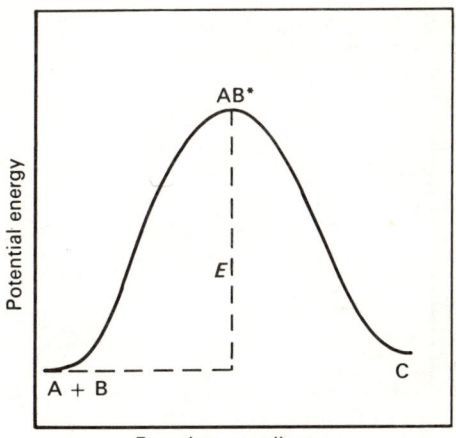

Fig 4.10. **Changes in potential energy for the reaction A + B ⇌ AB * ⇌ C. The difference in energy between AB* and A + B is the "energy of activation" (E) for the forward reaction.**

of the increased kinetic energy of the reactants coupled with the low energy of activation in the presence of enzyme. It is a general feature of catalysis, however, both enzymic and non-enzymic, that the factor by which the rate of a reaction is increased by a particular increase in temperature is less for the catalysed reaction than for the uncatalysed. Thus the lower energy of activation in the presence of the catalyst would appear to be the most important factor in the enhancement of velocity. Exactly how catalysts manage to lower E still cannot be fully explained. Factors that probably contribute to the lowering will be considered in Chapter 6.

The energy of activation for an enzymic reaction may be obtained from a study of the maximum velocity as a function of temperature. Integration of equation (20) gives:

$$\log k = \frac{-E}{2 \cdot 303 RT} + \text{constant} \qquad (21)$$

Since $V = ke_0$, at a constant enzyme concentration V is a measure of k and a plot of $\log V$ against $\frac{1}{T}$ should be linear with a slope equal to $\frac{-E}{2 \cdot 303 R}$ (Fig. 4.11). Most enzymic reactions do give linear plots when data are treated in this way, but occasionally discontinuities occur in

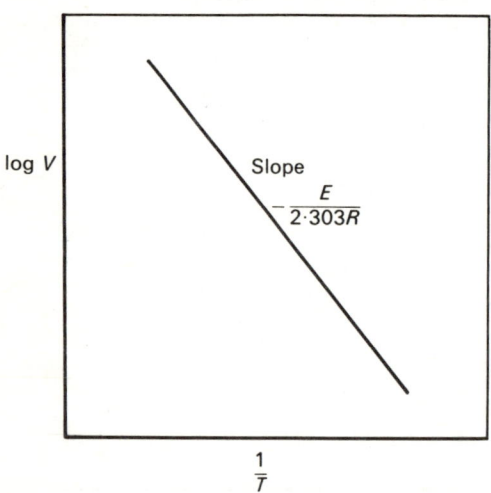

Fig 4.11.   Arrhenius plot for enzymic reaction. The energy of activation is obtained from the slope of the linear plot.

such plots, i.e. the energy of activation does not remain constant over the whole temperature range. The discontinuous plots are not easy to interpret and several causes probably contribute to the departure from ideal behaviour. Reversible inactivation of the enzyme by heat may explain the results in some cases.

Comparison of the values for the energy of activation (E) for a particular reaction in the presence and absence of enzyme is interesting from the physiological viewpoint. Most enzymic reactions, however, do not proceed at a measur-

able rate in the absence of the enzyme but the figures given in the table below for the hydrolysis of sucrose catalysed by H^+ ions and an enzyme obtained from yeast illustrate the greater efficiency, in terms of the lowering of E, of the enzymic process.

Reaction	Catalyst	E (cal/mole)
Hydrolysis of sucrose	H^+	26 000
Hydrolysis of sucrose	Yeast enzyme	13 000

(from Sizer)

THERMAL INACTIVATION OF ENZYMES

As mentioned previously, elevated temperatures generally lead to protein denaturation. The denaturation process can be studied by subjecting the enzyme to a particular temperature. Samples are withdrawn at timed intervals for determination of residual activity at a temperature at which the enzyme is quite stable. If the process is repeated at a number of different temperatures then the energy of activation E for the denaturation process can be determined using the integrated form of the Arrhenius equation. Generally, it is found that E is very large for enzymes, and other proteins, being 40 000 to 100 000 cal/mole. Such a large value would be expected to make denaturation difficult which is contrary to experimental findings. The anomaly stems from the unusual properties of proteins. In their native state proteins are maintained in a very special three-dimensional shape by interactions between various groups in the side chains of the constituent amino acids (Chapter 3). The native protein represents an ordered condition and disruption of the tertiary structure leads to a large entropy increase as the polypeptide chain assumes a less ordered arrangement. The entropy of activation on heating a protein is sufficient to overcome the large energy barrier represented by E. That the thermal inactivation of enzymes resembled the denaturation of proteins by heat was one of the early pieces of evidence put forward in support of the view that all enzymes are proteins.

The rate of inactivation of an enzyme by heat can be a useful means of characterization. Thus the different forms of lactate dehydrogenase present in human and other animal tissues differ in their susceptibility to heat as also do alkaline phosphatases originating from different organs. These differences can be made use of clinically to identify the source of an elevated level of enzyme circulating in the plasma (Chapter 9).

Very often the presence of substrate or an essential co-factor helps to stabilize an enzyme against denaturation. The remarkable chemical specificity of most enzymes for their substrates implies a high degree of organization in that region of the protein where the substrate binds. Denaturating agents may disorganize this region even when little change is brought about in the rest of the molecule unless substrate is present to aid the retention of the active conformation.

5 Inhibitors, Activators and Coenzymes

INHIBITION OF ENZYMIC ACTIVITY

It is a property of catalysts that they can be "poisoned" by certain chemical reagents. For example, platinum metal, which catalyses hydrogenations, has its catalytic function impaired by cyanide, arsenic and hydrogen sulphide. The activity of enzymes is also reduced (or inhibited) by a number of agents. An interesting similarity exists between the behaviour of a colloidal platinum solution and the enzyme catalase. Both these agents catalyse the decomposition of hydrogen peroxide and both are poisoned or inhibited by cyanide.

Enzyme inhibitors are important in medicine, pharmacy, veterinary science and agriculture, as well as sometimes serving as useful tools to the biochemist who is interested in investigating metabolic pathways and the mechanism of enzyme action. It is a feature of living systems that the breakdown or synthesis of important compounds proceeds by a sequence of small steps, each step being catalysed by an enzyme effecting a small chemical change. An inhibitor acting on just one of the enzymes in the sequence will effectively block the pathway and prevent the metabolic process often with fatal consequences. The very potent poison, cyanide, is an obvious example: the reason for its great toxicity is that it inhibits cytochrome oxidase in the mitochondrial respiratory chain and thus prevents aerobic respiration. The therapeutic actions of drugs and antibiotics can, in some cases, be traced to inhibition of enzymic processes and many pesticides kill because they affect enzymes in the nervous system of insects, rodents, and other pests. Inhibitors of enzymes essential for nervous function in man ("nerve gases") have also been developed as potential weapons for chemical warfare.

When an inhibitor specifically blocks a single step in a metabolic pathway, metabolites formed at stages prior to the point of blockage accumulate. Identification of the metabolites enables certain deductions to be made about the metabolic route. Inhibition of the glycolytic pathway by iodoacetate helped in the identification of intermediary metabolites and the effect of malonate on the respiration of tissue minces was used by Krebs in the elucidation of the sequence of reactions in the tricarboxylic acid cycle. Antibiotics provide the biochemist with a means of investigating aspects of protein and enzyme biosynthesis and it is the greater effect of antibiotics on these processes in bacteria than in the host which accounts for their effectiveness in therapy of infectious disease.

Inhibitors may be reversible or irreversible. Reversible inhibitors react rapidly with the enzyme and an equilibrium is set up between the inhibited enzyme, on the one hand, and the free enzyme plus inhibitor, on the other. The degree of inhibition in such a system depends on the concentration of the inhibitor and removal of the inhibitor by dialysis restores full enzymic activity. Irreversible inhibitors cannot be removed by dialysis, and the inhibition is usually progressive with time. If the concentration of the irreversible inhibitor is greater than that of the enzyme, total and permanent inactivation will result. The "nerve gas" types of inhibitor already mentioned are irreversible and are effective at very low concentrations.

REVERSIBLE INHIBITION

Reversible inhibitors have been grouped into three main classes depending on the effect the inhibitor has on the kinetics of the interaction between enzyme and substrate. The classes are (i) competitive, (ii) non-competitive and (iii) uncompetitive. The ways in which enzyme, substrate and inhibitor are assumed to interact in

(1)

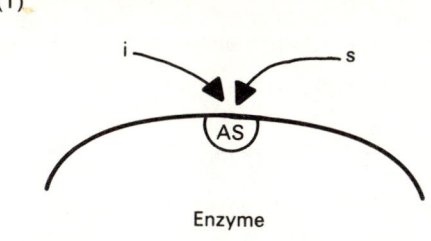

(2)

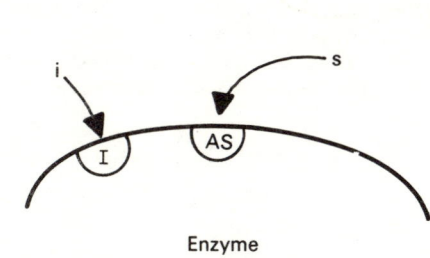

(3)

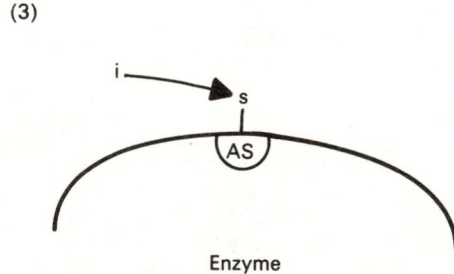

Fig 5.1. Diagrammatic representation of different types of reversible enzyme inhibition. A competitive inhibitor (1) can only combine with the enzyme molecule at the substrate-binding site (the active site, AS). A non-competitive inhibitor (2) combines with a separate inhibitor-binding site, I. An uncompetitive inhibitor (3) combines with the enzyme-substrate complex.

each of these classes are shown diagrammatically in Fig. 5.1.

(i) Competitive inhibition

The inhibitor is usually a structural analogue of the substrate and is assumed to bind to the enzyme at the normal substrate-binding site to form an enzyme-inhibitor complex. Although the inhibitor is able to combine with the enzyme and thus "compete" with the substrate, the specificity requirements of the enzyme are such that bond breakage in the inhibitor cannot occur. Hence, the presence of inhibitor effectively reduces the concentration of binding sites available to the true substrate and will appear to modify the binding constant of the substrate. Because binding of both the substrate and inhibitor is reversible, the fraction of the enzyme in the form of the inactive enzyme-inhibitor complex will depend on the relative concentrations of substrate and inhibitor. If the concentration of substrate is increased sufficiently, the inhibitor will be displaced from the enzyme and the inhibition overcome. Thus it is to be expected that the maximum velocity of the enzymic reaction (i.e. the velocity in the presence of a large excess of substrate) will not be affected by the presence of a competitive inhibitor.

The system may be represented by the following set of equations:

$$E + S \; \underset{k_{-1}}{\overset{k_{+1}}{\rightleftharpoons}} \; (ES) \xrightarrow{k_{+2}} E + Products$$

$$E + I \; \underset{k_{-3}}{\overset{k_{+3}}{\rightleftharpoons}} \; (EI)$$

E, S and I are free enzyme, substrate and inhibitor respectively. (ES) is the active complex that breaks down to form products while (EI) is the "dead-end" enzyme-inhibitor complex. The velocity constants apply to the reactions in the directions indicated. Let the total enzyme concentration be e_0, the concentration of ES be x and of EI be y. Provided that the concentrations of S and I are both large compared with e_0 then the amount of them contained in complexes with enzyme can be ignored since it will be negligible compared with the total concentrations. Let the total concentrations be s and i for substrate and inhibitor respectively. Rate equations can now be written for the reaction system.

The change in concentration of the (ES) complex with time is given by:

$$\frac{dx}{dt} = k_{+1} (e_0 - x - y)s - (k_{-1} + k_{+2})x \qquad (1)$$

At the steady state the rates of formation and breakdown of (ES) will be equal and $\frac{dx}{dt} = 0$.

Therefore,

$$k_{+1}(e_0 - x - y)s = (k_{-1} + k_{+2})x \qquad (2)$$

When equilibrium is attained between (EI) and E + I then

$$k_{+3}(e_0 - x - y)i = k_{-3}y \qquad (3)$$

The velocity of the reaction, v, depends on the concentration of (ES). (See equation (7) in Chapter 4). Therefore,

$$v = k_{+2}x$$

Using this relationship and equations (2) and (3) it can be shown that:

$$v = \frac{k_{+2}e_0 s}{K_m\left(1 + \frac{k_{+3}i}{k_{-3}}\right) + s} \qquad (4)$$

If the binding constant of inhibitor to the enzyme is K_i, then

$$K_i = \frac{k_{-3}}{k_{+3}}$$

Substituting in (4) and replacing $k_{+2}e_0$ by V and $\frac{k_{-1}+k_{+2}}{k_{+1}}$ by K_m, gives

$$v = \frac{Vs}{K_m\left(1 + \frac{i}{K_i}\right) + s} \qquad (5)$$

The rate equation predicts that the presence of a competitive inhibitor modifies K_m by the factor $\left(1 + \frac{i}{K_i}\right)$ but has no effect on the maximum velocity V.

Lineweaver-Burk plots in the presence and absence of inhibitor both intersect on the $\frac{1}{v}$ axis at $\frac{1}{V}$ whereas the inhibitor plot cuts the $\frac{1}{s}$ axis at $\frac{-1}{K_m\left(1+\frac{i}{K_i}\right)}$ (Fig. 5.2.1). Provided i is known and K_m can be derived from data taken in the absence of inhibitor, K_i can be calculated. Often

it is of interest to decide whether an inhibitor is acting in a competitive manner or not and the answer is immediately apparent from a comparison of the intercepts on the $\frac{1}{v}$ axis.

(ii) Non-competitive inhibition

The inhibitor is not usually structurally related to the substrate and is assumed to bind at a separate site on the enzyme distinct from the active site. There is no competition between substrate and inhibitor. Binding of the inhibitor does not modify the affinity of the enzyme for substrate but prevents the formation of products. The interactions between enzyme, substrate and inhibitor may be represented as follows:

$$E + S \underset{k_{-1}}{\overset{k_{+1}}{\rightleftharpoons}} (ES) \xrightarrow{k_{+2}} E + \text{Products}$$

$$E + I \underset{k_{-3}}{\overset{k_{+3}}{\rightleftharpoons}} (EI)$$

$$(EI) + S \underset{k_{-1}}{\overset{k_{+1}}{\rightleftharpoons}} (EIS)$$

$$(ES) + I \underset{k_{-3}}{\overset{k_{+3}}{\rightleftharpoons}} (EIS)$$

This time, because I and S bind at separate sites, a ternary complex (EIS) may be formed. (EIS) can arise by two routes: S may bind to (EI), or I to (ES). It is assumed that I has no effect on the binding of S, and *vice versa*. Thus the rate constants for the binding of S to E or (EI) will be identical and the same will hold true for the relationships of the rate constants for the binding of I to E or (ES). (EI) and (EIS) are both inactive complexes.

A rate equation may be derived for an enzyme reaction in the presence of a non-competitive inhibitor and it takes the form of equation (6)

$$v = \frac{Vs}{(K_m + s)\left(1 + \frac{i}{K_i}\right)} \qquad (6)$$

Where K_m, V, K_i, i and s have the same significance as before.

Comparison of equation (6) with the ordinary Michaelis equation (equation (10) in Chapter 4) shows that the effect of a non-competitive inhibitor is to divide V by the factor $\left(1 + \frac{i}{K_i}\right)$

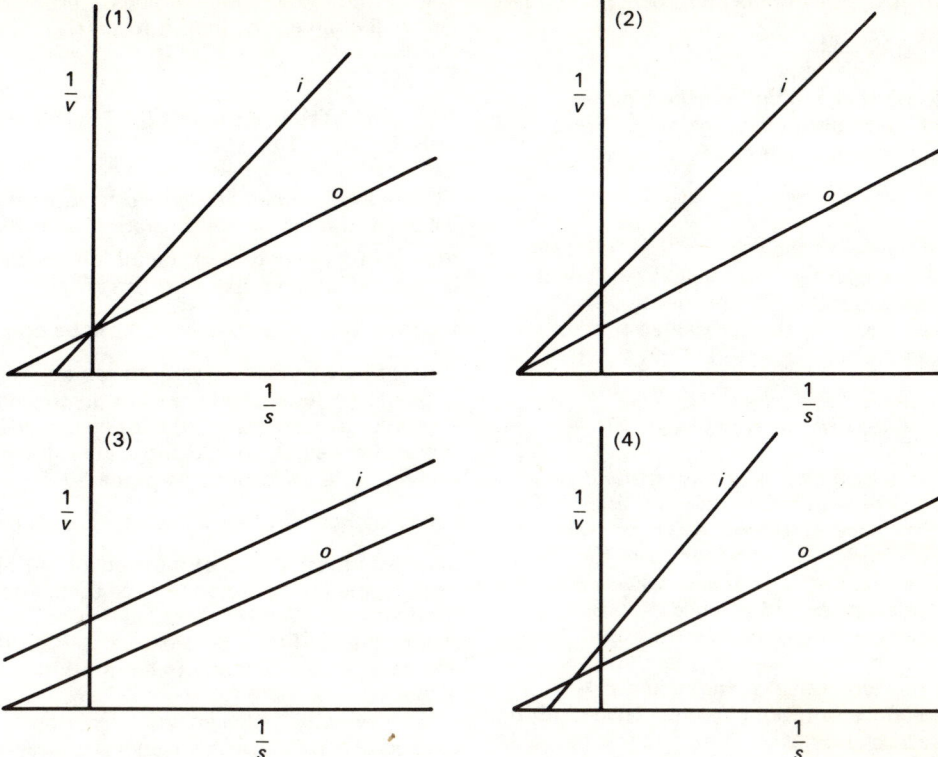

Fig 5.2. Effects of different types of inhibitors on the plot of $\frac{1}{v}$ against $\frac{1}{s}$ (1) Competitive, (2) non-competitive, (3) uncompetitive and (4) mixed inhibition. In each case line *"o"* is without inhibitor and line *"i"* in the presence of inhibitor.

Lineweaver-Burk plots with and without inhibitor intersect on the $\frac{1}{s}$ axis at $\frac{-1}{K_m}$ but the plot with inhibitor cuts the $\frac{1}{v}$ axis at $\frac{1+\frac{i}{K_i}}{V}$ (Fig. 5.2.2.).

(iii) Uncompetitive inhibition

This type of inhibition can be recognized from the appearance of Lineweaver-Burk plots. The inhibitor affects the intersections on both the vertical and horizontal axes by an equal factor and plots of $\frac{1}{v}$ against $\frac{1}{s}$ in the presence and absence of the inhibitor are parallel (Fig. 5.2.3.). This kind of inhibition behaviour is rare with single-substrate enzymic reactions but it can arise when the inhibitor binds to the enzyme sub-

strate complex (ES), but not to free E. The situation can be summarised by the following.

$$E + S \underset{k_{-1}}{\overset{k_{+1}}{\rightleftharpoons}} (ES) \overset{k_{+2}}{\longrightarrow} Products$$

$$(ES) + I \overset{K_i}{\rightleftharpoons} (ESI)$$

Note that (EI) does not form and there is only one route for the formation of the inactive ternary complex (ESI). Solution of the rate equation shows that:

$$v = \frac{\dfrac{Vs}{1 + i/K_i}}{\dfrac{K_m}{1 + i/K_i} + s}$$

The intercepts on a double reciprocal plot will be

$$\frac{1 + i/K_i}{V} \quad \text{and} \quad \frac{-(1 + i/K_i)}{K_m}.$$

Uncompetitive inhibition occurs more commonly with bisubstrate reactions. For a catalysed reaction of the type:

$$A + B \longrightarrow C + D$$

let A combine with the enzyme first followed by B to form a ternary complex (EAB) which breaks down to release the products C and D. The sequence can be represented by the following scheme:

$$E + A \rightleftharpoons (EA) \overset{+ B}{\rightleftharpoons} (EAB) \rightleftharpoons C + D$$

An inhibitor I which is a structural analogue of B can compete with B for a binding site on (EA). Thus if the concentration of A is kept constant, Lineweaver-Burk plots obtained from the plotting of data obtained at various concentrations of B in the presence and absence of inhibitor will be characteristic of competitive inhibition. When the same inhibitor is included in the reaction mixture, but A is varied and B kept at a fixed concentration, uncompetitive inhibition results because I binds to (EA) which is the enzyme-substrate complex formed from A. The example is thus analogous to the single substrate case already described. Competitive inhibitors are usually fairly easy to find because of their similarity in general structure to the substrate, so that with bisubstrate reactions it is also fairly easy to demonstrate uncompetitive inhibition.

"MIXED" INHIBITION

Many reagents that have been shown to inhibit enzymes reversibly give kinetic patterns that do not fit exactly into any of the three classes just discussed. The inhibition in such cases is sometimes referred to as "mixed" because there appears to be a mixture of competitive and non-competitive behaviour. The inhibitor appears both to modify the binding of substrate and to reduce the maximum velocity. Double-reciprocal plots do not intersect on either the vertical or horizontal axes but at a point to the left of the $\frac{1}{v}$ axis and above the $\frac{1}{s}$ axis (Fig. 5.2.4). Many inhibitory substances belong to the "mixed" group. Perhaps it is not surprising that a sub-stance that reduces the binding of a substrate can at the same time inhibit its catalytic break-down.

ALTERNATIVE METHOD OF PLOTTING INHIBITION DATA

The methods already described for obtaining a value for the inhibition constant K_i and for identifying the nature of the inhibition rely on data taken at a number of substrate concentrations so that $\frac{1}{v}$ *versus* $\frac{1}{s}$ plots can be constructed.

With some enzymes, those that can act upon large molecules such as starch or glycogen for example, the substrate does not have a definite molecular weight, so that substrate solutions of known molarity cannot be prepared. Thus values of $\frac{1}{s}$ cannot be calculated. In such cases an alternative method of plotting inhibition data suggested by Dixon may be used. Although particularly useful for these "awkward" substrates the method is of general applicability and can be used to study the inhibition of any enzyme.

The velocity of the enzymic reaction is studied at a fixed substrate concentration but with increasing concentrations of inhibitor. The substrate concentration is changed to another fixed value and the inhibition experiment repeated. The reciprocals of the velocities are plotted against the concentration of inhibitor, i. A straight line results (Fig. 5.3). With a non-competitive inhibitor, the plots intersect on the i axis at $-K_i$. For a competitive inhibitor, the point of intersection lies above the i axis but the value of i at this point again gives $-K_i$. Thus the actual concentrations of substrate used are not needed in order to estimate K_i. With this method an uncompetitive inhibitor gives parallel plots and although K_i cannot be obtained without further data, the class of inhibition can still be readily recognized. That Dixon's method will give the kind of results just outlined is easily verified from a consideration of the relationship between $\frac{1}{v}$ and i for each kind of inhibition.

First, for a competitive inhibitor, rearrangement of the rate equation (5) gives:

$$\frac{1}{v} = \frac{iK_m}{K_i Vs} + \frac{1}{V}\left(1 + \frac{K_m}{s}\right) \qquad (8)$$

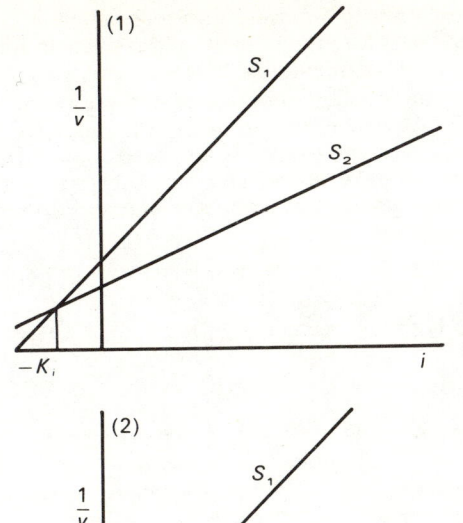

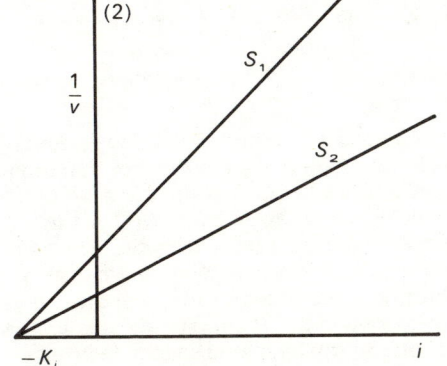

Fig 5.3. Dixon plots ($\frac{1}{v}$ against inhibitor concentration, i, at two substrate concentrations, s_1 and s_2) for (1) competitive and (2) non-competitive inhibition.

Equation (8) predicts a linear relationship between $\frac{1}{v}$ and i. Assume that two concentrations of substrate s_1 and s_2 were chosen. At the point of intersection of the two plots the following identity holds:

$$\frac{iK_m}{K_iVs_1} + \frac{1}{V}\left(1 + \frac{K_m}{s_1}\right) = \frac{iK_m}{K_iVs_2} + 1\frac{1}{V}\left(1 + \frac{K_m}{s_2}\right)$$

Rearrangement of the identity shows that

$$\frac{i}{K_i}\left(\frac{1}{s_1} - \frac{1}{s_2}\right) = \left(\frac{1}{s_2} - \frac{1}{s_1}\right) \qquad (9)$$

so that
$$i = -K_i$$

Second, the rate equation (6) for a non-competitive inhibitor rearranges to equation (10)

$$\frac{1}{v} = \frac{K_m + s}{Vs} + \frac{i(K_m + s)}{K_i \, Vs} \qquad (10)$$

Which shows that when $\frac{1}{v}$ is plotted against i the result is linear. For two substrate concentrations s_1 and s_2 it is easy to show from a consideration of identities that at the point of intersection of two plots:

$$\left(\frac{1}{s_1} - \frac{1}{s_2}\right) = \frac{i}{K_i}\left(\frac{1}{s_2} - \frac{1}{s_1}\right) \qquad (11)$$

and
$$i = -K_i$$

Third, the appropriate linear equation linking $\frac{1}{v}$ and i for an uncompetitive inhibitor is

$$\frac{1}{v} = \frac{K_m}{Vs} + \frac{1}{V} + \frac{i}{VK_i} \qquad (12)$$

Notice that the slope of the line is $\frac{1}{VK_i}$. V and K_i are both constants and so all plots will be parallel lines regardless of the values of s. In this case, K_i cannot be obtained from equation (12) without a knowledge of K_m and V.

INHIBITION BY REACTION PRODUCTS

One of the reasons for the fall in the rate of an enzymic reaction with time is the accumulation of products. If the reaction is freely reversible, the back reaction will begin once significant amounts of product are present and will cause a net decrease of velocity in the forward direction. Even with virtually irreversible reactions, however, inhibition by products can often be demonstrated. For example, various phosphatases that hydrolyse organic phosphate compounds liberating inorganic phosphate (Pi) are competitively inhibited by Pi. This is easy to understand because Pi binds to the catalytic site since it is a potential substrate. Thus it will be a competitive inhibitor of the organic phosphate substrate.

$$R{-}O{-}P + E \rightleftharpoons \left(E\begin{smallmatrix}R\\ \diagdown\\ P\end{smallmatrix}\right) \xrightarrow{\;H_2O\;} E + ROH + P$$

(R—O—P and P both bind to E)

With bisubstrate reactions the inhibition patterns obtained with the products can be complex. The order in which the substrates bind to the enzyme and the order in which the products are released affects the inhibition behaviour. Only one possible mechanism will be considered here as an example. This is the so called "ordered" system in which the reactants and products are assumed to bind to, and be released from, the enzyme in a definite order. Let A be the first substrate bound, and B the second, and let C be the first product and D the last product to be released. The situation can be described by the following reaction scheme.

$$E + A \rightleftharpoons (EA) \quad \ldots \ldots \text{1st substrate}$$
$$(EA) + B \rightleftharpoons (EAB) \quad \ldots \ldots \text{2nd substrate}$$
$$(EAB) \rightleftharpoons (ECD)$$
$$(ECD) \rightleftharpoons (ED) + C \quad \ldots \ldots \text{1st product}$$
$$(ED) \rightleftharpoons E + D \quad \ldots \ldots \text{2nd product}$$

A and D both combine with E, making D a competitive inhibitor of A. This is readily verified by experiments in which the concentration of A is varied while that of substrate B is held constant. If the concentration of A is raised sufficiently, the inhibitory effect of D can be totally overcome. On the other hand, B and D do not combine with the same form of enzyme so D will not be a competitive inhibitor of B, but will inhibit non-competitively because it effectively reduces the amount of enzyme with which B can combine. Raising the concentration of B will not overcome the inhibition. Similarly, C will be a non-competitive inhibitor of both A and B because it combines with a form of enzyme that is different from either of the forms binding A or B.

SOME INHIBITORS OF BIOCHEMICAL AND MEDICAL INTEREST

It is fairly easy to find examples of competitive inhibitors because of the close structural similarity that exists between inhibitor and substrate. Competitive inhibitors of many enzymes have been described, but a few examples are mentioned here which have application in investigative biochemistry and medicine.

The enzyme succinate dehydrogenase catalyses one of the steps in the sequence of reactions that make up the tricarboxylic acid cycle. The reaction converts succinate to fumarate and it is an important step because the hydrogen atoms that are removed from succinate are channelled into the mitochondrial electron transport chain for oxidation by respiratory oxygen with the production of energy in the form of ATP (adenosine triphosphate). The succinate dehydrogenase reaction is inhibited competitively by malonate. Comparison of the substrate and inhibitor shows that competitive inhibition is not unexpected.

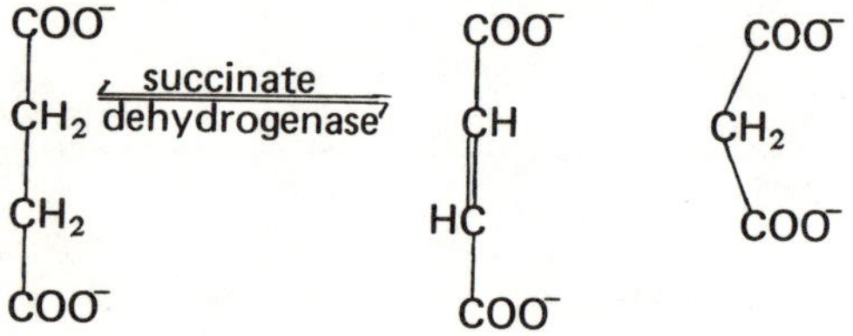

succinate   fumarate   malonate

Malonate is a 3-carbon dicarboxylic acid, whereas succinate is the next higher homologue. The inhibitory effects of malonate were part of the evidence obtained by Krebs for the reactions of the tricarboxylic cycle. When the tissue extracts studied by Krebs were respiring aerobically in the presence of malonate there was an increased production of succinate from oxaloacetate; the increased succinate levels resulted from the blockage of the cycle at the succinate dehydrogenase step so allowing the substrate to accumulate. The inhibitory effect of malonate is reversible and is ultimately overcome by raising the concentration of succinate.

Sulphonamide drugs have the general structure:

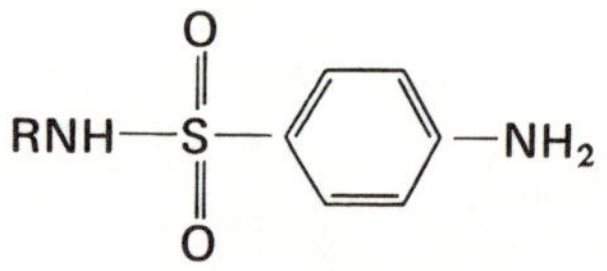

and have an important medical use as antibiotics. Studies have shown that the inhibition of growth of certain micro-organisms by sulphonamides arises from a block in the biosynthesis of folic acid which is a key cofactor in several important metabolic reactions. The biosynthesis of folic acid involves the incorporation of p-amino-benzoate whose structure is:

The similarity with the sulphonamides can be seen. Inhibition of folic acid synthesis will, in turn, lead to the inhibition of all folic acid dependent reactions in susceptible organisms. Human tissues do not possess the enzyme systems necessary for folic acid synthesis (folic acid is a vitamin for man) and are thus un-affected by sulphonamides.

Puromycin is an antibiotic substance synthe-sized by certain moulds and it has the structure:

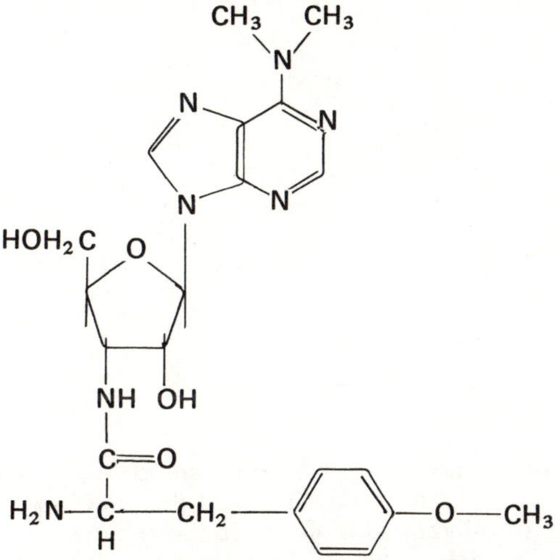

It inhibits protein synthesis, probably by acting as an analogue of an amino-acyl-tRNA complex (Chapter 3).

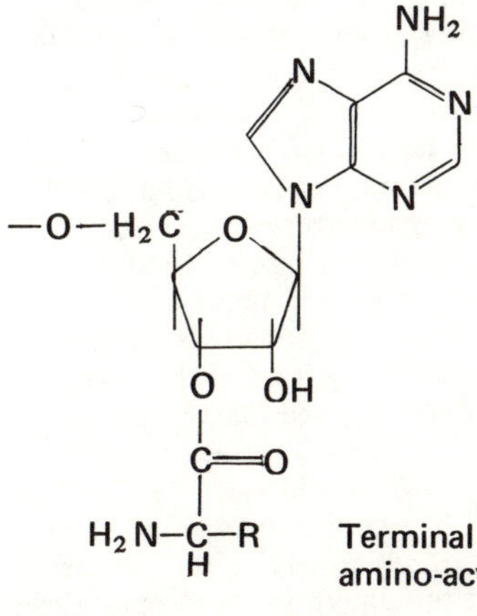

Terminal adenosine of an amino-acyl-tRNA complex

Puromycin would be expected to compete with amino-acyl-tRNAs for incorporation into the growing polypeptide chain, making further additions of amino acids impossible by blocking the transacylation reaction. The isolation from puromycin-treated ribosomal preparations of peptides of varying chain length carrying puromycin bound to the C-terminal end confirms that puromycin is a competitive inhibitor. The rapid protein syn-thesis necessary for bacterial growth makes the bacteria more susceptible to this action of puro-mycin than the cells of the host.

Eserine (physostigmine) and prostigmine are inhibitors of acetylcholinesterases and are impor-tant pharmacological agents. The formulae of these compounds are shown below.

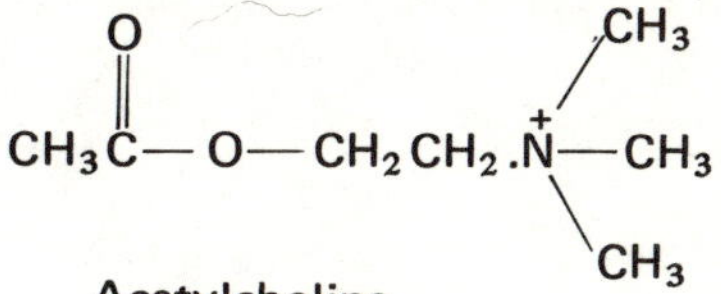

Acetylcholine

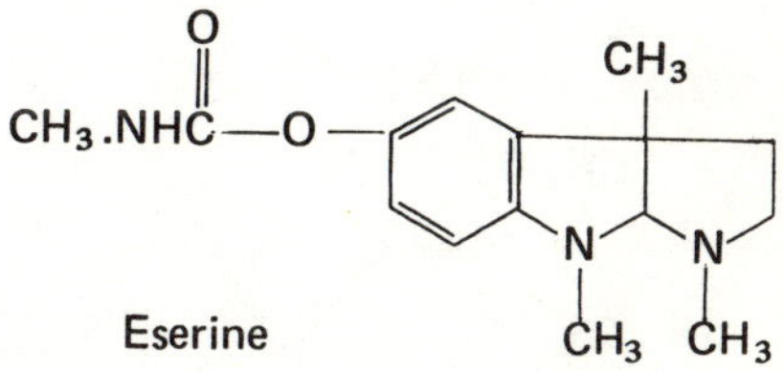

Eserine

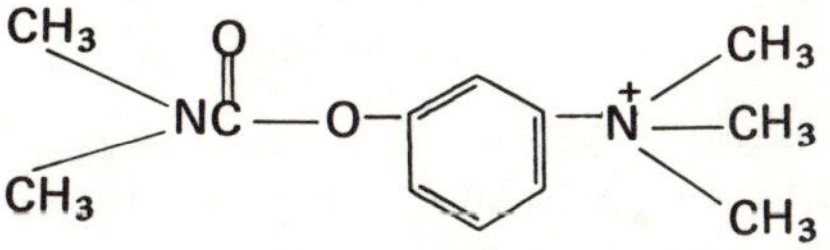

Prostigmine

The inhibitors resemble acetylcholine in containing methylated nitrogen atoms. Both eserine and prostigmine contain carbamoyl linkages (i.e. $-N-C-$ with $=O$) and on reaction with enzyme the carbamoyl group is transferred to the active centre.

The carbamoyl enzyme is hydrolysed slowly, so making the inhibition effectively irreversible. The structural similarity of acetylcholine and the inhibitor suggests that inhibition occurs by competition and that this is so is shown by the reduction of inhibition by the presence of substrate. The slow breakdown of the enzyme-inhibitor complex, however, does mean that the inhibition tends to be progressive with time.

A number of organophosphorus compounds are among the most potent of all known inhibitors. The compounds have the general formulae

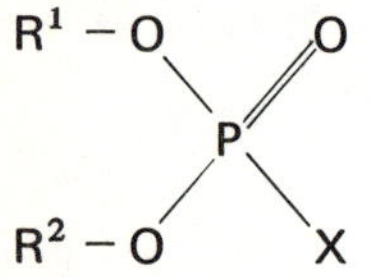

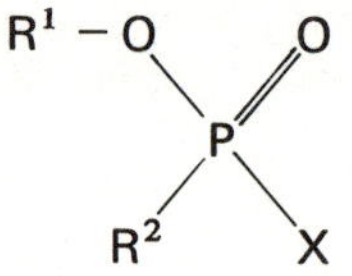

Where R^1 and R^2 are alkyl groups, X can be F, CN or —O—⬡—NO_2. An inhibitor developed as the war-gas Sarin has the structure

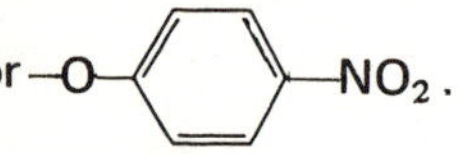

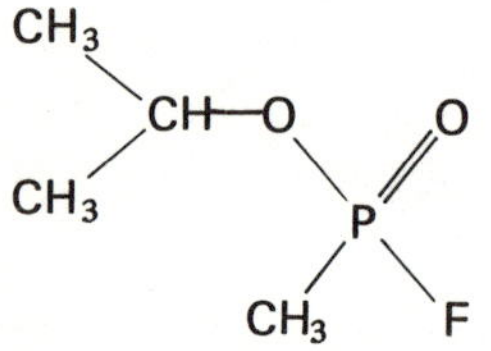

while another of the same group has been widely used in studies of enzyme action. This is di-isopropyl fluorophosphonate (DFP)

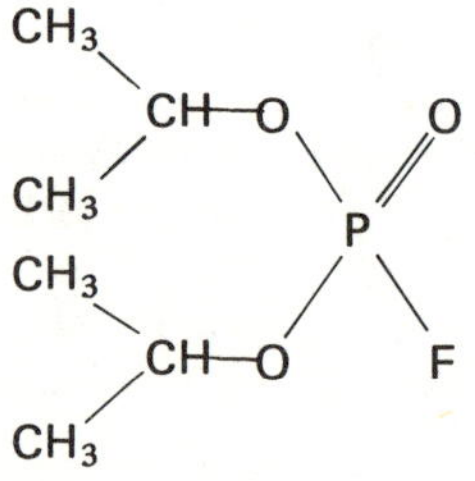

DFP and its relatives inhibit esterases, particularly cholinesterases and acetylcholinesterases. They have received the name of "nerve gas" because of the effects on the nervous system arising from the accumulation of acetylcholine when the acetylcholinesterase of nervous tissue is inhibited, resulting in the blockade of neuromuscular transmission and paralysis. The

inhibitors also act upon trypsin and chymotrypsin.

In the reaction of the inhibitor with the enzyme the breakage of the P — F bond is catalysed and the remainder of the molecule becomes bound to the active site.

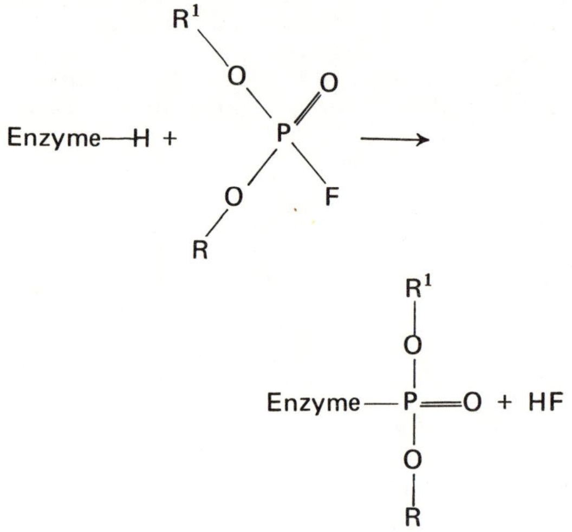

The phosphorylated enzyme can be very stable so explaining the irreversibility of the inhibition. Also, the binding of inhibitor to the enzyme is mole-for-mole and since the molar concentration of enzyme is usually very low it means that the inhibitor is very effective at similar low concentrations. Labelling of DFP with ^{32}P has enabled the amino acid residue to which the inhibitor binds to be identified after total hydrolysis of the inhibited enzyme since the phosphate bond survives hydrolysis. It has been found with cholinesterases, chymotrypsin, trypsin and other esterases that DFP is bound to a particular serine residue by an ester linkage through the oxygen of the OH group of the amino acid. This is evidence for the importance of this group at the active centre of these enzymes.

Many enzymes contain free —SH groups that appear to be important in their catalytic function. Thiol groups are generally quite reactive and substances that react with such groups usually inhibit this class of enzyme. Iodoacetate is of historical interest in this connection because of its inhibitory effect on glycolysis. Later investigations showed that the enzyme of the glycolytic sequence that was particularly susceptible to iodoacetate poisoning was that which converts D-glyceraldehyde-3-phosphate to D-1,3 diphosphoglycerate.

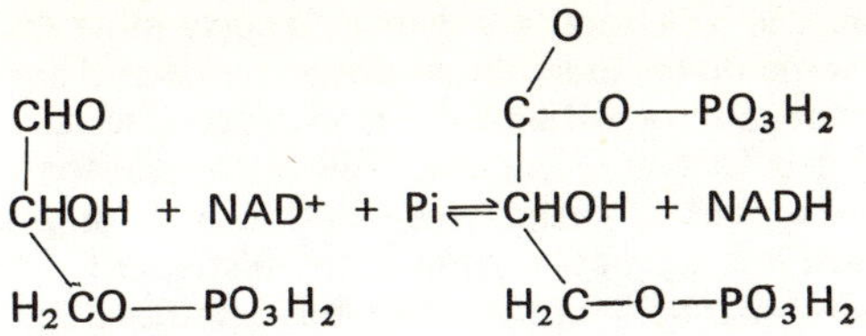

$$\begin{array}{c} CHO \\ | \\ CHOH \\ | \\ H_2CO-PO_3H_2 \end{array} + NAD^+ + Pi \rightleftharpoons \begin{array}{c} O \\ \| \\ C-O-PO_3H_2 \\ | \\ CHOH \\ | \\ H_2C-O-PO_3H_2 \end{array} + NADH$$

The enzyme contains numerous —SH groups that can react with iodoacetate to form a stable S— alkyl derivative of cysteine.

$$Enz{-}SH + ICH_2COONa$$
$$\rightarrow HI + Enz{-}S{-}CH_2COONa$$

Thiol groups form inhibitory mercaptides with organic compounds of heavy metals, e.g. p-chloromercuribenzoate (PCMB):

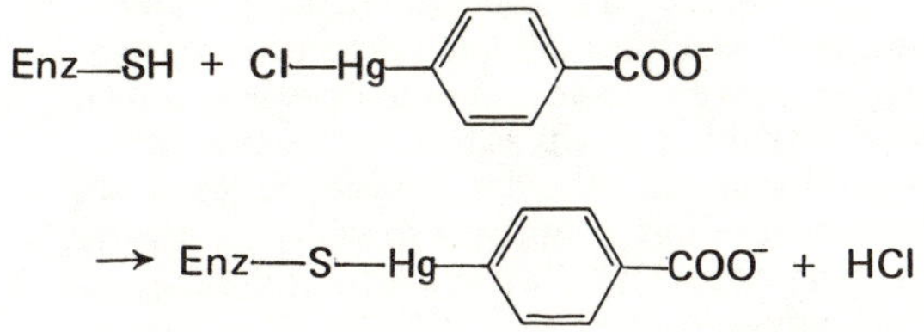

$$Enz{-}SH + Cl{-}Hg{-}\langle\bigcirc\rangle{-}COO^-$$

$$\rightarrow Enz{-}S{-}Hg{-}\langle\bigcirc\rangle{-}COO^- + HCl$$

Titration with PCMB can be used to determine the number of thiol groups in enzymes. The total number of reactive thiols in glyceraldehyde phosphate dehydrogenase varies from species to species, but enzyme prepared from human tissues contains seven groups that react with PCMB within 2 min of mixing.

Some enzymes require cofactors in order to function catalytically. One such factor is lipoic acid (p. 66) which in its reduced form possesses two thiol groups.

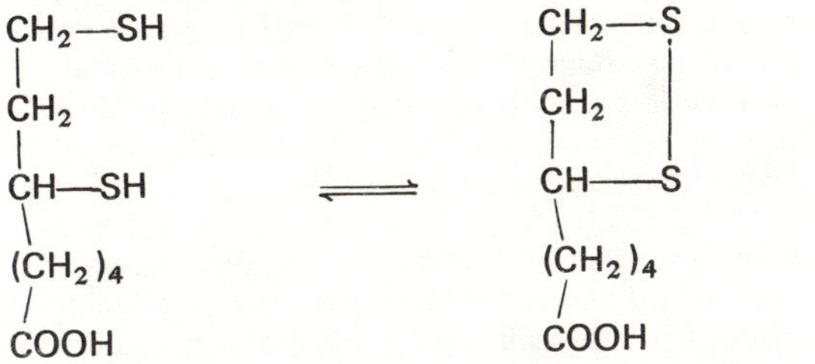

$$\begin{array}{c} CH_2-SH \\ | \\ CH_2 \\ | \\ CH-SH \\ | \\ (CH_2)_4 \\ | \\ COOH \end{array} \rightleftharpoons \begin{array}{c} CH_2-S \\ | \quad | \\ CH_2 \quad | \\ | \quad | \\ CH-S \\ | \\ (CH_2)_4 \\ | \\ COOH \end{array}$$

Reduced lipoate   Oxidized lipoate

A compound that reacts readily with the thiols of lipoate will inhibit an enzymic reaction which depends on the participation of lipoate although it is not a true enzyme inhibitor in that it does not combine with the enzyme itself. Certain compounds of arsenic are very inhibitory in this respect, e.g. the war gas, Lewisite.

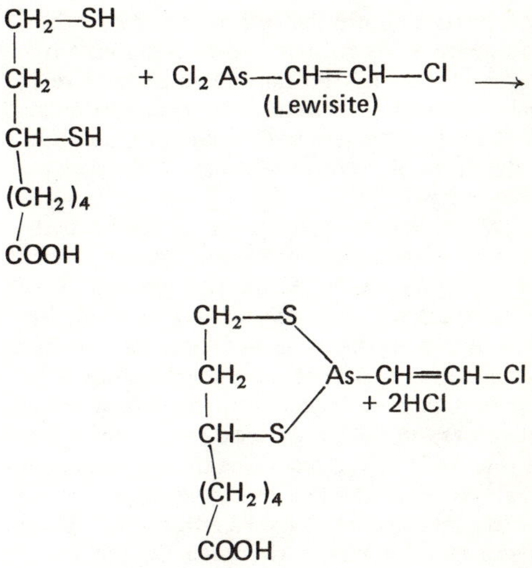

$$\begin{array}{c} CH_2-SH \\ | \\ CH_2 \\ | \\ CH-SH \\ | \\ (CH_2)_4 \\ | \\ COOH \end{array} + Cl_2As{-}CH{=}CH{-}Cl \longrightarrow$$
$$\text{(Lewisite)}$$

$$\begin{array}{c} CH_2-S \\ | \quad \backslash \\ CH_2 \quad As{-}CH{=}CH{-}Cl \\ | \quad / \\ CH-S \quad\quad + 2HCl \\ | \\ (CH_2)_4 \\ | \\ COOH \end{array}$$

The pyruvate dehydrogenase and oxoglutarate dehydrogenase complexes of enzymes rely on lipoate as a cofactor and are inhibited by Lewisite and related reagents. The compounds formed between lipoate and arsenicals are usually fairly stable so that the inhibition is irreversible. Treatment with an excess of another dithiol compound can often overcome the inhibition, however. A compound originally developed as an antidote to Lewisite (British Anti-Lewisite or BAL) is effective in reversing the inhibitory action of arsenicals as is dithiothreitol (Cleland's reagent).

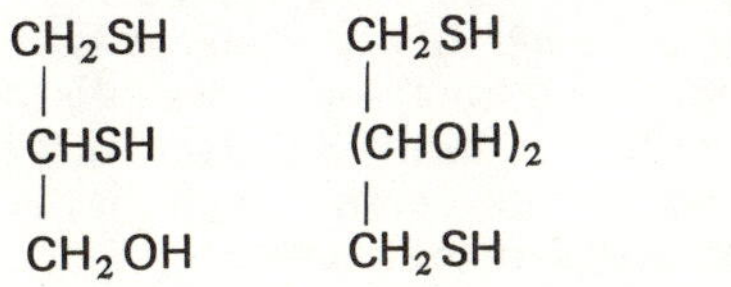

$$\begin{array}{cc} CH_2SH & CH_2SH \\ | & | \\ CHSH & (CHOH)_2 \\ | & | \\ CH_2OH & CH_2SH \\ \\ BAL & \text{Cleland's reagent} \end{array}$$

Dithiothreitol is a useful reagent for the protection of reactive thiol groups against reaction with inhibitors or oxidation during enzyme isolation.

ACTIVATORS AND COENZYMES

Many enzymes need a non-protein cofactor in the reaction medium before any breakdown of substrate can be demonstrated. The existence of essential cofactors was first suggested to account for the loss of the ability of yeast extracts to ferment glucose to alcohol after dialysis of the extracts. Fermentation could be restarted by adding the diffusate back to the yeast extract. As cofactors were separable by dialysis it seemed that they were of relatively small molecular size compared with proteins, and their non-protein nature was emphasized by the demonstration that many cofactors were stable to heat.

Dialysable cofactors can be divided broadly into *activators* and *coenzymes*. Coenzymes bind to the enzyme at the active site where they take a direct part in catalysis. A chemical change is impossible in their absence because coenzymes are obligatory reactants and are the donors or recipients of groups from appropriate substrates. Activators, on the other hand, generally increase the rate of enzymic reactions by bringing enzymes or substrates into a more reactive state, rather than by taking part in the reaction itself. Sometimes a reaction can be demonstrated in the total absence of activator, albeit at a rate much below the maximum. Activators are usually simpler substances than coenzymes, and activation by bivalent metal ions, e.g. Mg^{2+}, seems to be particularly common.

An absolute distinction between coenzymes and activators cannot be made because substances such as metal ions, which are usually regarded as activators, are important constituents of the active site of certain enzymes and according to the broad definition given above could be considered to be coenzymes. Again, some enzymes collectively referred to as allosteric or regulatory (Chapter 7) bind coenzyme-like molecules at points remote from the active site and, although they have no catalytic function, they bring the enzymes into their most active conformations. It is correct to call such molecules activators even though they are not necessarily of simple structure.

The distinction between coenzymes and true substrates is difficult to make from a purely chemical point of view: both are bound at the active site and the coenzyme as well as the substrate may become chemically modified during the catalysed reaction. A physiological distinction can usually be made, however,

because substrates are usually recognizable as intermediates in a metabolic sequence and are chemically modified further in biosynthetic or degradative pathways. Coenzymes, on the other hand, are usually restored to their original form in a separate enzymic step instead of being channelled into a metabolic pathway. Thus there is usually a recycling of coenzymes and very often they function as carriers of groups from one substrate to another.

Although coenzymes are smaller molecules than proteins, many of them nevertheless have complex chemical structures and cannot be fully synthesized by man and other higher animals. Thus, a dietary source is obligatory and many vitamins, particularly those belonging to the B group, are known to contain the components of coenzymes which are not synthesized by man. The symptoms of vitamin deficiencies presumably arise because metabolism is blocked when there are insufficient amounts of essential coenzymes.

The following discussion illustrates types of enzymic reactions in which activators and coenzymes are involved and includes accounts of the Vitamin B group as well as of some other vitamins and coenzymes. The examples of activation are limited to those in which inorganic ions are involved; allosteric effects are considered in Chapter 7.

ACTIVATION

It has been mentioned already that activators are considered to increase the rate of enzymic reactions by promoting the formation of the most active state of the enzyme or reactants. The exact mechanisms by which this is achieved are not fully understood, but it is possible to speculate on several general schemes by which activation may take place. Some of the possible mechanisms are considered below. In general, it is a little easier to account for the activating effects of cations than for that of anions because the former have been studied more extensively and anionic effects are often relatively unspecific.

(i) Structural role of activators

Many enzymes (metalloenzymes) are known to contain metal ions as an integral part of their structure. The function of the metal may be to stabilize tertiary and quaternary protein structure, but in some cases the metal ion seems to provide an important reactive group at the catalytic (or active) site. When the activator is a constituent of the enzyme molecule, it is usually bound very firmly and its removal requires the use of strong metal-chelating agents, e.g. ethylenediamine tetra acetic acid (EDTA). After such treatment, activation is usually demonstrable when the enzyme is dialysed against a solution containing the important metal ion. If the removal of the metal is accompanied by significant changes of tertiary or quaternary protein structure, restoration of full activity on dialysis against metal ions may take some time, because the refolding of polypeptides is frequently found to be time-dependent.

Several different metals including zinc, iron, copper and molybdenum have been detected in enzymes. Zinc has been found in carbonic anhydrase, alcohol dehydrogenase, alkaline phosphatase (from *E.coli* and human placenta) and in bovine carboxypeptidase A. The zinc can be replaced experimentally by cobalt in *E.coli* phosphatase and carboxypeptidase A and catalytic activity is retained.

When the enzyme metal ion functions directly in catalysis it is probably providing an electropositive centre to which chemical groups in the substrates are coordinated. In the carbonic anhydrase reaction, for example, it is likely that the bicarbonate ion is coordinated to zinc and that after C—O bond cleavage the zinc is left with a coordinated OH^- ion.

(ii) Combination of activator with substrate or product

For some enzymes it appears that a preliminary combination of substrate molecules with metal ions to form substrate-metal complexes is necessary before the substrate can be bound to the enzyme and broken down. In such cases, the "true" substrate is the substrate-metal complex and it is the concentration of the complex rather than the total substrate concentration which determines the rate of catalysis. If the reactions forming the "true" substrate and the enzyme-substrate complex are represented as follows:

$$M + S \underset{}{\overset{K}{\rightleftharpoons}} MS$$

(metal) (substrate)   (metal substrate complex)

$$E + MS \rightleftharpoons (E-MS) \longrightarrow Products$$

It can be seen that the concentration of MS for any concentrations of M and S is dependent on the equilibrium constant K. When K is known, the concentration of MS may be calculated and used in the rate equation for the enzyme reaction in order to determine K_m and the maximum velocity.

$$v = \frac{[MS] V}{[MS] + K_m}$$

In most reactions involving adenosine triphosphate (ATP) there seems to be obligatory formation of a complex of ATP with magnesium.

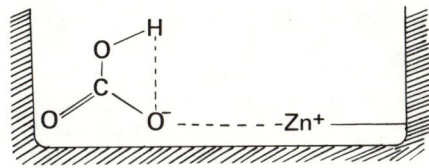

HCO_3^- ion coordinated to zinc atom

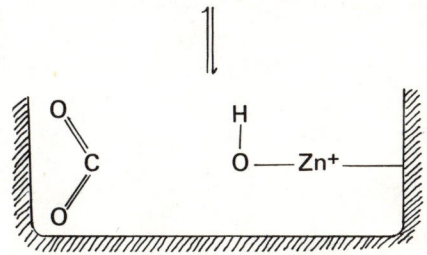

OH^- ion coordinated to zinc atom

$MgATP^{2-}$ complex

The reasons why complexes are preferred by some enzymes are unknown, but it has been suggested that the metal ion may act as a bridge between substrate and enzyme, i.e. the metal portion of the complex binds to the active site and holds the substrate in a favourable position for catalytic breakdown. Alternatively, the metal may facilitate substrate binding by affecting the molecular conformation of the substrate or by neutralizing negative charges which would tend to be repulsed by negatively-charged enzyme catalytic groups. Also it has been suggested that, with ATP, neutralization of charged phosphate groups in the bivalent metal ion complex may make it easier for a nucleophilic OH^-, itself negatively charged, to approach the terminal P atoms during P—O bond breakage.

For a freely reversible reaction, or one in which the product is a potent inhibitor of catalysis of the forward reaction, removal of the product by reacting it with a suitable sub-stance may effect a considerable activation of the forward reaction. Thus, interaction with a product is one way in which an activator could favour catalysis. The effect of Ca^{2+} ions on lipases may be explained in this way. Lipases, which act on triglycerides to produce glycerol and free fatty acids, are known to be inhibited by fatty acids. Ca^{2+} ions probably remove free fatty acids from solution as in-soluble calcium-soaps, so reducing the inhibition by these products.

(iii) Reversible combination with enzyme

Some enzymes may possess specific binding sites for molecules of activator, the sites being separate from the active site. When the acti-vator site is filled, effects are transmitted through the protein to the active site which is then induced either to accept and bind substrate more readily (i.e. there is an effect on K_m), or allow substrate molecules already bound to the active site to be broken down more rapidly. When there is an accelerated rate of breakdown of the *ES* complex the maximum velocity is affected, but it is also quite possible for an acti-vator to change both K_m and V_{max}.

The existence of an independent activator site on the enzyme is a characteristic of allo-steric enzymes (see Chapter 7) but in the absence of the cooperative effects which are common with allosteric enzymes, "non-competitive activation" is probably a better description of activation which follows the binding of a molecule to the activator site. Kinetically, the equations describing such activation are analogous to non-competitive inhibition except that the bound activator increases V_{max}, rather than decreasing it. In some cases occupancy of the activator site may be a prerequisite of catalysis, i.e. the activator is essential and the enzyme is totally inactive in its absence.

The activation of salivary and pancreatic amylases by chloride ions probably involves combination of the enzyme with activator. The addition of 5 mM chloride almost trebles the enzymic activity of salivary amylase and, at the same time, displaces the pH optimum from pH 6·0 to pH 7·0. Cl^- ions may combine with a positively charged group on the enzyme close to the active centre thus increasing the pK value of a group important in catalysis (see discussion on effects of pH in Chapter 4). Other univalent anions also shift the pH optimum of the enzyme but are, in general, less effective at increasing activity. Br^- and I^- ions, which are chemically related to Cl^-, can activate the enzyme to some extent, but nitrate does not. Thus there seems to be a degree of specificity with regard to the activation of amylase by anions. With many other enzymes, however, the effect of anions on activity can probably be attributed almost entirely to changes in pH optima.

COENZYMES

Thiamine — Vitamin B$_1$

Thiamine is present in all living organisms but animals depend on plants and micro-organisms for their supply. In man, the disease beri-beri is associated with thiamine deficiency. Thiamine occurs in cereal grains in the free vitamin form whose structure is shown below. The molecule contains a pyrimidine derivative joined by a

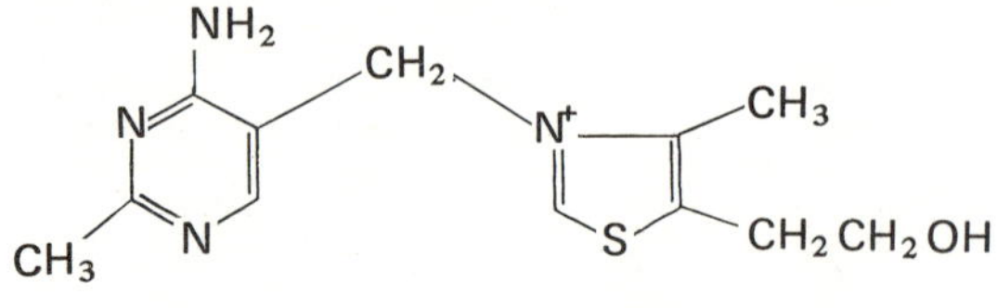

Pyrimidine derivative   Thiazole derivative

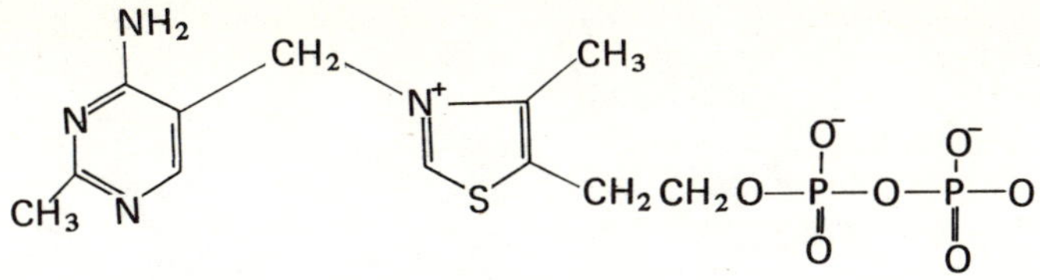

Thiamine pyrophosphate

methylene bridge to a sulphur-containing thiazole derivative. Living tissues can phosphorylate thiamine to form thiamine pyrophosphate (TPP), the structure of which was established by K. Lohmann in 1937.

TPP is the active coenzyme form of Vitamin B_1, and it is important in several reactions of carbohydrate metabolism including the oxidative decarboxylation of pyruvate and α-oxoglutarate and also the transketolase reactions that are a feature of the hexose monophosphate shunt pathway of glucose utilization. The reactive portion of the coenzyme molecule seems to be the thiazole ring. The positive charge on the quaternary nitrogen atom is believed to aid the loss of a proton from C2 of the thiazole ring by repulsion, leaving a negatively charged carbanion:

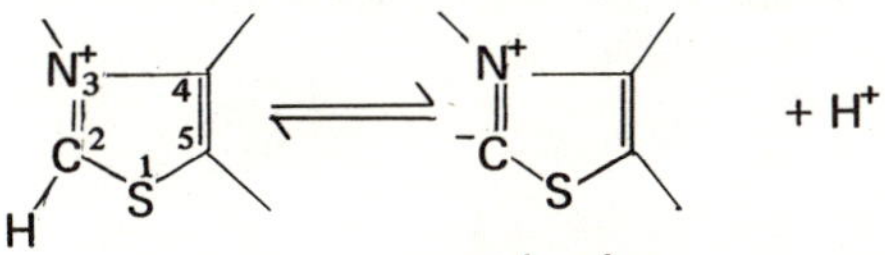

carbanion

The carbanion is a nucleophilic centre which can readily form an addition compound with a keto acid, e.g. pyruvate:

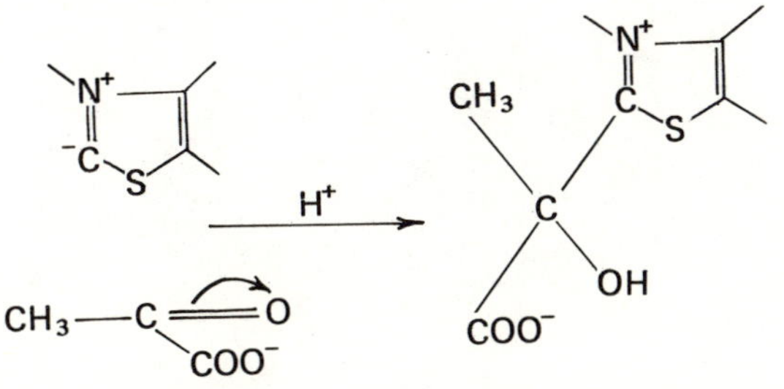

Pyruvate   Pyruvate-TPP complex

The addition compound decarboxylates very readily leaving an aldehyde-TPP compound which is very reactive. The aldehyde moiety can be transferred to a suitable acceptor such as lipoic acid (see below)

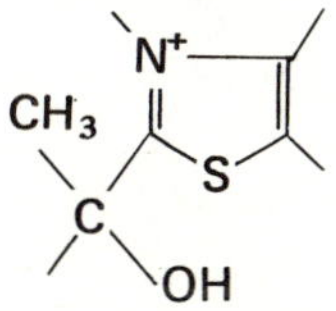

Aldehyde-TPP complex

In a transketolase reaction a ketol-TPP compound reacts with an aldehyde forming a new ketol, e.g.

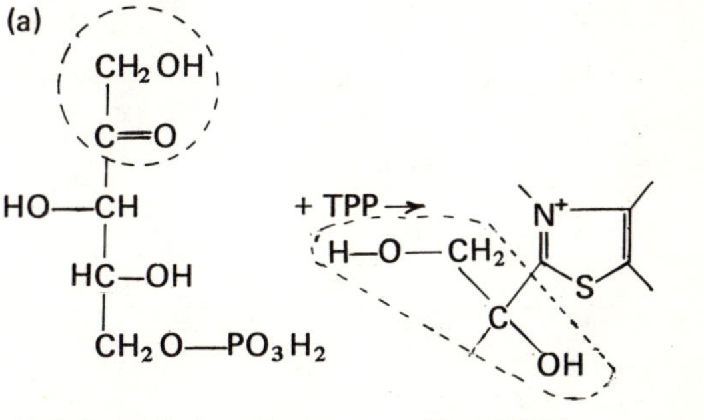

(a)

Xylulose-5-phosphate   Ketol-TPP

Glyceraldehyde-3-phosphate

(b)

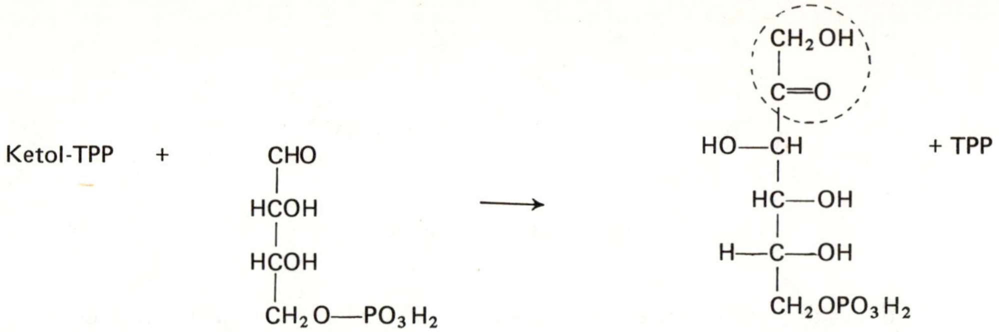

$$\text{Ketol-TPP} \quad + \quad \begin{array}{c} \text{CHO} \\ | \\ \text{HCOH} \\ | \\ \text{HCOH} \\ | \\ \text{CH}_2\text{O—PO}_3\text{H}_2 \end{array} \quad \longrightarrow \quad \begin{array}{c} \text{CH}_2\text{OH} \\ | \\ \text{C}\text{=}\text{O} \\ | \\ \text{HO—CH} \\ | \\ \text{HC—OH} \\ | \\ \text{H—C—OH} \\ | \\ \text{CH}_2\text{OPO}_3\text{H}_2 \end{array} \quad + \text{TPP}$$

Erythrose-4-phosphate   Fructose-6-phosphate

The importance of the removable proton on C2 of the thiazole ring is shown by experiments in which a methyl group is substituted for the H atom. The methylated thiamine derivative has no coenzyme activity.

Niacin

This name is used for two compounds, nicotinic acid and nicotinamide.

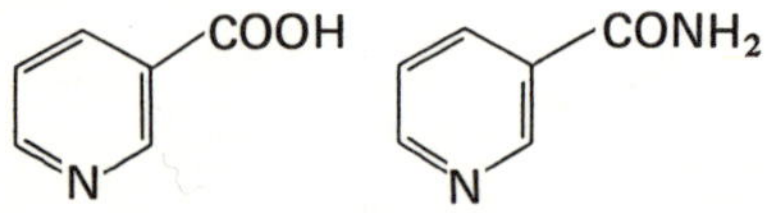

nicotinic acid   nicotinamide

Niacin is a specific curative agent for pellagra, a disease in humans with the symptoms of dermatitis and dementia, and a condition in dogs known as black tongue.

Nicotinamide is found in two very similar coenzymes which differ from each other only in the number of phosphate groups they contain. Both coenzymes function similarly as hydrogen carriers in dehydrogenatin reactions. Nicotin-amide is linked to a ribose phosphate molecule forming a typical nucleotide. The nucleotide is joined through a pyrophosphate linkage to adenosine. The coenzymes are now usually referred to as nicotinamide adenine dinucleo-tide (NAD$^+$) and nicotinamide adenine

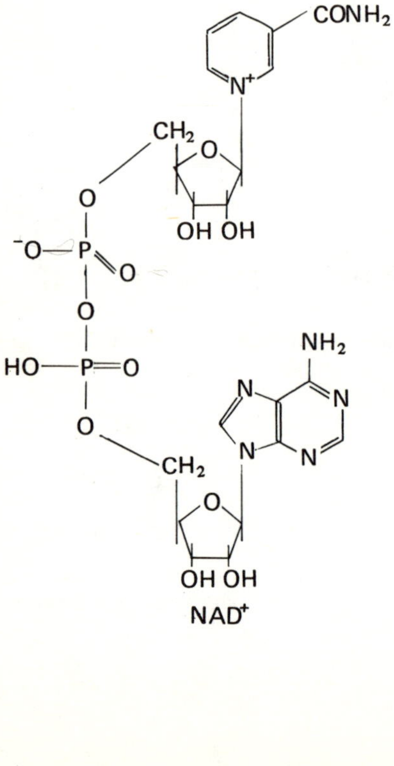

NAD$^+$

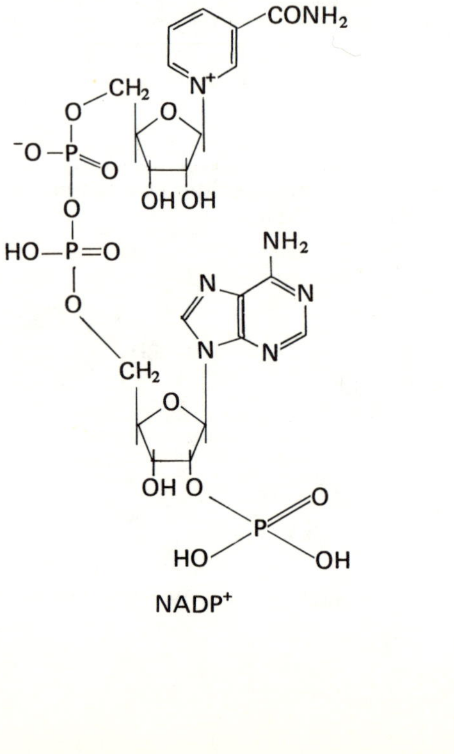

NADP$^+$

dinucleotide phosphate (NADP$^+$). An older terminology calls them diphosphopyridine nucleotide (DPN) and triphosphopyridine nucleotide (TPN) respectively.

In the structures given above note that the N atom of the nicotinamide ring is quaternary and carries a positive charge, and that NADP$^+$ has an extra phosphate group at the 2' position of the adenosine part of the molecule.

NAD$^+$ and NADP$^+$ are involved in a large number of oxidation-reduction reactions which are usually reversible. Most dehydrogenase enzymes show a marked specificity for NAD$^+$ or NADP$^+$ but a few are known that will function equally well with either. The lactate dehydrogenase reaction is a typical example of an enzyme linked to NAD$^+$.

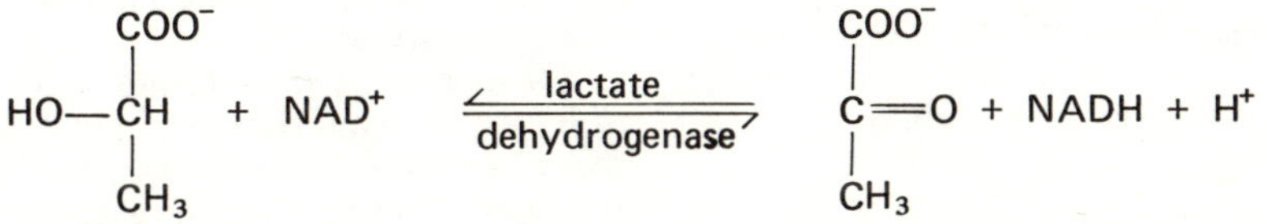

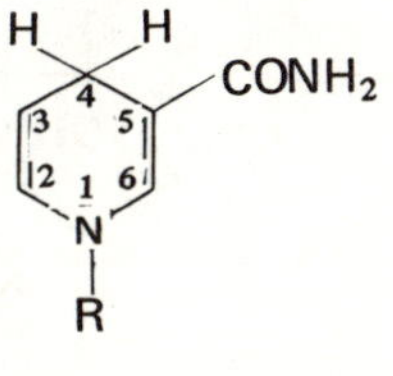

nicotinamide ring at C4. The other hydrogen atom removed from the substrate is released as a proton into the solvent. Thus reduction of NAD$^+$ is correctly written as:

$$NAD^+ + 2H \rightleftharpoons NADH + H^+$$

However, the less strictly correct abbreviations NAD(P) for the oxidized and NAD(P)H$_2$ for the reduced forms are often used.

In NADH and NADPH the hydrogens at C4 project on opposite sides of the plane of the nicotinamide ring. By appropriately labelling a substrate with deuterium it is possible to convert NAD$^+$ to the deuterated form at C4, making the latter asymmetric and allowing two possible isomers denoted A and B.

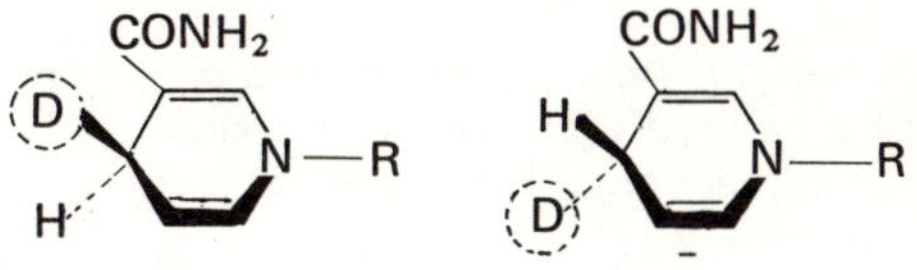

L-lactate   pyruvate

In general, NAD$^+$-linked dehydrogenases are usually associated with catabolic energy-producing pathways whereas NADP$^+$ seems to be more often associated with reductive anabolic reactions.

Reactions coupled to the nicotinamide coenzymes involve the removal or addition of the equivalent of two hydrogen atoms. The reduced forms of NAD$^+$ and NADP$^+$ have the formula shown below, where R represents the remainder of the coenzyme molecule. Reduction converts the quaternary N atom to the neutral, tervalent state and one of the H atoms removed from the substrate is incorporated into the

A form   B form

Such studies have indicated that the addition and subtraction of H from C4 is stereospecific, i.e. dehydrogenases distinguish between the two H atoms attached to C4. Some dehydrogenases add and subtract hydrogens from the A position, whereas others react with B. Muscle lactate dehydrogenase forms or dehydrogenates only the A form, but muscle glyceraldehyde-3-phosphate dehydrogenase reacts with form B. The exclusive nature of the stereospecificity emphasises the subtlety of the orientation of substrates and coenzymes at the active sites of dehydrogenases.

When the coenzymes are reduced there is a change in their absorption spectra and a strong absorption band appears at 340 nm. Following the change in extinction at 340 nm with time in a spectrophotometer is a very convenient method of assaying dehydrogenases (Chapter 2; Fig. 2.3). The oxidation-reduction reaction can be followed in either direction since the conversion of NADH to NAD^+ is accompanied by a decrease in absorbance and NAD^+ to NADH leads to an increase. The spectra of oxidized and reduced NADP are identical with those of NAD^+ and NADH respectively.

Riboflavin — Vitamin B$_2$

Riboflavin is composed of a substituted iso-alloxazine ring linked to a straight chain alcohol, D-ribitol.

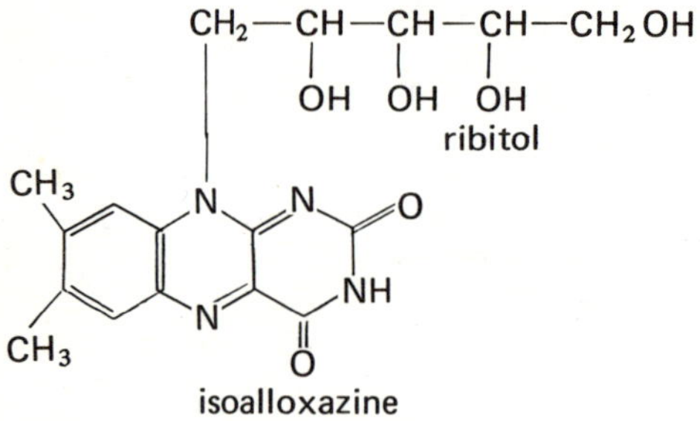

Riboflavin

This vitamin is a component of two coenzymes which act as hydrogen carriers in oxidation-reduction reactions. One of the coenzymes is formed by phosphorylation of the terminal 5' OH group of the ribitol moiety and is usually called flavin mononucleotide (FMN) to distinguish it from a more common flavin coenzyme first demonstrated as important for D-amino acid oxidase. This second flavin is flavin adenine dinucleotide (FAD). The structure of FAD is shown below and, as its name suggests, it contains an adenosine residue linked through the 5'-phosphate of the FMN structure.

Flavin coenzymes act as hydrogen carriers in oxidation-reduction systems but differ from NAD^+ and $NADP^+$ in being more strongly bound to enzymes. Coenzymes which are firmly bound to enzymes are often known as prosthetic groups. Such groups do not act as carriers of groups from one enzyme to another, but seem to undergo a complete catalytic cycle while attached to the same enzyme protein. Flavins, together with biotin and pyridoxal phosphate are examples of prosthetic groups.

Flavoproteins are involved in the transfer of hydrogen from NADH to the carriers of the mitochondrial electron transport chain so that a system exists for the eventual oxidation of hydrogen removed from a substrate by atmospheric oxygen with the formation of water. In addition, flavin coenzymes take part in metabolic oxidations which involve the removal of two hydrogens on adjacent carbon atoms of a substrate to form a double bond. Succinate dehydrogenase, an enzyme of the tricarboxylic acid cycle, and acyl dehydrogenases of the β-oxidation pathway of fatty acid degradation catalyse FAD linked reactions in which a double bond is formed.

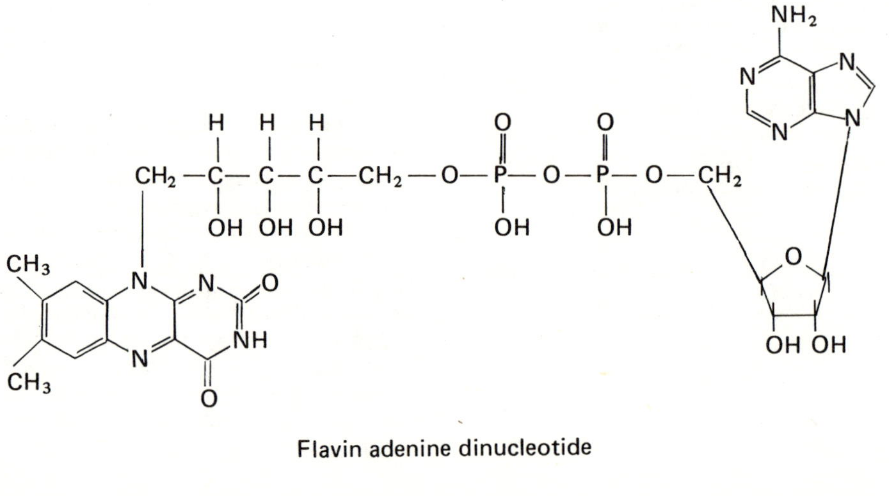

Flavin adenine dinucleotide

(a)

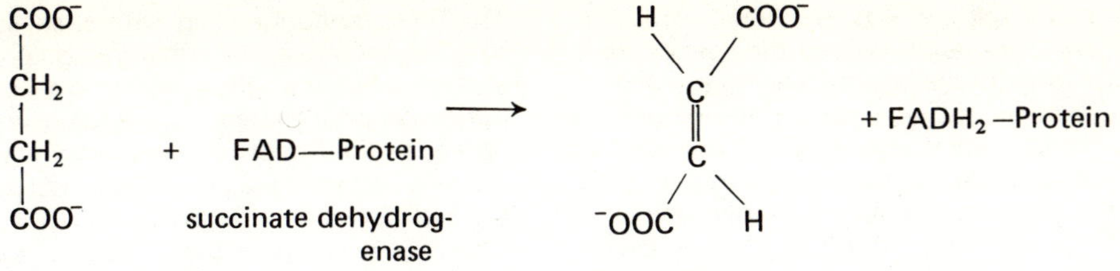

$$\begin{array}{c} COO^- \\ | \\ CH_2 \\ | \\ CH_2 \\ | \\ COO^- \end{array} \quad + \quad FAD\text{---}Protein \\ \text{succinate dehydrog-} \\ \text{enase}$$

succinate

$\longrightarrow$

$$\begin{array}{c} H \quad\quad COO^- \\ \diagdown \quad \diagup \\ C \\ \| \\ C \\ \diagup \quad \diagdown \\ {}^-OOC \quad\quad H \end{array} \quad + \quad FADH_2\text{---}Protein$$

fumarate

(b)

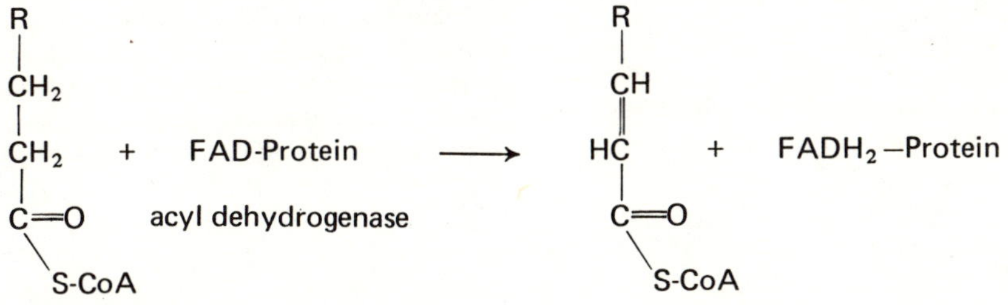

$$\begin{array}{c} R \\ | \\ CH_2 \\ | \\ CH_2 \\ | \\ C\!\!=\!\!O \\ | \\ S\text{-CoA} \end{array} \quad + \quad FAD\text{-Protein} \\ \text{acyl dehydrogenase}$$

Fatty-acyl CoA

$\longrightarrow$

$$\begin{array}{c} R \\ | \\ CH \\ \| \\ HC \\ | \\ C\!\!=\!\!O \\ | \\ S\text{-CoA} \end{array} \quad + \quad FADH_2\text{---}Protein$$

The reversible oxidation and reduction of flavin coenzymes involves the isoalloxazine ring. Reduction seems to occur in two steps with the formation of a semiquinone form of riboflavin as an intermediate.

The semiquinone containing an unpaired electron is likely to be fairly stable because numerous resonance forms of the isoalloxazine structure are possible. Also, many flavoproteins contain a metal ion such as iron or molybdenum which may help to stabilize the semiquinone.

Flavoproteins are abundant in both plants

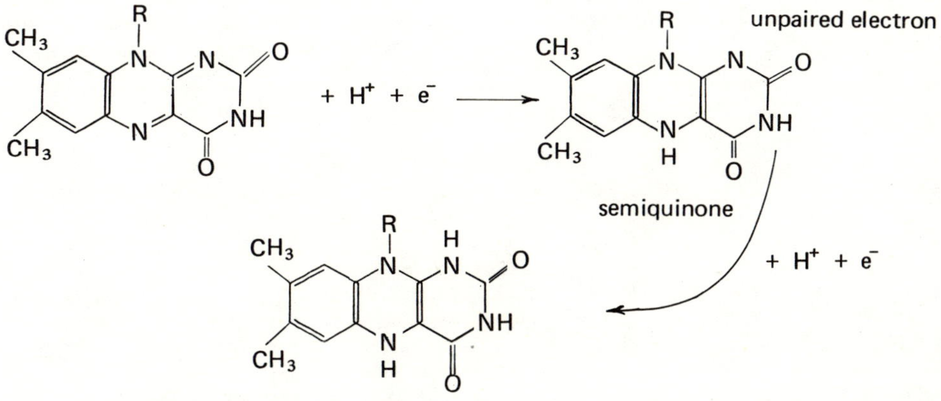

semiquinone

FADH$_2$

and animals which are thus the dietary source of riboflavin for man. Provided the coenzyme can be freed from the accompanying protein during digestion, a deficiency is not encountered. Riboflavin deficiency in man is characterized by an inflammation of the tongue and impairment of vision with photophobia. A simple deficiency of riboflavin is unlikely to occur because there will almost certainly be shortages of other B vitamins in a diet where amounts of riboflavin are small. Thus, as with most other vitamins it is not possible to relate exactly the biochemical functions of riboflavin to the symptoms of deficiency.

Pyridoxine — Vitamin B_6

Pyridoxine was the first compound with Vitamin B_6 activity to be isolated and identified, but

The three compounds are interchangeable in the diet and are approximately equally effective in supporting growth, but it is now known that pyridoxal phosphate and pyridoxamine phosphate are the important coenzyme forms of the vitamin. Vitamin B_6 can be synthesized by the flora of the human large intestine but it is not known for certain whether man can derive all his B_6 requirements from this source. The vitamin is abundantly distributed in animal tissues and plants particularly cereal grains.

Pyridoxal phosphate-dependent enzymes catalyse a large number of important reactions of amino acid metabolism, such as transamination and decarboxylation. In transamination an amino group is transferred from an amino acid to a keto acid. The transfer generates a new amino acid and a new keto acid. Glutamic acid donates the amino group

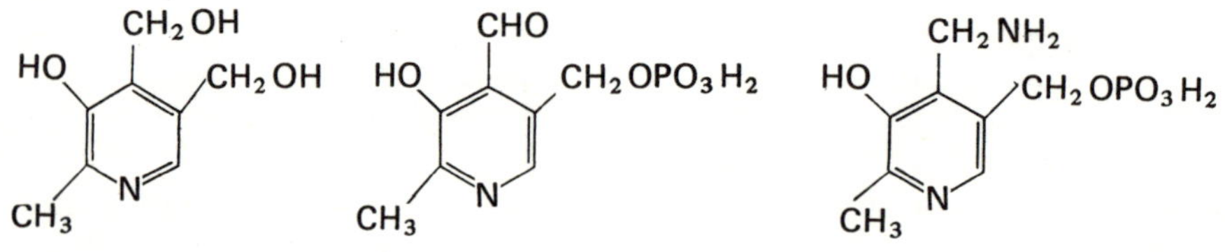

Pyridoxine   Pyridoxal phosphate   Pyridoxamine phosphate

soon after it was shown that two other compounds, pyridoxal and pyridoxamine, which occur naturally as phosphates, also possessed vitamin-like properties.

in many transaminations. Many tissues contain an enzyme catalysing an amino transfer from glutamate to oxaloacetate which is of interest to the clinical biochemist (Chapter 9).

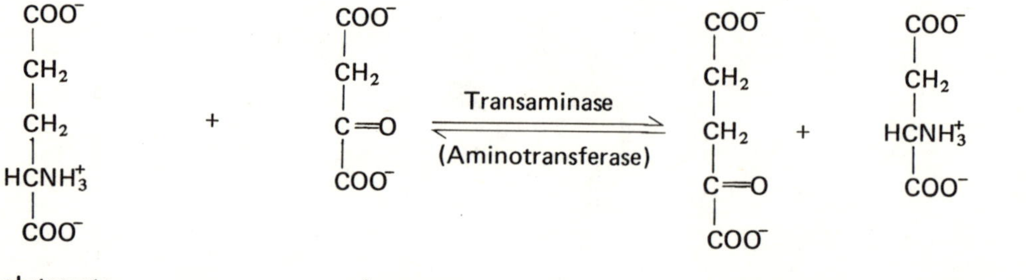

glutamate   oxaloacetate   α-oxoglutarate   aspartate

Braunstein and Kritzmann suggested that a condensation of the amino acid and pyridoxal phosphate occurs to form a Schiff's base.

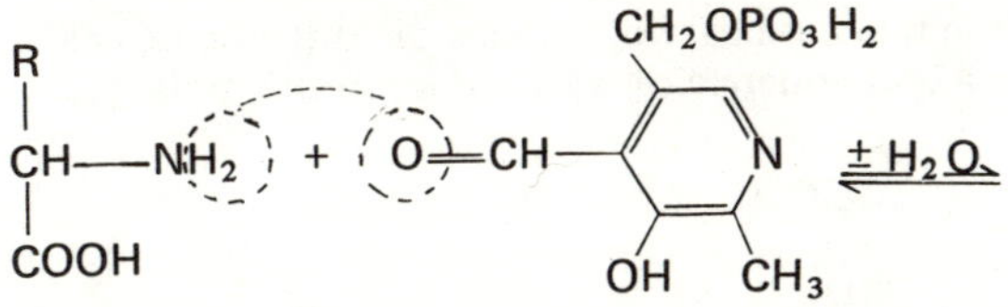

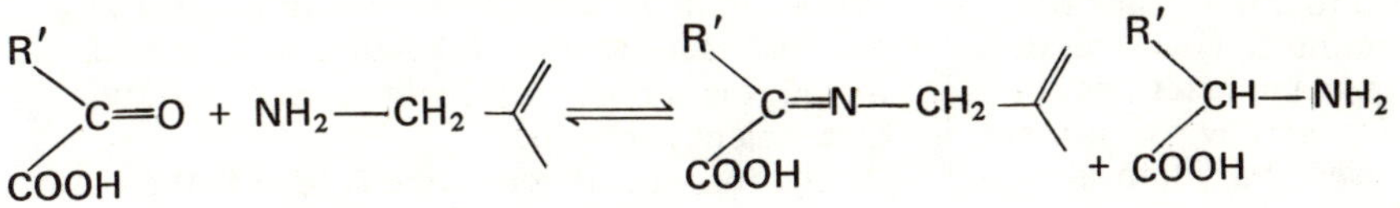

By retracing the steps given above a new amino acid is formed and the pyridoxal form of the coenzyme is regenerated for recycling.

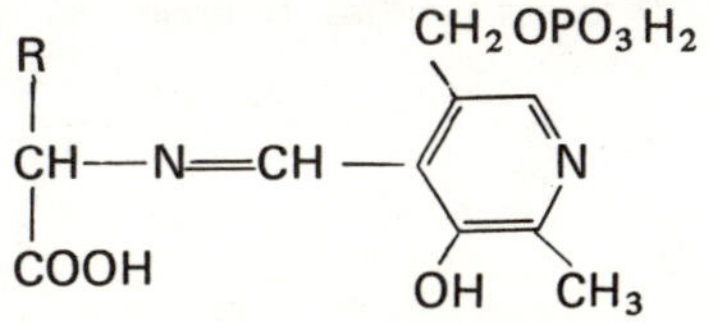

$$\underset{\substack{|\\COOH}}{\overset{\substack{R\\|}}{CH}}-NH_2 \;+\; O=CH-\!\!\!\underset{\substack{|\\OH\;\;CH_3}}{\overset{CH_2OPO_3H_2}{\bigotimes}}\!\!\!N \quad \xrightarrow{\pm H_2O} \quad \underset{\substack{|\\COOH}}{\overset{\substack{R\\|}}{CH}}-N=CH-\!\!\!\underset{\substack{|\\OH\;\;CH_3}}{\overset{CH_2OPO_3H_2}{\bigotimes}}\!\!\!N$$

Rearrangement of the Schiff's base followed by hydrolysis liberates a keto acid and leaves the amino group attached to the coenzyme, i.e. the coenzyme has become pyridoxamine phosphate.

There is good evidence to support Braunstein's scheme and it appears that formation of a Schiff's base is the primary reaction in all pyridoxal phosphate dependent reactions. The coenzyme

$$\underset{\substack{|\\COOH}}{\overset{\substack{R\\|}}{HC}}-N=CH-\!\!\!\Big\langle \xrightarrow{\text{Rearrangement}} \underset{\substack{|\\COOH}}{\overset{\substack{R\\|}}{C}}=N-CH_2-\!\!\!\Big\langle \xrightarrow{\pm H_2O} \underset{\substack{|\\COOH}}{\overset{\substack{R\\|}}{C}}=O \;+\; \underset{\substack{\text{Pyridoxamine}\\\text{phosphate}}}{H_2N-CH_2-\!\!\!\Big\langle}$$

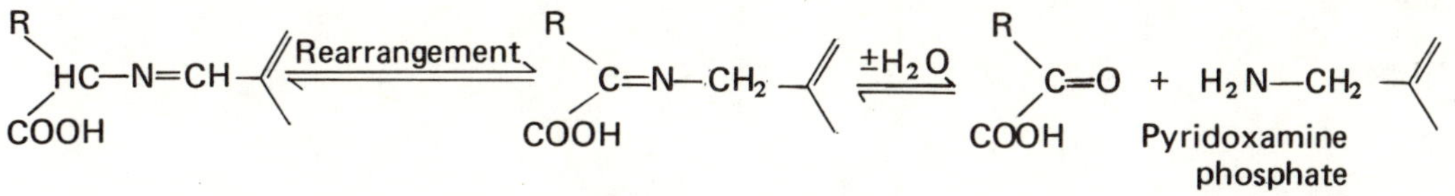

If all the steps are assumed to be reversible, the pyridoxamine form of the coenzyme can react with the donor keto acid to form another Schiff's base.

is bound to the enzyme through an $\epsilon-NH_2$ group of lysine until the amino group of the approaching substrate enters the active site of the enzyme and displaces the $\epsilon-NH_2$ of lysine.

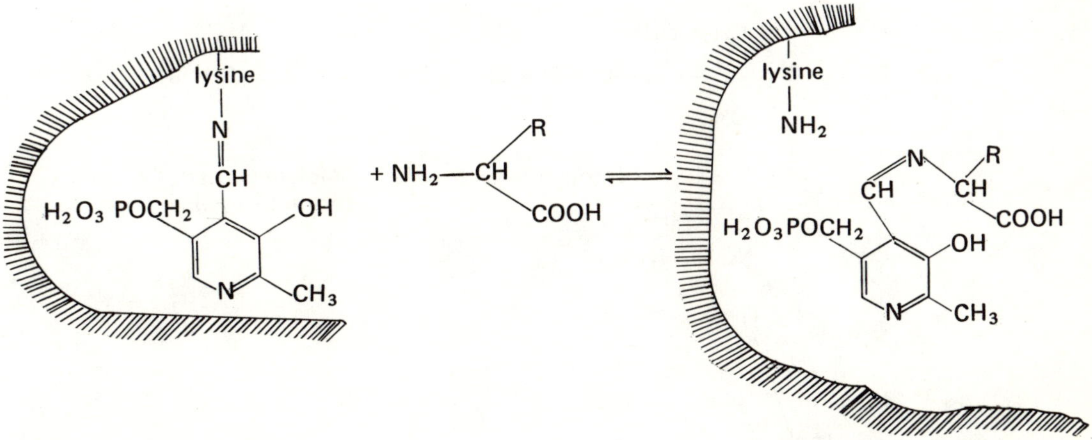

$$\underset{\substack{|\\COOH}}{\overset{\substack{R'\\}}{C}}=O \;+\; NH_2-CH_2-\!\!\!\Big\langle \quad\rightleftharpoons\quad \underset{\substack{|\\COOH}}{\overset{\substack{R'\\}}{C}}=N-CH_2-\!\!\!\Big\langle \quad \underset{\substack{|\\+\;COOH}}{\overset{\substack{R'\\}}{CH}}-NH_2$$

Pyridoxal phosphate bound to enzyme at active site

Schiff's base of substrate and pyridoxal phosphate at active site

The formation of a Schiff's base appears to facilitate the removal of CO_2 from an amino acid in reactions catalysed by decarboxylases. Brain tissue contains an important glutamate decarboxylase which forms γ-amino butyric acid.

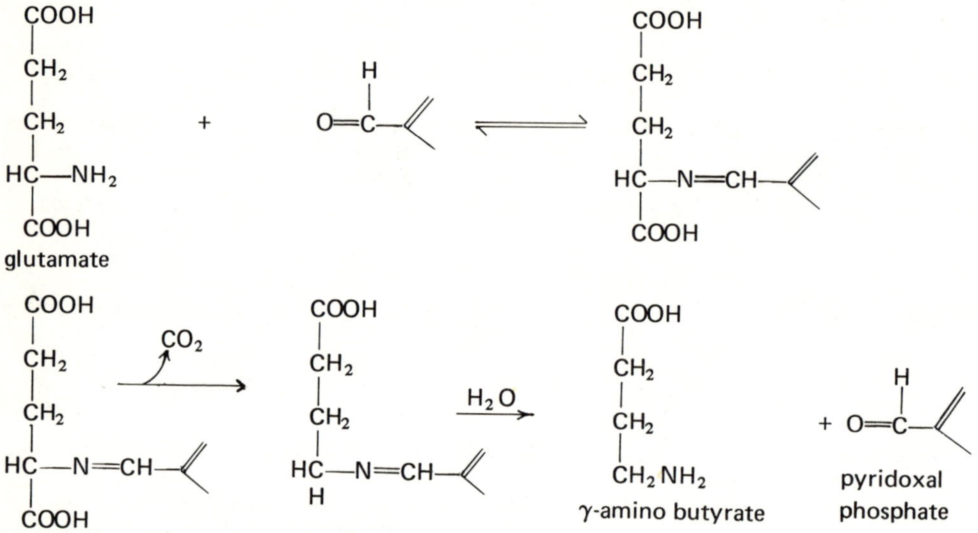

enzyme is obviously essential for full catalytic activity.

Pantothenic acid

The biochemical importance of pantothenic acid is a consequence of its presence in the molecule

glutamate

γ-amino butyrate

pyridoxal phosphate

Pyridoxal phosphate is found in plant and animal phosphorylases bound to lysine residues but the role of the coenzyme in reactions catalysed by phosphorylases is unknown at present. In has been shown, however, that removal of pyridoxal inactivates phosphorylase so the co-

of coenzyme A. Most of the vitamin occurs in this form and because its occurrence is so widespread it is probably difficult to reach a state of deficiency in man.

The formula of coenzyme A (CoASH) is shown below.

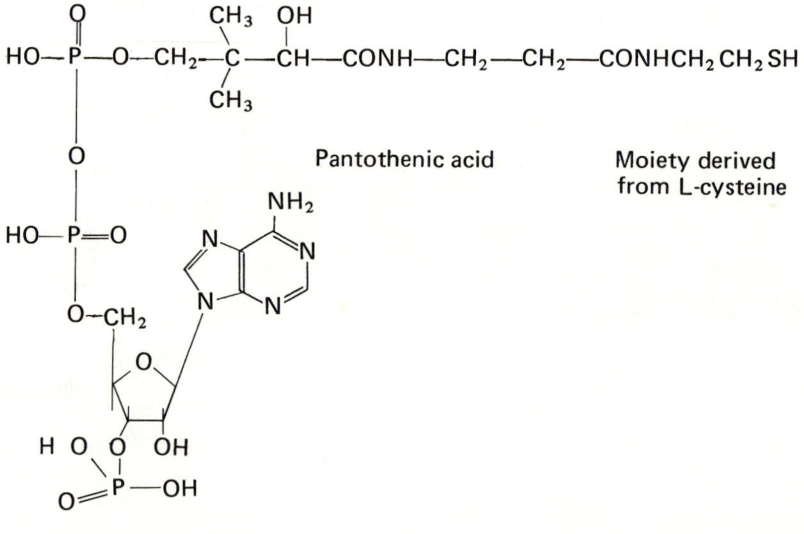

Pantothenic acid

Moiety derived from L-cysteine

It is a complex molecule containing an adenine nucleotide, but the chemically reactive part is the thiol group which is derived from L-cysteine.

Many metabolic reactions generate carbon compounds with —COO⁻ groups. Carboxylic acids are relatively inert and before they can be metabolized further they must be converted to a more reactive derivative. Very often the derivative is a thiol ester formed between the carboxyl group and CoA-SH.

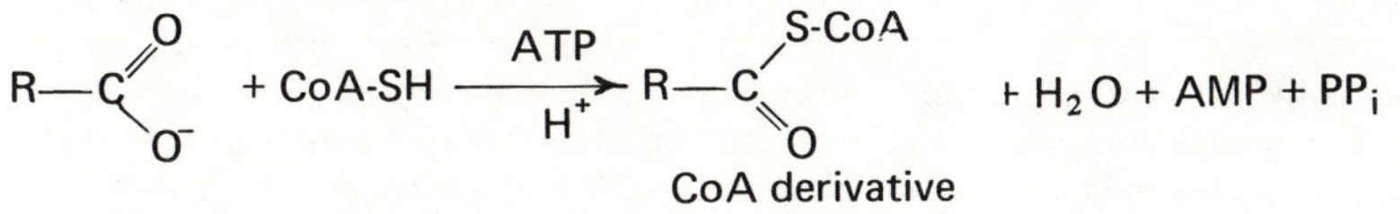

CoA derivative

This ester formation is normally linked to ATP hydrolysis since it is an endergonic reaction.

Thiol esters are more reactive than their oxygen counterparts. The C—S bond is longer and weaker than the C—O bond of esters making nucleophilic displacement at the C atom easier with the release of CoA-SH.

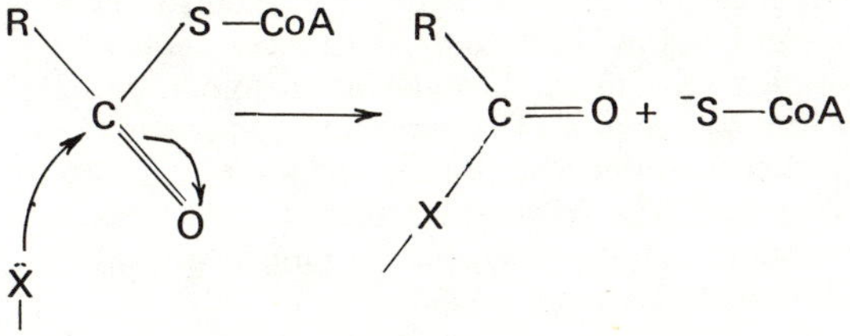

An example of such a reaction is the biosynthesis of the transmitter substance acetylcholine from choline and acetyl-CoA.

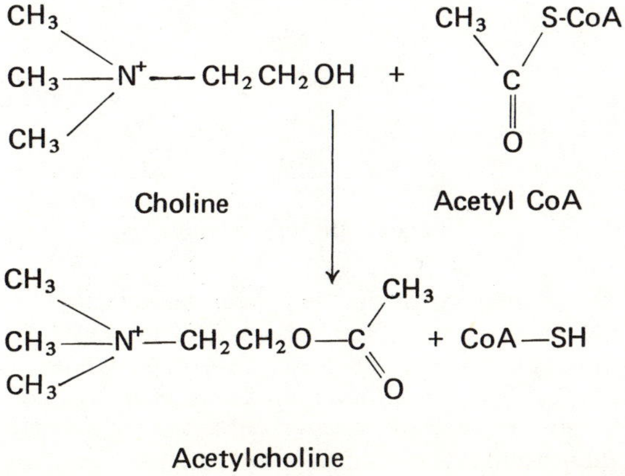

Whereas O esters are stabilized by resonance, sulphur does not release its electrons for double bond formation so readily. As a result the —C=O group of thiol esters is much more like the carbonyl groups of aldehydes and ketones in properties. The effects of polarization of the carbonyl group can be transmitted to an adjacent C atom making a hydrogen on the latter more likely to dissociate in the form of a proton. In so doing, the α-C atom becomes a nucleo-

$+ H_2O + AMP + PP_i$

philic centre and very reactive. Claisen condensations, that are a feature of the reactions of aldehydes and ketones, can then occur at the α-C atom.

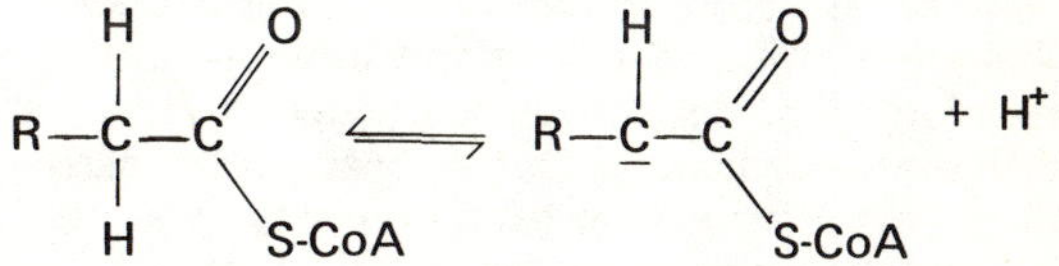

The biosynthesis of citrate from acetyl-CoA and oxaloacetate is an example from metabolic pathways of a condensation occurring at the α-C atom.

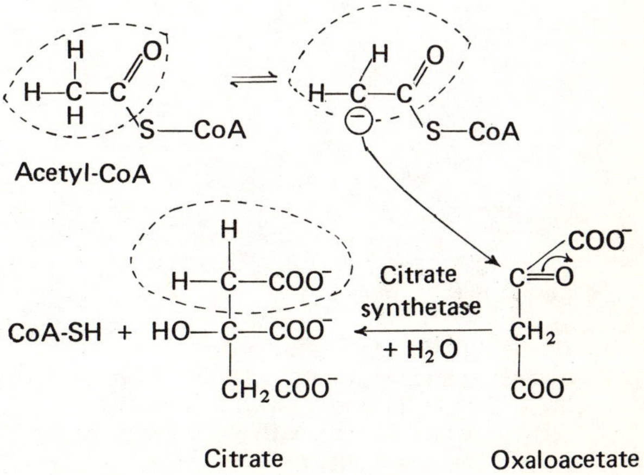

Acetylcholine

Citrate   Oxaloacetate

The increased reactivity of the α-C atom also allows carboxylation reactions to occur, such as the formation of malonyl-CoA from acetyl-CoA and CO_2, and facilitates the dehydrogenation steps in the β-oxidation of long chain fatty acids.

$$CH_3CO\text{-}SCoA \ + \ CO_2 \longrightarrow \underset{CH_2}{\overset{COO^-}{}}\text{—CO—SCoA}$$

acetyl-CoA   malonyl-CoA

Many enzymes have reactive thiol groups which form thiol esters with their substrates and the question arises of why coenzyme A is needed. The properties of thiol esters formed between enzyme and substrate would be very similar to those of CoA derivatives so that it could be argued that the binding of a carboxylic acid to a thiol enzyme would automatically increase its reactivity sufficiently to allow metabolism to proceed. The importance of CoA may be to act as a store for active acyl intermediates. Acylation of the coenzyme and subsequent reaction of the derivative usually occur on different enzymes and CoA may act as the carrier of an acyl intermediate from one enzyme active site to another. In the fatty acid synthetase complex of enzymes an acetyl residue from acetyl-CoA is transferred to a thiol group of the complex, and malonyl is transferred from malonyl-CoA to a second thiol. In both these processes CoA is liberated, presumably to act as carrier for more acyl residues.

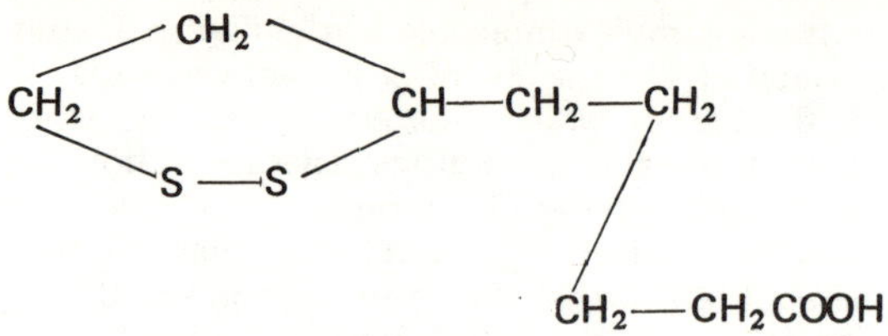

lipoic acid

The coenzyme is bound to protein by a peptide bond formed between the carboxyl group and an ϵ-amino group of a lysine residue. Lipoyl enzymes are important constituents of the multi-enzyme complexes that catalyse the oxidative decarboxylations of some 2-oxoacids, e.g. the formation of acetyl-CoA from pyruvate, and the conversion of oxoglutarate to succinyl-CoA. To complete the two reactions just quoted, CoASH is obviously required, but in addition thiamine pyrophosphate is necessary to function as a carrier of an intermediate which is eventually passed to lipoate.

The sequence of reactions by which pyruvate is converted to acetyl-CoA is believed to be as set out below. An exactly analogous sequence applies to the decarboxylation of oxoglutarate. First, a pyruvate-TPP compound is formed which then decarboxylates, leaving a two-carbon fragment. The fragment is transferred to one of the S atoms of lipoic acid forming S-acetyl-dihydrolipoic acid.

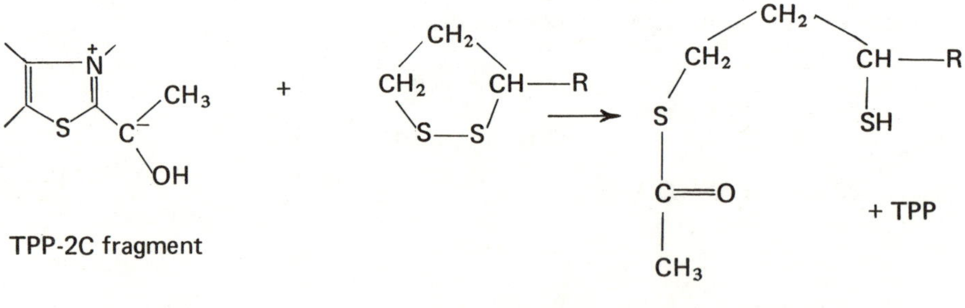

TPP-2C fragment

S-acetyl dihydrolipoate

Lipoic acid

Lipoic acid is a growth factor for some micro-organisms and was first crystallized from beef liver. Chemically, the coenzyme is an eight-carbon carboxylic acid with a disulphide bridge linking C atoms 6 and 8.

An interchange of the acyl group now occurs between lipoate and coenzyme A. This leaves lipoate in the dihydro form. Before reuse the disulphide form of lipoate must be produced by a dehydrogenation reaction catalysed by a specific dihydrolipoate dehydrogenase.

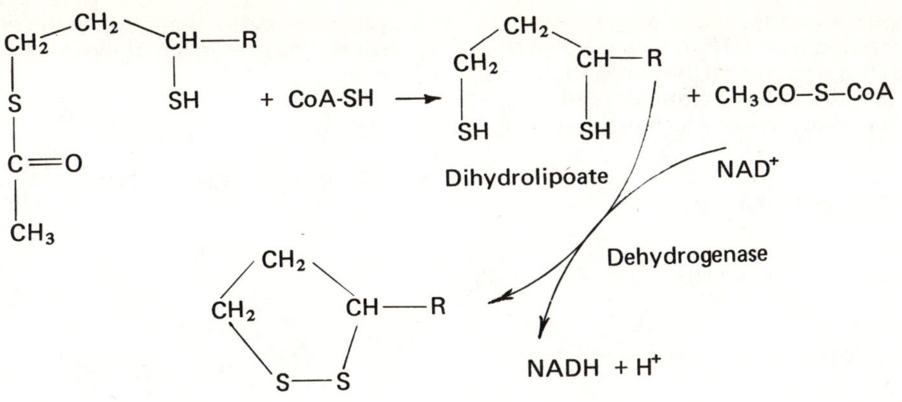

The overall scheme shows how lipoate acts as a bridge between TPP and coenzyme A. The TPP derivative needs to be formed to facilitate decarboxylation and acetate is eventually captured as the CoA derivative ready for reaction with oxaloacetate, to form citrate for example. The enzymes catalysing the decarboxylation, transfer to lipoate, formation of acetyl CoA and regeneration of lipoate are components of a multi-enzyme complex, but the components can be separated under the right conditions and shown to catalyse the reactions outlined in the scheme. This evidence lends support to the proposed mechanism.

Biotin — Vitamin H

Feeding animals with large amounts of egg white leads to symptoms of nutritional deficiency. Studies of the deficiency disease showed that egg yolk, milk, yeast and liver were all very effective in reversing the symptoms and in the years 1936—46 the growth factor was isolated from egg yolk and liver. That isolated from liver was originally named Vitamin H but it is now usually called biotin.

Biotin deficiency arises from ingestion of egg white because the latter contains a protein, avidin, which binds the growth factor very tightly. Avidin is a potent inhibitor of biotin-dependent enzymes. The structure of biotin was determined in 1942 and is shown below.

The vitamin occurs mainly combined with protein. Binding to protein is by a peptide link formed between the COOH of the vitamin and an ϵ-amino group of a lysine residue in the protein. Both in structure and in its mode of binding to enzymes, biotin resembles lipoic

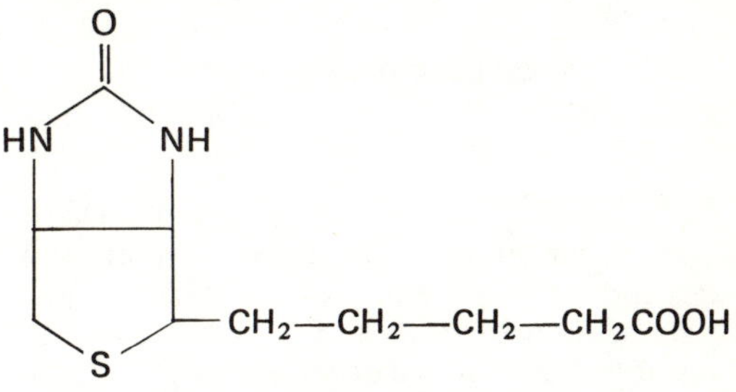

Biotin

acid. As biotin and lipoic are covalently bound to enzymes, dialysis is insufficient to separate enzyme and coenzyme and both may be classed as prosthetic groups.

That biotin might be important in "CO_2-fixation" reactions was suggested by the finding, in 1942, that a dicarboxylic acid, aspartic acid, could partially replace biotin as a growth factor for yeast. A few years later, the incorporation of ^{14}C labelled bicarbonate into aspartate was proved to be biotin-dependent. Before the role of biotin was understood it was believed that "activated-CO_2" was required for CO_2-fixation.

It was suggested that activation of CO_2 might occur by reaction with ATP to form carbamyl phosphate. It is now known, however, that activated-CO_2 is a carboxy-biotin-enzyme complex, the formation of which requires ATP.

$$ATP + HCO_3^- + \text{biotin-enzyme} \underset{}{\overset{Mg^{2+}}{\rightleftharpoons}} CO_2 -$$

$$\text{biotin-enzyme} + ADP + P_i$$

A molecule of CO_2 is incorporated into biotin at the 1' position.

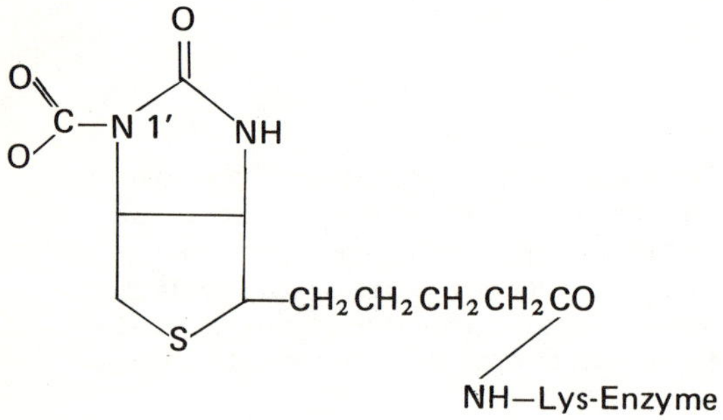

Carboxy-biotin enzyme

Carboxylation reactions are used biologically to increase the length of the carbon chain in substrates and are important in anabolic pathways. In fatty acid synthesis a key priming reaction incorporates CO_2 into acetyl-CoA to give a derivative of the 3-carbon acid, malonic acid. Biotin is an essential cofactor in the reaction.

$$ATP + HCO_3^- + CH_3CO\text{-}SCoA \xrightarrow[Mg^{2+}]{Biotin}$$

(reactive form   acetyl-CoA
of CO_2)

$$\begin{array}{c} COO^- \\ | \\ CH_2CO\text{---}SCoA + ADP + P_i \end{array}$$

malonyl-CoA

Pyruvate carboxylase is another biotin enzyme. It catalyses the conversion of pyruvate to oxaloacetate.

$$ATP + HCO_3^- + CH_3CO.COO^- \underset{Mg^{2+}}{\overset{Biotin}{\rightleftharpoons}}$$

pyruvate

$$\begin{array}{c} COO^- \\ | \\ CH_2\text{---}CO.COO^- + ADP + P_i \end{array}$$

oxaloacetate

The enzyme probably performs a useful physiological function in the reversal of glycolysis by overcoming the irreversible step by which phosphoenolpyruvate is converted to pyruvate. The oxaloacetate formed in the carboxylase reaction can give rise to phosphoenolpyruvate by reaction with guanosine triphosphate (GTP).

$$\begin{array}{c} COO^- \\ | \\ CH_2CO.COO^- + GTP \end{array} \underset{\text{carboxy kinase}}{\overset{\text{phosphoenolpyruvate}}{\rightleftharpoons}}$$

$$\begin{array}{c} COO^- \\ | \\ CH_2{=}C\text{---}OPO_3H_2 + CO_2 + GDP \end{array}$$

phosphoenolypruvate

Folic acid

The structure of folic acid is based on a group of compounds known as pterins which contain a ring system that was unknown to organic chemists until the pigments of butterfly wings were isolated and identified. The vitamin contains a p-aminobenzoic acid residue covalently bound to L-glutamic acid.

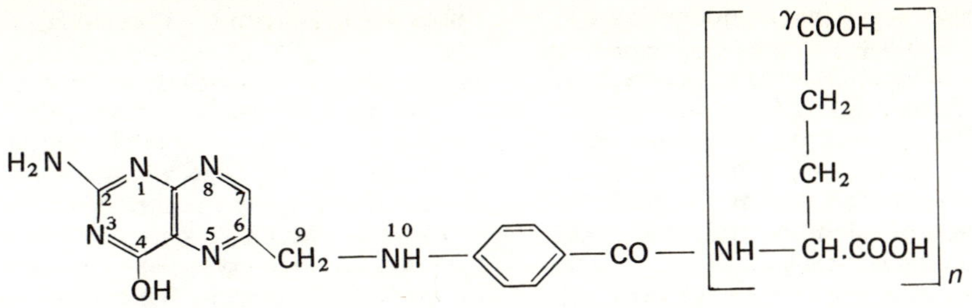

pterin derivative   p-aminobenzoic acid

Polyglutamic acid derivatives are known with up to seven glutamic molecules joined to each other through γ-glutamyl peptide bonds.

An understanding of the biochemical function of folic acid derived from studies of bacteria for which the vitamin is an essential nutritional factor. Deficiency in man is associated with megaloblastic anaemia in which changes in the blood and bone marrow identical with those of pernicious anaemia caused by Vitamin B_{12} deficiency are seen. Apart from unsuitable diets or methods of cooking which destroy the vitamins, folate deficiency can result from an increased requirement for it (e.g. in pregnancy),

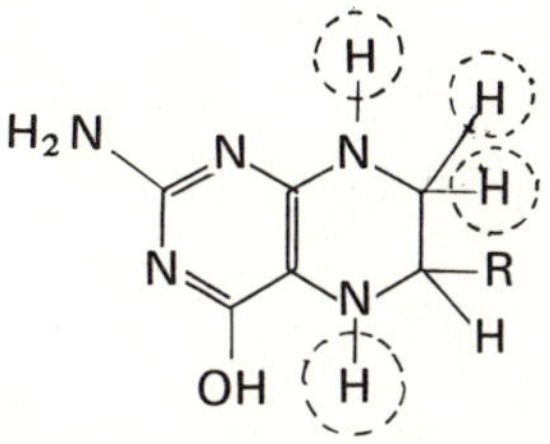

Tetrahydrofolate (FH_4)
Four H-atoms added on
reduction are ringed.

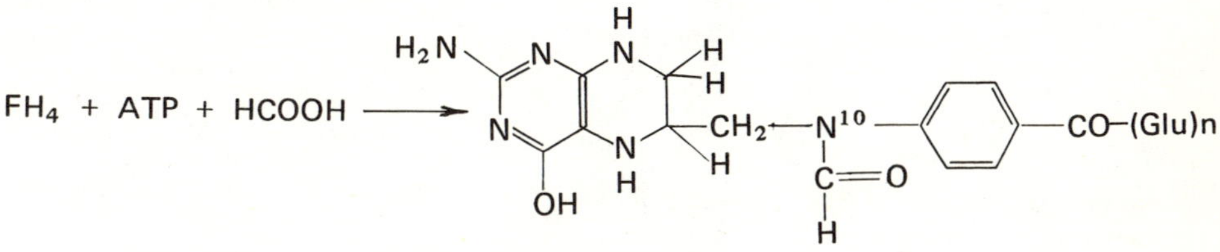

$$FH_4 + ATP + HCOOH \longrightarrow$$

Formyl N^{10}-FH_4

or from malabsorption from the intestine as a result of coeliac disease or tropical sprue or, to a less severe degree, from surgical removal of part of the jejunum.

There are numerous reactive coenzyme forms of the vitamin and all of them are formed from folate by reduction. Reductase enzymes linked to NADPH convert folic acid to tetrahydrofolate (FH_4).

FH_4 and related derivatives act as carriers of 1-C units, particularly formate. Formyl-FH_4 derivatives are required for the biosynthesis of purine nucleotides, serine and glycine. Formyl N^{10}-FH_4 is formed from FH_4 and formate by a reaction involving ATP.

Formyl N^{10}-NH_4 can undergo ring closure between the C atom of the formyl group and N^5 of the pterin ring. The derivative thus formed is called methenyl N^{5-10}-FH_4.

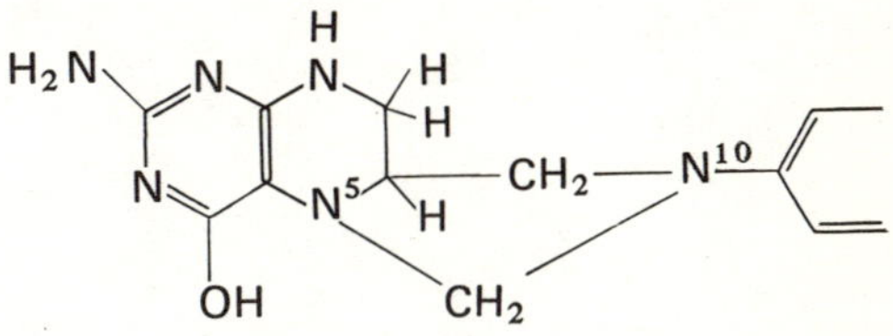

Methenyl N^{5-10}-FH_4

Both of these formylated compounds are important in the biosynthesis of purine nucleotides. Earlier in this chapter mention was made of the competitive inhibition by sulphonamide of enzymes that metabolize p-aminobenzoic acid. When it became apparent that folate was very important in nucleic acid metabolism the antibiotic effects of sulphonamide drugs could be interpreted biochemically.

Other forms of the vitamin carrying 1-C groups (e.g. methyl; CH_3) are known. All these derivatives can arise from the two examples described here by interactions with appropriate enzymes.

Cobamide coenzymes — Vitamin B_{12}

A variety of compounds with Vitamin B_{12} activity have been isolated from animal and microbial sources. The vitamin contains an atom of tervalent cobalt and forms of the vitamin differ in the anion that is coordinated to the cobalt atom. Vitamin B_{12} can be isolated from liver as a red crystalline substance containing a CN^- or OH^- anion. The cyanide derivative is known as cyanocobalamin (formula below, left).

The coenzyme form of the vitamin has the CN^- group replaced by 5'-deoxy adenosine (shown below, right).

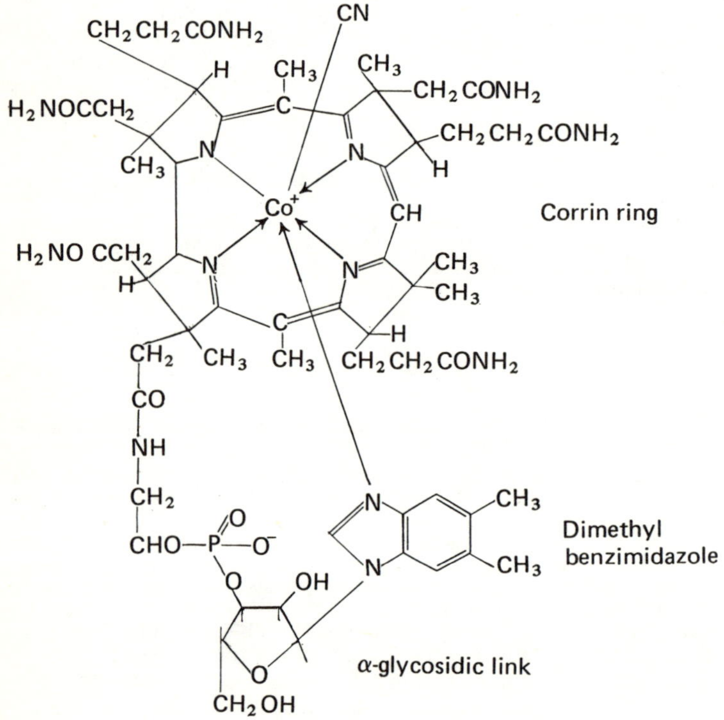

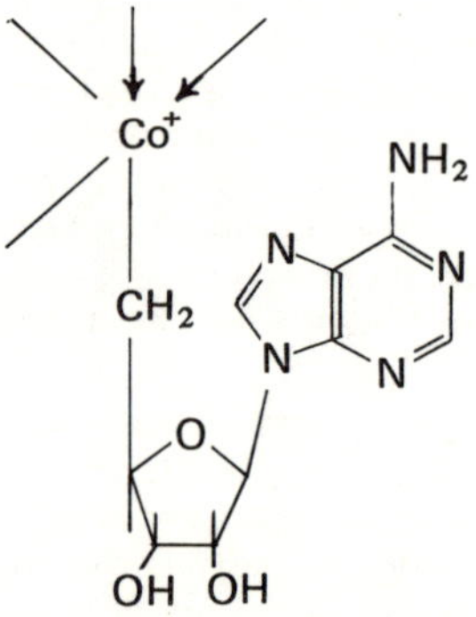

Corrin ring

Dimethyl benzimidazole

α-glycosidic link

5'-deoxy adenosine

The 5'-deoxy adenosine portion of the molecule is easily replaced by CN^- or OH^- and it is likely that cyanide and hydroxy derivatives are artefacts of the isolation procedure and that the vitamin exists in nature always as the coenzyme form.

The coenzyme is more complex in structure than any other coenzyme known. A nucleotide-like group composed of dimethyl benzimidazole joined by an α-glycosidic link to a ribose-3-phosphate residue, is coordinated to the cobalt ion. Naturally occurring analogues have been discovered in which dimethyl benzimidazole is replaced by adenine or guanine.

Several reactions are known for which the coenzyme is essential including the isomerization of methyl malonyl-CoA to succinyl-CoA.

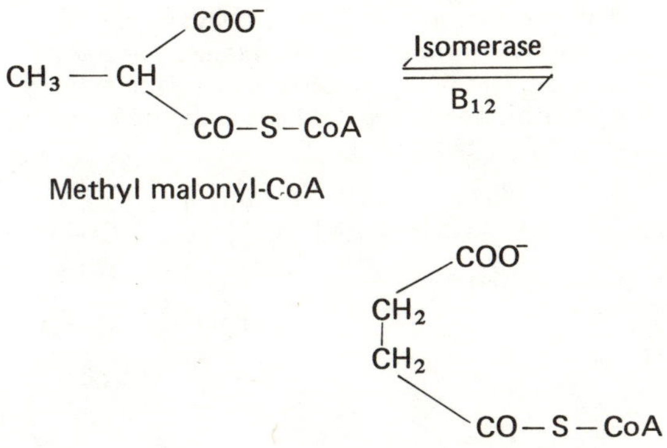

Methyl malonyl-CoA

succinyl-CoA

The isomerization allows entry into the tricarboxylic acid cycle of the terminal metabolite of the β—oxidation of fatty acids containing an odd number of carbon atoms. The reaction involves the transfer of hydrogens to and from the $-CH_2$ group of the 5' deoxy adenosine portion of B_{12}. Other B_{12}-linked reactions may proceed by a somewhat different mechanism however, involving other parts of the coenzyme molecule.

Pernicious anaemia results from a deficiency of Vitamin B_{12}. Failure to absorb the vitamin from the intestine classically follows atrophy of the glandular mucosa of the fundus of the stomach, which produces the intrinsic factor necessary for absorption of Vitamin B_{12}. The vitamin is effective in treatment if administered parenterally. The biochemical relationship between anaemia and coenzyme function is unknown but as more understanding of the mechanisms of B_{12}-dependent enzymes is reached it may become possible to suggest causes of the anaemia accompanying deficiency.

Ascorbic acid — Vitamin C

A deficiency of ascorbic acid in man and guinea pigs causes scurvy. Other mammals, the rat for example, can synthesize the vitamin from carbohydrate. It is also synthesized by higher plants, and fresh fruits and vegetables are the main dietary source of the vitamin. The antiscurvy factor was first isolated in crystalline form in 1928 and the structure of the vitamin was worked out by degradation and synthesis just four years later. Vitamin C is alternatively called L-ascorbic acid.

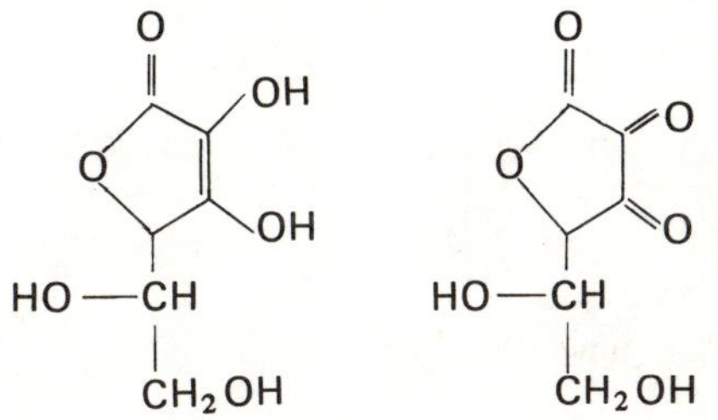

L-Ascorbic acid   L-Dehydroascorbic acid

Ascorbic acid is a strong reducing agent and is readily converted to L-dehydroascorbic acid. Both forms of the acid are lactones and the acidic properties arise from the tendency of the ene-diol structure of the reduced form to ionize.

In spite of the fact that the effects of a deficiency of ascorbic acid have been known for many years, the biochemical function of the vitamin is still very obscure. It seems likely that the interconversion of the oxidized and reduced forms could make ascorbate an effective hydrogen carrier in biological systems. Plants contain an ascorbate oxidase enzyme which converts ascorbic acid to the dehydro form but animal tissues lack this enzyme and it is possible that conversion can be brought about by the electron transport chain. In scurvy there is disturbance of carbohydrate, fat and amino acid metabolism which suggests a key role for the vitamin in these processes. The metabolism of the aromatic amino acid tyrosine seems particularly affected. It has been suggested that the reducing properties of ascorbic acid are useful for generating "active-SH" groups from disulphide linkages in enzymes and coenzymes and, while this is a possible function of the vitamin, it seems almost certain that other more direct actions of ascorbate will be discovered eventually.

Glutathione — GSH

This compound is present in all living cells and was first discovered in 1888. It remained forgotten, however, until 1921 when F. G. Hopkins rediscovered it. Finding that the substance contained glutamic acid and cysteine, Hopkins named it glutathione. By the middle of the 1930s the structure of glutathione (GSH) had been established by degradation and synthesis.

GSH is a tripeptide, γ-L-glutamyl-L-cysteinyl-glycine.

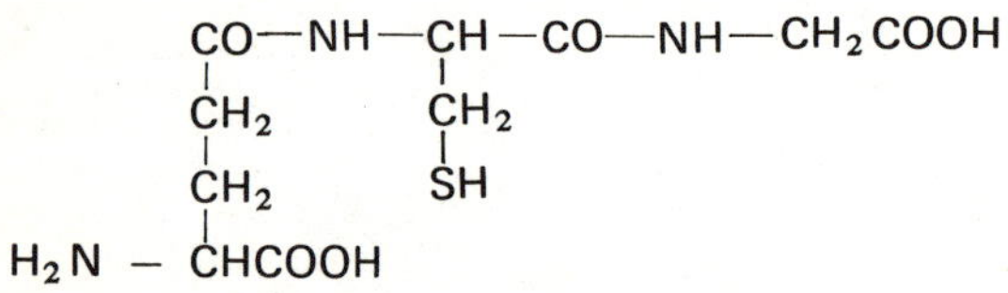

$$CO-NH-CH-CO-NH-CH_2COOH$$
$$CH_2 \qquad CH_2$$
$$CH_2 \qquad SH$$
$$H_2N - CHCOOH$$

Glutathione

Oxidation of two molecules of GSH leads to a disulphide compound, oxidized glutathione (GSSG).

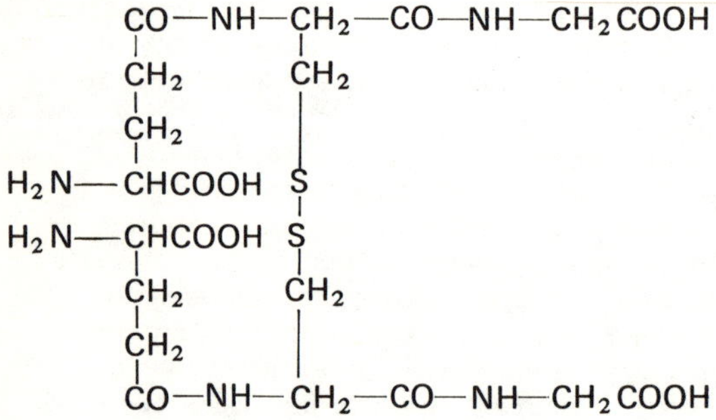

$$CO-NH-CH_2-CO-NH-CH_2COOH$$
$$CH_2 \qquad CH_2$$
$$CH_2 \qquad |$$
$$H_2N-CHCOOH \quad S$$
$$H_2N-CHCOOH \quad S$$
$$CH_2 \qquad CH_2$$
$$CH_2 \qquad |$$
$$CO-NH-CH_2-CO-NH-CH_2COOH$$

Oxidized glutathione

The oxidation by molecular oxygen was shown to be catalysed by tissue extracts and it was believed at first that glutathione had a respiratory function. The growth of knowledge of the cytochromes led to the abandonment of the respiratory theory and subsequently numerous

coenzyme roles have been suggested for glutathione in which it functions in an analogous way to lipoic acid. However, only in very few instances have specific coenzyme requirements for glutathione been demonstrated. A specific involvement of glutathione has been shown for the glyoxalase reaction. This is a reaction that converts methyl glyoxal to lactate.

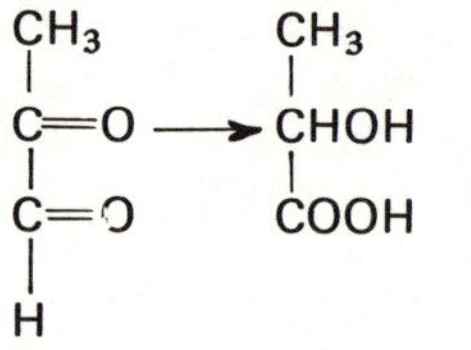

$$CH_3 \qquad CH_3$$
$$C=O \longrightarrow CHOH$$
$$C=O \qquad COOH$$
$$H$$

methyl glyoxal   lactate

The conversion occurs in two steps catalysed by two components, glyoxalase I and glyoxalase II. The first step results in a thiol ester of gluthathione with lactate which is then cleaved by glyoxalase II.

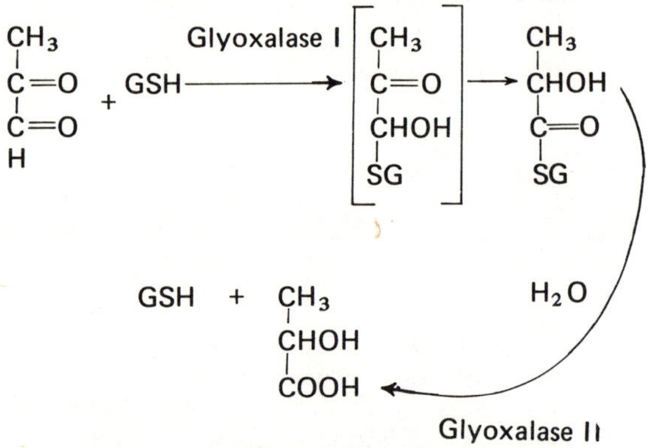

$$CH_3 \quad \text{Glyoxalase I} \quad \begin{bmatrix} CH_3 \\ C=O \\ CHOH \\ SG \end{bmatrix} \longrightarrow \begin{array}{c} CH_3 \\ CHOH \\ C=O \\ SG \end{array}$$
$$C=O \; + \; GSH \longrightarrow$$
$$C=O$$
$$H$$

$$GSH \; + \; \begin{array}{c} CH_3 \\ CHOH \\ COOH \end{array} \qquad\qquad H_2O$$

Glyoxalase II

The very widespread occurrence of glyoxalase in living tissues led investigators into a search for a reaction forming methyl glyoxal so that glyoxalase could be fitted into the glycolytic sequence. When it became clear that methylglyoxal was remote from glycolysis there was a loss of interest in the physiological importance of the reaction. A. Szent-Gyorgyi has reawakened interest by the hypothesis that methyl glyoxal is a growth-inhibiting substance and that it and glyoxalase are present in the tissues to control cell division. During periods of no growth glyoxalase is unable to break down methyl glyoxal, but when tissue growth is required the enzyme is allowed to come into contact with the substrate to break it down and release the inhibition. Szent-Gyorgyi believes that cancer results from the loss of the normal ability of cells to prevent the action of glyoxalase on methyl glyoxal during quiescent periods.

A non-specific role suggested for glutathione is that it protects essential thiol groups on enzymes from oxidation or generates thiols from disulphides. Disulphide exchange reactions have been demonstrated with glutathione, and an NADPH-linked enzyme that reduces oxidized glutathione is of widespread occurrence. The sequence of reactions involved is outlined below.

$$
\text{(i)} \quad
\begin{array}{c} \top \\ S \\ | \\ S \\ \bot \end{array}
+ \text{GSH} \rightarrow
\begin{array}{c} \top \\ SH \\ \\ S\!-\!SG \\ \bot \end{array}
+ \text{GSH} \rightarrow
\begin{array}{c} \top \\ SH \\ \\ SH \\ \bot \end{array}
+ \text{GSSG}
$$

Inactive enzyme   Active enzyme

$$
\text{(ii)} \quad \text{GSSG} + \text{NADPH} + \text{H}^+ \xrightarrow{\text{Reductase}} 2\text{GSH} + \text{NADP}^+
$$

The reductase regenerates glutathione for recycling. In living cells practically all of the glutathione is present in the thiol form so the reductase must catalyse the immediate reduction of any oxidized glutathione formed by disulphide exchange.

GENERAL CONCLUSIONS

One of the most interesting and noticeable features of coenzyme structure is the tendency for them to contain nucleotide or nucleotide-like groups, especially the adenine-ribose-phosphate (AMP) complex. From the foregoing accounts of the part played by coenzymes in the chemical events during catalysis, it is seen that the non-nucleotide portion of the molecule usually carries the functional groups. This immediately suggests the possibility that the nucleotide fragment is chiefly concerned with the binding of the coenzyme to protein. Evidence that perhaps supports this suggestion is the finding that coenzymes like biotin, pyridoxal

phosphate and lipoic acid, which do not possess nucleotide groups, need to be covalently bound to protein through a lysine residue whereas nucleotide coenzymes such as NAD^+ do not.

Is the role of the nucleotide fragment then just to act as an anchor so that the reactive portion of the coenzyme is brought into proximity with the enzyme active site? Recent evidence indicates that the function may be more than this. When NAD^+ binds to dehydrogenases the AMP portion probably induces the correct fit at the active site for the nicotinamide end of the coenzyme, which is involved in the oxidation-reduction reaction. It has been shown that nicotinamide itself is unable to bind to dogfish lactate dehydrogenase but the addition of AMP to a mixture of enzyme plus nicotinamide brings about a change in protein conformation and nicotinamide can then bind. The AMP fragment of NAD^+ therefore seems to perform at least two functions. First, it does seem to anchor the coenzyme by interactions of the purine ring with the side chains of protein amino acids and, second, it facilitates the binding of the reactive part of the coenzyme to the active site close to catalytically important groups of the protein. If nucleotide-induced conformational changes are of general occurrence then the significance of the nucleotide fragment is clear.

The choice of nucleotides to act as anchors for coenzymes could have occurred purely by chance very early in the evolutionary time-scale. Presumably an alternative choice could have resulted in amino acids carrying functional coenzyme groups with binding properties analogous to those of nucleotides, i.e. capable of binding non-covalently so that shuttling from one enzyme to another can occur fairly easily. The demands on the available pool of amino acids for protein synthesis may, however, have necessitated a selection of other kinds of molecules for building into coenzymes. An alternative theory that accounts for the pre-valence of AMP in coenzymes stems from a knowledge of protein synthesis. Completion of a polypeptide chain leaves a coiled poly-peptide attached to a molecule of tRNA through adenosine (Chapter 3). It is possible that an "imprint" of the nucleotide is left on the protein which may then always permit the binding of a nucleotide structure. If cleavage of the poly-peptide tRNA complex sometimes left the adenosine still bound to the polypeptide then not only an imprint would be left but a structure analogous to an enzyme with bound coenzyme. The imprint theory implies that all proteins should be capable of binding nucleotides which does not seem to fit the facts as we known them. The *potential* for binding may be a general property, however. That it does not occur with all proteins may be a reflection of differences in primary structure since this dictates tertiary structure and the latter in turn determines the specificity for binding of any ligand, nucleotides included. Thus although an imprint may be left, there are no suitable groups on the "non-binding" protein to hold a nucleotide in place.

Assuming, then, that certain polypeptides evolved which could bind nucleotide structures, a way was open for natural selection to operate along two parallel paths. First, changes in enzyme primary structure that facilitated and strengthened nucleotide binding could occur and, second, the development and diversification of true coenzymes by the addition of one or more groups to nucleotides would be possible. The end result is an enzyme with a specific coenzyme requirement. In some proteins the potential for nucleotide binding seems to have been put to a different use. These are the enzymes whose activities are regulated by the binding of nucleotides such as AMP. The nucleotides probably trigger off conformational changes that affect the binding of substrate so that the actions of nucleotides both as regulatory modifiers and as components of coenzymes seem to be promoted in identical ways.

6 Specificity and the Active Site

When the kinetics of enzyme action were discussed the assumption was made that an enzyme-substrate complex is formed during enzymic catalysis. The molecular weights of enzymes range from about 10 000 to a few million. Taking the average molecular weight of the amino acids which are the constituents of proteins as 100 indicates that even the smallest known enzyme (with a molecular weight of 9000) contains approximately 90 amino acids. Most substrates are very much smaller molecules than the enzymes that act upon them; many substrates are of similar molecular size to a single amino acid. The use of reagents which chemically modify only those amino acids in the regions of the protein where substrate is assumed to bind, and kinetic binding studies, have shown that enzymes contain very few of these regions. Many enzymes, in fact, seem to possess just one binding site for substrate. Thus only a small area of the enzyme protein can be involved in any interactions that lead to the formation of an enzyme-substrate compound. Large molecules such as proteins, nucleic acids and poly-saccharides are substrates for some enzymes and in such cases a dissimilarity in molecular size between enzyme and substrate may not exist. Even in these cases, however, a very limited region of the substrate, perhaps a single chemical bond, is involved in the catalysed reaction. This suggests that even large substrates are bound to a very small region of the protein. The region where substrate is assumed to bind is called the active site or centre of the enzyme.

That the active site constitutes such a small proportion of the enzyme molecule raises the interesting question of why enzymes are such large molecules. Since any treatment that affects protein tertiary structure usually results in a loss of activity, the spatial alignment of potentially reactive groups is obviously very important. D. E. Koshland suggested that the amino acids in enzyme proteins could be divided into three classes according to function. Certain amino acids carry reactive groups that act on the substrate during catalysis; others, called contact amino acids, might be important in binding the substrate to the enzyme in such a way that the bond to be broken is brought to a favourable position for attack by the catalytic amino acids; while the third class of amino acid side chains might play a purely structural role in maintaining the three dimensional shape characteristic of the native enzyme. The three classes of amino acids are illustrated diagrammatically in Fig. 6.1 which represents a portion of the enzyme molecule including the active site.

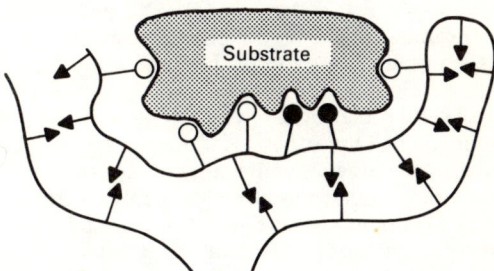

Fig 6.1. Binding of substrate to the active site region of the enzyme. The "contact" amino acids (represented by O) determine the specificity of the enzyme by inter-acting with groups on the substrate while the "catalytic" amino acids (represented by ●) act on a particular chemical bond in the substrate. The active site is a cleft formed by folding of the polypeptide chain and the majority of amino acids (represented by ▲) are responsible for maintaining the three-dimensional shape of the protein, and hence the cleft, so that substrate molecules can be correctly bound and acted upon.

The contact amino acids are assumed to contribute to the specificity of the enzyme for its substrate. Only a limited change in the structure of the substrate could be tolerated before the ability of the groups on the contact amino acids to hold the substrate in place is lost. The forces involved in the binding of substrate would be expected to include those associated with the maintenance of protein secondary and tertiary structure, e.g. hydrogen bonding, ion pair formation and hydrophobic interactions. It is of interest in this connection that for the enzyme ribonuclease which hydrolyses the phosphodiester links of pyrimidine nucleotides in RNA there is evidence that the specificity for pyrimidine nucleotides resides in their ability to form hydrogen bonds with groups on the enzyme. The groups on the substrate which participate in hydrogen bonding to the enzyme are the same as those that are involved in base-pairing of double-stranded polynucleotide structures.

To explain specificity, Emil Fischer in 1894 suggested that the absorption of substrate on to the enzyme is analogous to a key fitting into a lock. The matching fit of substrate and enzyme thus brings the bond to be broken close to the reactive groups. From studies of the kinetics of the binding of substrates and inhibitors to a number of enzymes, Koshland concluded that the rigidity of structure implied in the lock and key theory was unlikely to be met in enzymes. Koshland's results suggested that there was a certain degree of flexibility in the arrangement of the amino acid side chains at the active site. The approach of substrate may itself produce the arrangement at the active site most favourable for catalysis to ensue. This "induced fit" hypothesis is illustrated in Fig. 6.2. Substrate analogues which are competitive inhibitors may also induce changes of conformation on binding but fail to bring about the proper alignment of the catalytic groups.

Direct proof of the induced fit hypothesis is difficult to obtain. As more data become available on the three dimensional structure of enzymes from X-ray diffraction studies, it may be possible to decide whether a conformational change does occur when substrate binds. In the few cases for which it has been possible to examine protein crystals into which the substrate or an analogue has been allowed to diffuse, the X-ray data sometimes indicate a conformational change in the region of the active site accompanying the binding of a substrate analogue. A change is particularly noticeable in carboxypeptidase A where a tyrosine residue moves about 14 Å when an analogue enters the active site. Such large changes have not

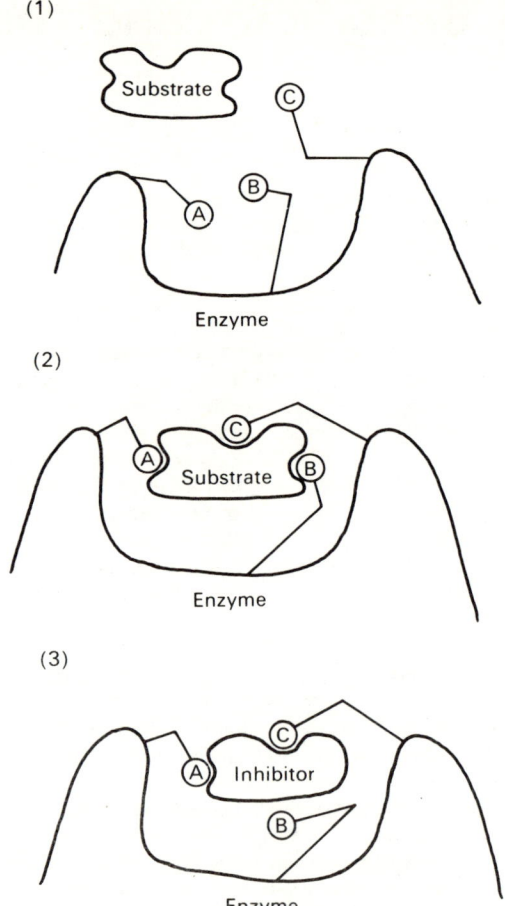

Fig 6.2. Representation of "induced fit" hypothesis. In (1) the active site groups A, B and C are not in the best position for interaction with substrate. Close approach of the substrate leads to a change in the orientation of the groups (as in 2) and sets up a favourable active site. The binding of a competitive inhibitor is shown in (3). The resemblance to the substrate is sufficient to cause some alignment of the important groups but the fit is imperfect and catalysis does not occur.

been detected in lysozyme or chymotrypsin, however.

THE IDENTIFICATION OF REACTIVE GROUPS AT THE ACTIVE SITE OF ENZYMES

Mapping the region that constitutes the active site has been attempted, with varying amounts of success, for many enzymes. The more that can be learned about the nature and arrangement of groups at the active site, the easier it may become to understand how the enzyme works. A full understanding of this problem is, at present, a remote goal because even with the much less effective catalysis brought about by much simpler substances such as platinum, it is not yet possible to describe the exact mechanism in many cases, but the identification of a particular group at the active centre at least provides a starting point for the formulation of a mechanism. Subsequently it may be possible to test the proposed mechanism by a different kind of experiment.

Some of the methods available for probing the active centre of enzymes have already been touched upon in other sections of this book. For instance, a study of the variation of K_m and maximum velocity with pH often provides the first indirect clues about ionizable groups involved in catalysis. As discussed in the section on pH, the pK_a for an ionizable group in a protein may differ significantly from the expected value in free solution because of interactions with neighbouring groups on the protein or with constituents of the buffer. Further difficulties arise because the pK_a values for several of the titratable groups found in proteins overlap to some extent. For example, phenolic hydroxyl groups of tyrosine and amino groups both have pK_a values close to 10. The method can be refined to some extent by repeating the experiments at several temperatures. The change in pK_a with temperature is often characteristic and may help in the identification of a particular group.

The surest way of mapping the active site is to react the enzyme with substrate, separate the enzyme-substrate complex and degrade the protein to determine the amino acid residue that is combined with the substrate. This method requires that conditions can be found under which the enzyme-substrate complex is very stable. Usually the intermediate complex cannot be isolated but in a few cases it has been possible to study it. Phosphoglucomutase is an enzyme of the glycolytic sequence which catalyses the interconversion of glucose-l-phosphate and glucose-6-phosphate. A phosphorylated enzyme intermediate is formed during the reaction.

Koshland was able to separate the phospho-enzyme and after degradation of the protein showed that the phosphate group was bound to a serine residue adjacent to a histidine residue. A reactive serine residue has been detected in a number of hydrolytic enzymes and the amino acid composition in the neighbourhood of the serine is very similar in many of these enzymes. Another example in which an enzyme substrate complex can be isolated is provided by alkaline phosphatase. Inorganic phosphate is not a true substrate for this enzyme but presumably it binds to the active site since it is competitive inhibitor of the enzyme and a product of its action. The phosphorylated enzyme is stable under acid conditions and, if a radioactive isotope of phosphorus is used in the phosphorylation process, the isotopic label can be found in phosphoryl serine when the protein is degraded.

When conditions cannot be found where the intermediate is stable it is sometimes possible, by chemical modification, to "fix" the intermediate as it is formed at the active site. For example, Westheimer using acetoacetate labelled with ^{14}C studied acetoacetate decarboxylase in the presence of the strong reducing agent sodium borohydride. The enzyme was inactivated but became labelled with the radioisotope. After hydrolysis of the labelled protein, N-ϵ-isopropyllysine was identified in the hydrolysate. Thus the intermediate must be formed by a reaction between the keto group of acetoacetate and the $-NH_2$ of a lysine residue and the intermediate could then be reduced by the borohydride to form a stable amine (*see* p. 78).

Many enzymes have been investigated by methods that rely upon labelling the active site with compounds that resemble the substrate to the extent that a covalent enzyme-substrate intermediate is formed, but differ from the true substrate in that the complex is stable and cannot react further. These compounds are sometimes called pseudo-substrates. The effect of DFP on esterase enzymes has already been described (Chapter 5). However, compounds such as DFP are chemically quite reactive and can react with groups in proteins remote from the active site. If the inhibition is rapid and stoichiometric amounts of inhibitor bring about almost total inhibition of the enzyme then

Glucose-1-phosphate + phosphoenzyme $\rightleftharpoons$ Enzyme + Glucose-1,6-diphosphate
Enzyme + Glucose-1,6-diphosphate $\rightleftharpoons$ phosphoenzyme + glucose-6-phosphate

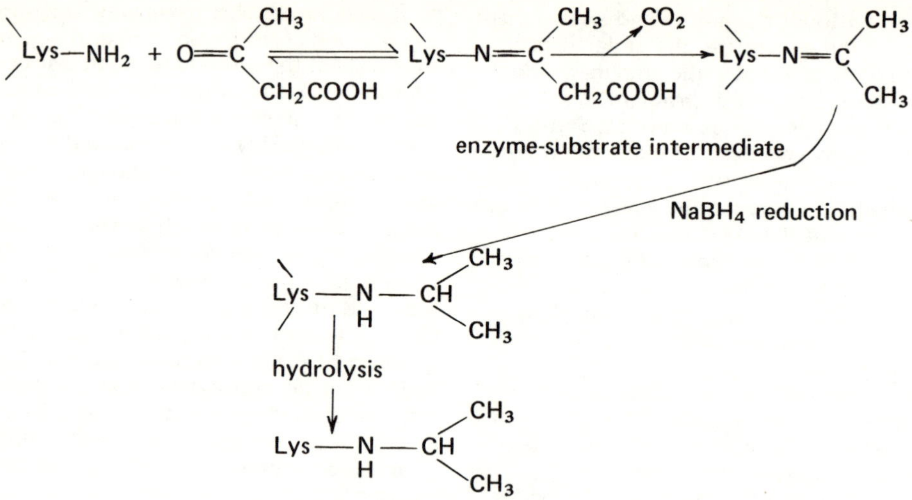

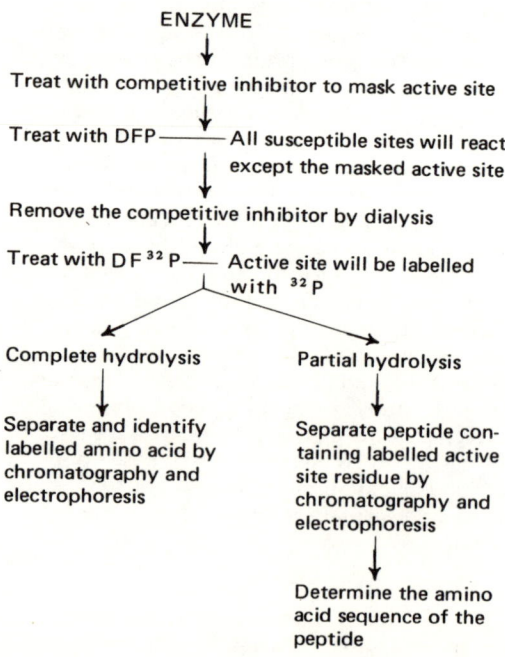

$$\text{Lys---NH}_2 \; + \; \text{O=C} \underset{\text{CH}_2\text{COOH}}{\overset{\text{CH}_3}{}} \rightleftharpoons \text{Lys---N=C} \underset{\text{CH}_2\text{COOH}}{\overset{\text{CH}_3}{}} \xrightarrow{\text{CO}_2} \text{Lys---N=C} \underset{\text{CH}_3}{\overset{\text{CH}_3}{}}$$

enzyme-substrate intermediate

NaBH$_4$ reduction

$$\text{Lys---N---CH} \underset{\text{CH}_3}{\overset{\text{CH}_3}{}} \quad \overset{\text{H}}{}$$

hydrolysis

$$\text{Lys---N---CH} \underset{\text{CH}_3}{\overset{\text{CH}_3}{}} \quad \overset{\text{H}}{}$$

Nϵ-isopropyllysine

there are good grounds for supposing that the label has indeed become attached to the active site. Complete hydrolysis of the inactive enzyme-inhibitor complex may then allow the identification of the phosphorylated amino acid residue. Partial hydrolysis to peptides followed by sequence determination may provide information about the residues adjacent to the labelled one. When there are doubts about the specificity of the reagent for the active site the latter may be protected by the presence of a competitive inhibitor while the enzyme is treated with an excess of DFP to modify all the reactive groups on the protein. Removal of the competitive inhibitor followed by treatment with DF^{32}P will result in the incorporation of the radioactive label into the active site only. The labelled group can then be separated after hydrolysis (Table 6.1). If a particular enzyme can be reacted with a number of pseudo-substrates and the same residue is modified in each case, then there is good circumstantial evidence that the residue is important in substrate binding.

In nearly all of the esterases and proteases that react stoichiometrically with DFP a serine hydroxyl group is phosphorylated and the serine occurs in a common or closely related sequence of amino acids (Table 6.2). This may indicate a common evolutionary precursor of all esterases. It is interesting to note that where the sequences are different the substitutions involve amino

acids of similar character, e.g. aspartic acid may be replaced by glutamic.

TABLE 6.1. EXAMPLE OF THE IDENTIFICATION OF ACTIVE SITE RESIDUES BY CHEMICAL MODIFICATION TECHNIQUES

ENZYME

↓

Treat with competitive inhibitor to mask active site

↓

Treat with DFP ——— All susceptible sites will react except the masked active site

↓

Remove the competitive inhibitor by dialysis

↓

Treat with DF^{32}P ——— Active site will be labelled with ^{32}P

Complete hydrolysis / Partial hydrolysis

Separate and identify labelled amino acid by chromatography and electrophoresis

Separate peptide containing labelled active site residue by chromatography and electrophoresis

↓

Determine the amino acid sequence of the peptide

TABLE 6.2

Enzyme	Sequence around reactive serine residue
Trypsin	— Asp — Ser — Gly —
Chymotrypsin	— Asp — Ser — Gly —
Thrombin	— Asp — Ser — Gly —
Alkaline phosphatase	— Asp — Ser — Ala —
Butyryl cholinesterase	— Glu — Ser — Ala —

For those enzymes where it is difficult to find a suitable pseudo-substrate a technique known as affinity labelling may give some clue to the identity of the amino acid residues at the active site. From specificity studies with substrates and competitive inhibitors it should be possible to deduce which portion of the substrate is important for binding to the enzyme, i.e. to identify the structures in the substrate that presumably interact with "contact" amino acid residues rather than the "catalytic" residues on the enzyme. Thus, a synthetic compound containing the appropriate group or groups should fit into the active site. If in addition the compound contains a chemically reactive centre capable of linking covalently to one or more amino acid residues, then it is possible that after the group has entered the active site region it will react to form a stable derivative with a group at or near the active centre. Subsequent analysis of an enzyme hydrolysate should make clear which residue has been modified. Perhaps the best known example of this technique is that of Schoellman and Shaw in their study of chymotrypsin. They synthesized the compound shown below:

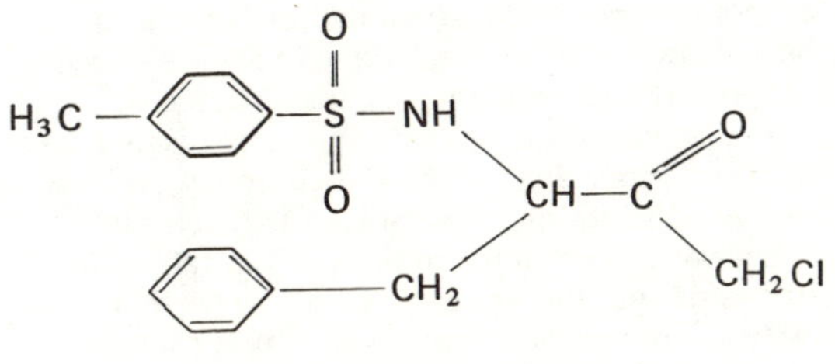

N-Tosyl-L-phenylalanyl chlormethyl ketone (TPCK)

α-Chymotrypsin exhibits a specificity for linkages involving L-phenylalanine and so TPCK is directed towards the active centre of the enzyme. The —CH$_2$Cl group confers strong alkylating properties on this compound. α-Chymotrypsin is alkylated by TPCK with complete loss of catalytic activity and examination of the alkylated enzyme has shown that the amino acid residue that is alkylated is a histidine.

Affinity labelling has not been widely used because of the difficulty in finding suitable labels but the method is likely to become more and more popular. There is the disadvantage that there is no proof that the modified residue is directly involved in catalysis, but it can be concluded that it is very close to the active centre in the native enzyme.

When it is impossible to find any means of fixing a label to catalytic residues, the enzymologist is forced to rely on less specific methods of identification. Practically all amino acid residues can be modified by reaction with suitable chemical reagents. If a certain compound is known to react with only one kind of amino acid residue and this compound inactivates the enzyme then it may be supposed that the residue is of catalytic importance. The more specific the reaction between the modifier and the amino acid residue the more certain one can be that only one kind of residue has reacted and the easier it is to identify active site residues. Unfortunately in most instances the useful chemical reagents used in this kind of study can react with several different amino acids. Also the three dimensional structure of the protein can affect the rate at which a particular residue reacts with the modifying agent.

Inhibition by iodoacetate is often taken to be indicative of involvement of a thiol group in catalysis for as we have seen already thiols are readily alkylated by this reagent (Chapter 5). However, other residues may be modified. For example, treatment of ribonuclease at pH 5·5 with iodoacetate inactivates the enzyme and investigation of digests of the inactivated enzyme shows that two histidine residues are alkylated (*see* top of p. 80). Provided the modified residues can be identified then the method is valuable but when predictions are made about the nature of catalytic groups only on the basis of whether the enzyme is inactivated or not in the presence of a certain agent too much reliance cannot be placed upon them.

Where a reagent is known to react with more than one kind of amino acid a method developed by Koshland and Ray can help in deciding which modifications affect the enzyme activity to the greatest extent. In this method the enzyme is

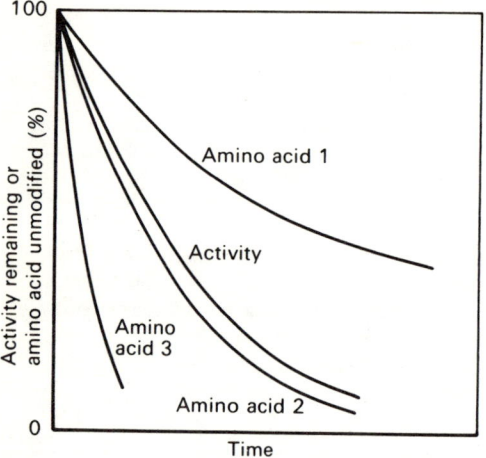

treated with an excess of reagent and the decline in activity is followed over a timed period. At intervals throughout this period samples are withdrawn and investigated for degree of amino acid modification. When the results of loss of activity and degree of modification are plotted against time (Fig. 6.3) it should be possible to tell which are the most important residues from the slope of the curves. An amino acid that is modified much faster than the enzyme is

Fig 6.3. The identification of active site residues by chemical modification. The enzyme is treated with an excess of the reagent and the reaction mixture sampled from time to time in order to follow the decline in enzymic activity and the degree of modification of amino acids of particular types. An amino acid that is modified at a rate which closely follows the decline in activity is probably important in catalysis (amino acid 2 in this example).

inactivated is unlikely to be important for catalysis. Those which are modified at a comparable rate with the inactivation are good candidates for catalytic involvement. For phosphogluco-mutase it was found that a methionine residue was important for catalysis and a histidine was needed for full catalytic activity but total modification of this residue did not inactivate the enzyme completely.

From the application of several of the methods that have been outlined it should be possible to obtain quite a lot of information about the nature of the active site. No information can be gained however about the three dimensional arrangement of the important amino acid residues. It may be possible to speculate about the distances between reactive residues from studies of accurately constructed models of substrates and inhibitors but the complete tertiary structure of the enzyme needs to be known before any assertions can be verified. Enzymologists who have studied the active centre of enzymes by chemical modification techniques have been pleased to find that in those enzymes where the tertiary structure is known, amino acids that have been shown to be important in catalysis are clustered together in space although they may be widely spaced linearly along the polypeptide chain, or may be located on different chains.

MECHANISMS OF CATALYSIS

It is very difficult to provide a full explanation of how any catalyst, enzymic or non-enzymic, increases the rate of a chemical reaction. Several factors that could conceivably contribute to the enhancement of reaction velocity have been recognized and their relative importance in particular cases has been a subject of discussion for some time. Because of the chemical complexity of proteins, enzymologists often seek an answer to problems of enzymic catalysis in studies of the non-enzymic catalysis of relatively simple organic reactions. These model systems have given some insight into the catalytic properties which certain of the groups present as the side chains of the amino acids of proteins may exhibit under particular conditions, but the models often seem remote from enzymic processes. As more is learned about the structure of enzymes and their important catalytic groups it becomes easier to piece together the evidence taken from model systems and from enzymic studies.

The chemist recognizes two main classes of catalysis: homogeneous and heterogeneous. Heterogeneous processes involve more than one phase: hydrogenation in the presence of a finely divided metal catalyst is an example of a heterogeneous process with a gaseous phase (hydrogen) and a solid phase (metal catalyst). Since enzymic catalysis occurs in solution, i.e. in an apparently single phase system, it is usually regarded as homogeneous, but before dismissing heterogeneous catalysis as being totally inappropriate to enzymic processes it is as well to remember that an enzyme in aqueous solution may perhaps be considered to constitute a separate phase from the aqueous environment. The formation of an enzyme-substrate complex is analogous to adsorption of a substrate on to a metal surface and the mathematical equation describing surface adsorption (the Langmuir isotherm) is of similar form to the Michaelis-Menten equation. The catalytic effects of surfaces may arise from the provision of a platform where reactant molecules can be brought close together thus facilitating a reaction between them. Where a gas is adsorbed by a surface, a layer of one molecule thickness is formed. There is a relatively strong attachment between the reactants and the molecules which form the surface and this could result in loosening of bonds in other parts of the reactants. Thus less energy may then be required to raise the reactants to the activated state which is necessary for reaction to take place (Chapter 4). How applicable surface phenomena are to enzyme action is impossible to assess at the present time, but the effect of the proximity of reacting molecules on the rate at which they react is probably important and will be discussed more fully in a later section.

GENERAL ACID-BASE CATALYSIS

A large number of homogeneous reactions are catalysed by acids or bases. In this context an acid is defined as any substance that can either donate a proton to the solvent or substrate or accept a pair of electrons to form a chemical bond. A base is any substance that can either accept a proton from the solvent or substrate or donate a pair of electrons to form a bond.

For a dissociation reaction that can be written:

$$HA \rightleftharpoons A^- + H^+$$

it follows from the definitions given above that HA is an acid since it can release a proton. Note that in the reaction in the reverse direction, however, A^- can accept a proton. Thus A^- is a base and it becomes apparent that all acids and bases exist in related or conjugated pairs. For acetic acid the conjugate acid can be written as CH_3COOH while CH_3COO^- represents the conjugate base. Although it is convenient to regard the dissociation simply as a release of H^+, it is likely that protons do not exist freely in aqueous solution but are solvated. Therefore they are usually represented as H_3O^+. Hence the dissociation of acetic acid in water is more correctly written:

$$CH_3COOH + H_2O \rightleftharpoons CH_3COO^- + H_3O^+$$

By writing out the equation in this way it can be seen that H_2O itself can act as a base. For the discussion in this chapter, dissociations will be written as though a free proton can exist in solution, although it should be remembered that this is a shorthand notation.

In catalysis by an acid, the addition of a proton to the substrate makes the latter very reactive and it then breaks down very rapidly forming the products. The rate of the decomposition of the substrate will depend largely on the rate of the proton transfer from the catalyst to the substrate, i.e. this step will be rate-limiting or rate determining. Consider an acid-catalysed reaction in which a substrate X is converted to product Y. The steps in the process may be represented:

$$HA + X \xrightleftharpoons{\text{slow}} XH^+ + A^- \qquad \text{(a)}$$
$$XH^+ \xrightleftharpoons{\text{fast}} Y + H^+ \qquad \text{(b)}$$

The reaction depicted in equation (a) is the rate limiting step; XH^+ decomposes very rapidly.

A base catalyst removes a proton from the substrate which then reacts more readily. The rate determining step is the abstraction of the H^+. Many reactions are catalysed by acids and bases working in concert, i.e. the reaction sequence of the catalysed process includes both proton donation and proton abstraction.

82

Enzymology and Medicine

For example the conversion of X to Y may proceed according to the scheme:

$$HA + X \rightleftharpoons AXH$$
(acid)

$$AXH + B^- \rightleftharpoons Y + BH + A^-$$
(base)

Concerted general acid-base catalysis may be especially important in enzyme action.

The effectiveness of an acid or base as a catalyst depends on its strength, i.e. the ease with which it dissociates. The rate of the catalysed process is related to the dissociation constant K_a by the Brønsted Catalysis Law which may be represented by the equation:

$$\log k_{cat} = a\log K_a + b \qquad (1)$$

where k_{cat} is the catalytic rate constant and a and b are constants. A plot of $\log k_{cat}$ against the negative log of K_a is linear. Figure 6.4. shows the results obtained for the dehydration of

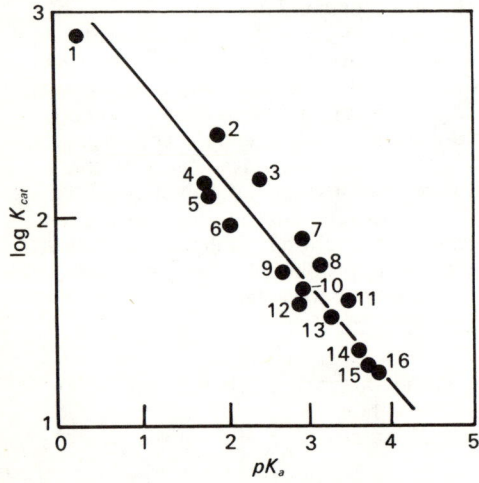

Fig 6.4. The dehydration of acetaldehyde hydrate catalysed by carboxylic acids. The log of the rate of reaction is plotted against the pK_a of the acid. 1, Dichloroacetic; 2, salicylic; 3, metanitrobenzoic; 4, chloroacetic; 5, bromoacetic; 6, phenoxyacetic; 7, p-chlorobenzoic; 8, benzoic; 9, formic; 10, diphenylacetic; 11, p-hydroxybenzoic; 12, o-toluic; 13, phenylacetic; 14, phenylpropionic; 15, acetic; 16, propionic. (Data from Bell and Higginson (1949) *Proc. R. Soc.* A, 197, 141.)

acetaldehyde hydrate catalysed by carboxylic acids and it is seen that the experimental points lie close to the fitted straight line.

General acid and base catalytic processes are dependent on pH. For a base catalysed reaction represented by

$$B^- + HX \rightleftharpoons BH + Y$$
(base) (substrate) (product)

the rate of the forward reaction v, is a function of the concentrations of B^- and HX, i.e.

$$v = k_{cat}[B^-][HX] \qquad (2)$$

Let the total concentration of base in the system be $[B_0]$ then

$$[B_0] = [B^-] + [BH] \qquad (3)$$

If the dissociation constant of BH is K_a, then the concentration of B^- at any pH is given by the equation:

$$[B^-] = \frac{K_a[B_o]}{K_a + [H^+]} \qquad (4)$$

Substituting for $[B^-]$ in the rate equation (equation 2) gives

$$v = \frac{kK_a[B_o][HX]}{K_a + [H^+]}$$
$$= \frac{k'[B_o][HX]}{K_a + [H^+]} \qquad (5)$$

The appearance of $[H^+]$ in the denominator of the rate equation indicates that as $[H^+]$ becomes large compared to K_a the velocity of the reaction decreases.

By the application of similar reasoning it can be shown that the rate equation for an acid catalysed process is

$$v = \frac{k''[B_o][HX][H^+]}{K_a + [H^+]} \qquad (6)$$

$[H^+]$ appears in the numerator of equation (6) so that a decrease in $[H^+]$ is accompanied by a fall in the rate of the catalysed reaction.

In Fig. 6.5.1 typical plots of the variance in

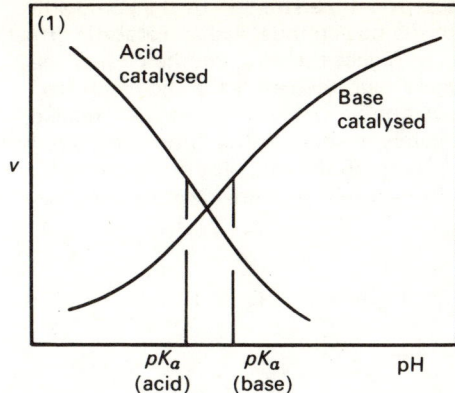

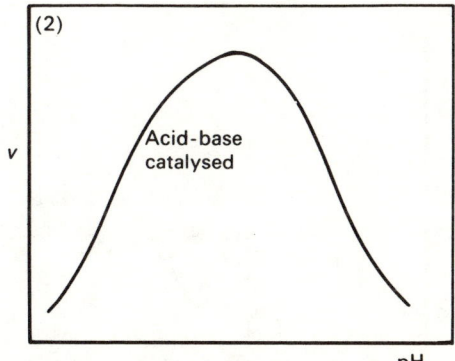

Fig 6.5. Effect of pH on rate of catalysis. (1) Typical rate profiles for acid and base-catalysed processes. The pHs at which the inflexions occur correspond to the respective pK_a values. (2) Bell-shaped profile for acid-base catalysed process. The shape of the bell is determined by the pK_a values of the acid and base species involved.

the rate of acid and base catalysed processes with pH are shown. Where a concerted acid-base process is operative the rate-pH curve is bell-shaped (Fig. 6.5.2). Enzymic activity-pH curves are often bell-shaped (Chapter 4) and it is very likely that some enzymes act as acid-base catalysts.

NUCLEOPHILIC AND ELECTROPHILIC CATALYSIS

A *nucleophilic reagent* is one that contains an atomic centre with a strong tendency to donate

an electron pair to form a covalent bond. Certain functional groups in proteins such as:

$$H \quad H \qquad H \qquad O \qquad H$$
$$\backslash\;/ \qquad | \qquad \| \quad \bar{O}: \qquad |$$
$$N: \qquad O: \qquad C \qquad S:$$
$$| \qquad | \qquad | \qquad |$$

are potential nucleophiles. An *electrophilic reagent* possesses an atomic centre with a strong affinity for electrons. Positively charged ions, e.g. Mg^{2+} and NH_4^+, are electrophiles, while compounds containing groups such as $C{=}O$ and $C{=}N{-}$ are electrophilic because the strong electron withdrawing tendencies of O and N atoms renders the neighbouring C atom electron-deficient (usually indicated by $\delta+$ written above the chemical symbol). The electron displacements are often indicated by arrows drawn in the direction of the displacement.

$$\delta+ \quad \delta- \qquad \delta+ \quad \delta-$$
$$C{=}O \qquad\qquad C{=}N$$

$$\delta+ \qquad\qquad\qquad \delta-$$
$$-CH{=}CH{-}C{=}O$$

A nucleophilic catalyst donates an electron pair to the substrate in a sequence that is partially or completely rate determining, i.e. in a multistep reaction the nucleophilic attack is the slowest step. By the donation of the electrons, the catalyst forms a covalent link with the substrate and the covalent intermediate then breaks down very rapidly. Electrophilic catalysis is the converse of nucleophilic and the removal of electrons from the substrate is the rate determining step. In fact, nucleophilic and electrophilic processes always occur together; when the catalyst is a nucleophile it attacks an electrophilic centre in the substrate and *vice versa*. In enzymic catalysis, an attack by enzyme nucleophilic groups on an electrophilic centre in the substrate is probably more usual than an attack by electrophilic enzyme groups on a substrate nucleophile. The formation of a covalent bond between nucleophile and substrate requires the displacement of a group already attached to the substrate, i.e. a nucleophilic reaction involves substitution of one group by another. The chemist distinguishes between substitutions that occur by direct displacement and those that occur in two steps; in the latter the group that is to be displaced is expelled from the substrate before the new bond is formed with the nucleophilic substitute.

DIRECT DISPLACEMENT

In such reactions, the attacking group "moves in" as the displaced group "moves out". The

exists between the strength of the conjugate acid of the nucleophile and its catalytic effectiveness. The higher the pK_a, i.e. the weaker the conjugate acid, the better the agent is suited for nucleophilic catalysis. As a result, nucleophilic catalysts should obey Brønsted's Catalysis Law. Figure 6.6 shows a plot of the rate of hydrolysis of p-nitrophenyl acetate catalysed by

$$OH^- + CH_3I \longrightarrow \left[HO \cdots\cdots > \overset{\overset{H}{|}}{\underset{\underset{H}{|}\ H}{C}} \cdots\cdots > I \right] \longrightarrow HO-CH_3 + I^-$$

rate of the reaction depends on the concentration of both the nucleophile and the substrate, i.e. the reaction is bimolecular. The hydrolysis of methyl iodide is an example of a bimolecular nucleophilic substitution.

   If the substrate contains an asymmetric C atom there is an inversion of configuration about the C atom accompanying the substitution. The C atom remains asymmetric, however, because there is a limited direction of approach of the substituent. These reactions are often referred to as Walden Inversions and described by analogy to the turning inside out of an umbrella.

HETEROLYSIS MECHANISM OF DISPLACEMENT

A substitution reaction of this kind is unimolecular with respect to the substrate. A group is expelled from the substrate leaving a planar ion in which the configuration about an asymmetric C atom is lost. The substituent then becomes attached to the planar ion. The complete bond breakage that occurs before the addition of new species results in racemization of an optically active substrate because there is no restriction on the direction of approach of the new substituent (*see* below).

   Bases, as well as nucleophiles can be defined as substances with tendencies to donate electrons. Similarly electrophiles can be classed as acids. Thus it is to be expected that a relationship

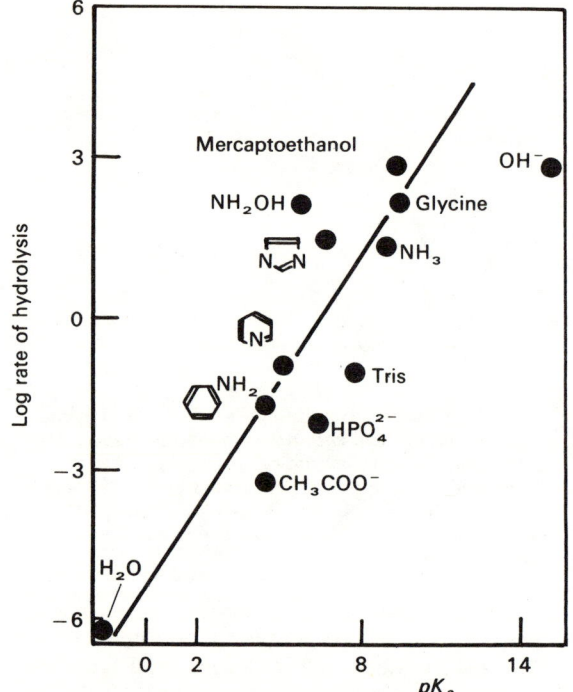

Fig 6.6. **Relationship between the rate of the catalysed breakdown of p-nitrophenyl acetate and** pK_a **values of substances which act on the substrate.** (Data from Jencks and Carriulo (1960) *J. Am. Chem. Soc.* **82**, 1778.)

$$Y-\overset{\overset{X}{|}}{\underset{\underset{Z}{|}}{C}}-W \longrightarrow \left[\overset{Y}{\underset{\underset{Z}{|}}{\overset{\diagdown}{C^+}}}\overset{X}{\diagup} \right] + W^- \xrightarrow{\ OH^-\ } Y-\overset{\overset{X}{|}}{\underset{\underset{Z}{|}}{C}}-OH + HO-\overset{\overset{X}{|}}{\underset{\underset{Z}{|}}{C}}-Y$$

a number of nucleophilic substances as a function of the basicity of the attacking reagent. For nucleophiles of similar molecular geometry, such plots are linear showing that the catalysis law holds. Because of the dependence of the reaction rate on the pK_a of the attacking reagent, the rate variance with pH is of the same general form as that of base catalysis. The distinction between whether a catalyst is acting as a nucleophile or general base is difficult to make because of the duality of nature of catalysts. The essential difference between the alternative mechanisms is that a covalent intermediate is formed in nucleophilic (or electrophilic) reactions. If the existence of an intermediate can be proved then there is good reason to suspect that a nucleophilic mechanism is involved. Comparison of the rates of the reactions in H_2O and D_2O can sometimes provide information of the catalytic mechanism. D_2O has relatively little effect on the rate of nucleophilic processes but the rate of donation or removal of a deuteron in acid-base reactions is different from the rate for a proton. Thus acid-base catalysis is characterized by a relatively large "isotope effect".

FUNCTIONAL GROUPS IN ENZYMES

Several groups present in the side chains of protein amino acids are known from studies with model systems to possess catalytic properties, and chemical modification studies (p. 79) usually provide evidence for the localization of one or more of such groups at the active centre of enzymes. This section discusses the nature of active groups in enzymes and the reaction mechanisms proposed for certain enzymes on the basis of the known reactivity and properties of the groups.

(i) Carboxyl

Free carboxyl residues are present in the side chains of glutamic and aspartic acid residues and at the C-terminal amino acids of polypeptide chains, of which there may be more than one in many enzymes. Carboxyl groups can act as an acid-base catalyst.

$$-COOH \rightleftharpoons -COO^- + H^+$$

In addition, the carboxylate anion is nucleophilic. The pK_a values for carboxyl groups in proteins lie between 3 and 4 so that at physiological pH values, it is the conjugate base or nucleophilic form of the group that predominates. Thus if a carboxyl group is important, the dissociated form is the likely catalyst. Having stated this, it must be remembered that in some proteins the group may be located in an apolar environment having the effect of raising the pK_a so that the $-COOH$ form may predominate close to pH 7.

Evidence of nucleophilic catalysis by $-COO^-$ is obtained from a study of the hydrolysis of acid anhydrides. With carboxylate as catalyst a "mixed" anhydride is formed in accordance with a nucleophilic mechanism.

catalyst

acetic anhydride

slow

mixed anhydride

H_2O

fast

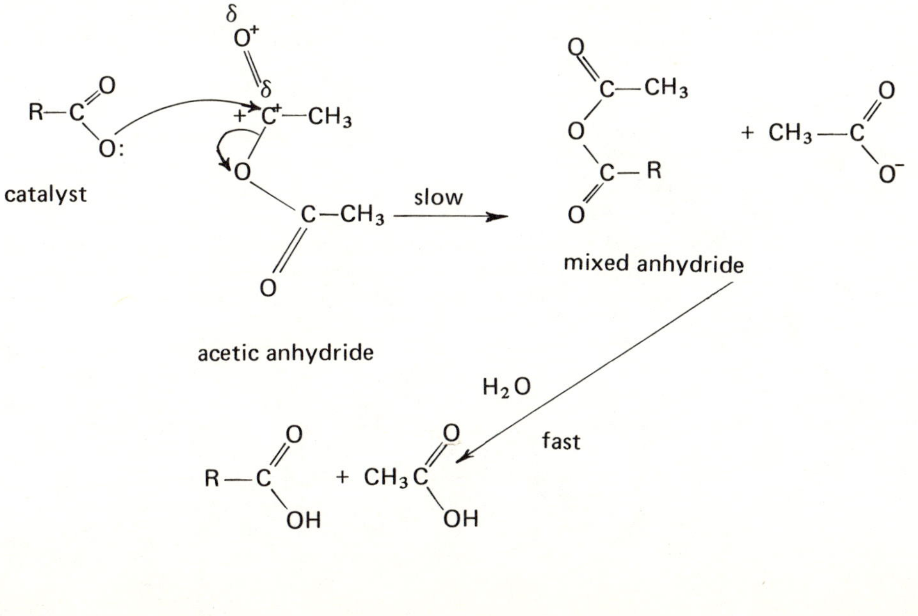

Pepsin shows a bell-shaped pH-activity curve with pK_a values for the functional groups of 1·0 and 4·8. It is very likely that at least one of the catalytic groups is a carboxyl. The higher pK_a may represent the ionization of a histidine residue but could be a carboxyl also. Considering that pepsin is optimally active at very low pH, it is quite probable that the enzyme functions as an acid catalyst.

The elucidation of the three dimensional structure of lysozyme has provided an example of how carboxyl groups may function in an enzyme. Lysozyme breaks down bacterial cell walls composed of a heteropolysaccharide in which alternating residues of N-acetylmuramic acid and N-acetylglucosamine (NAG) are linked together through β-glycosidic links to form a chain. The enzyme hydrolyses the link between the two components. Studies of model substrates have shown that a hexamer of NAG is a good substrate also and that it is cleaved between the fourth and fifth residues only. Other links in the chain of the hexamer are not attacked. Molecular models show that the hexamer can fit into a cleft on the enzyme,

identified as the "substrate cleft", and in the region of the bond between residues 4 and 5 are two carboxyl residues. One carboxyl is part of a glutamate residue, the other part of aspartate. The glutamate carboxyl is located in an apolar environment and is likely to be undissociated but the aspartate is almost certainly ionized and hydrogen bonded to neighbouring residues. When the NAG hexamer enters the substrate cleft, residue 4 is probably twisted into an unfavourable conformation while the other residues retain their usual stable conformation. The distance between the aspartate residue and the bond between residues 4 and 5 is probably too great for nucleophilic attack by $-COO^-$ and it is believed that the $-COOH$ of glutamate acts as an acid catalyst and transfers a proton to the bridge oxygen. The formation of the positively charged bridge aids the breaking of the bond between C_1 of residue 4 and oxygen. The breakage leaves C_1 with a positive charge which is stabilized by the negatively charged $-COO^-$ of aspartate. The addition of OH^- to C-1 completes the proposed sequence.

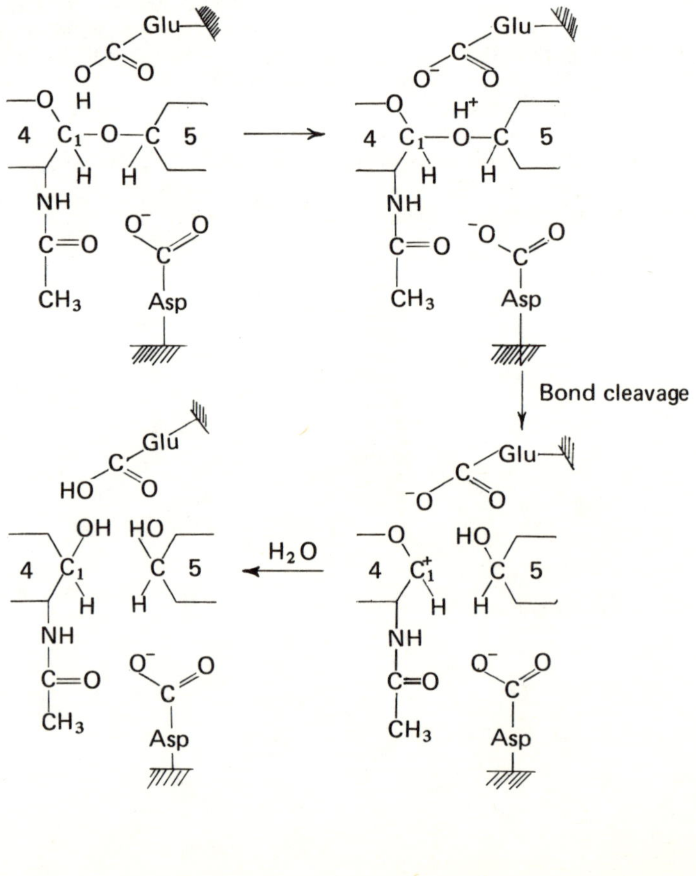

(ii) Amino

Primary amino groups are present in the side chains of lysine residues and at the N-termini of polypeptide chains. Amino groups can be shown to function, in model systems, as general acid, general base and nucleophilic catalysts. At pH 7 amino groups in proteins that are exposed to solvent will be in the conjugate acid form $-NH_3^+$ so that examples of general base or nucleophilic action in enzymic catalysis are unlikely to be numerous. For a few enzymes, however, lysine has been shown to be important for the formation of a reactive intermediate by a nucleophilic mechanism. In the discussion of methods used for exploring the active site of enzymes it was described how chemical reduction of the enzyme-substrate intermediate in the acetoacetate decarboxylase reaction led to the identification of an important lysine group in the enzyme, while studies of the non-enzymic decarboxylation of β-keto acids has shown that primary amines are effective catalysts for the reaction. The decarboxylation of acetoisobutyrate is ten times faster in 0·05 M aniline than in water. The reaction probably involves the formation of an imine intermediate following a nucleophilic attack by the amino group. It is very likely that the enzymic reaction follows a similar course.

Lysine residues have a special role in the binding of some coenzymes. Pyridoxal phosphate, a very important cofactor in enzymic transformations of amino acids (Chapter 5), is bound to lysine through the $-NH_2$ group forming a ketimine (Schiff's base).

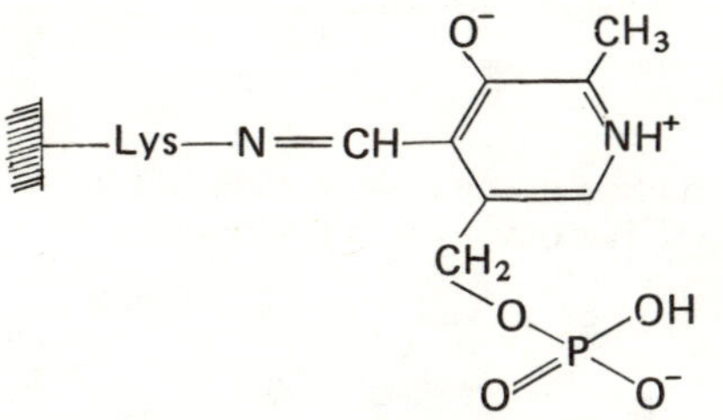

Ketimines are very reactive compounds and the transfer of enzyme-bound cofactor to the substrate may be made easier by ketimine formation. Phosphorylases also contain pyridoxal phosphate bound in a similar manner. Biotin is a cofactor in many decarboxylation reactions and is linked by a peptide bond to the ϵ-NH_2 group of a lysine residue in the decarboxylase.

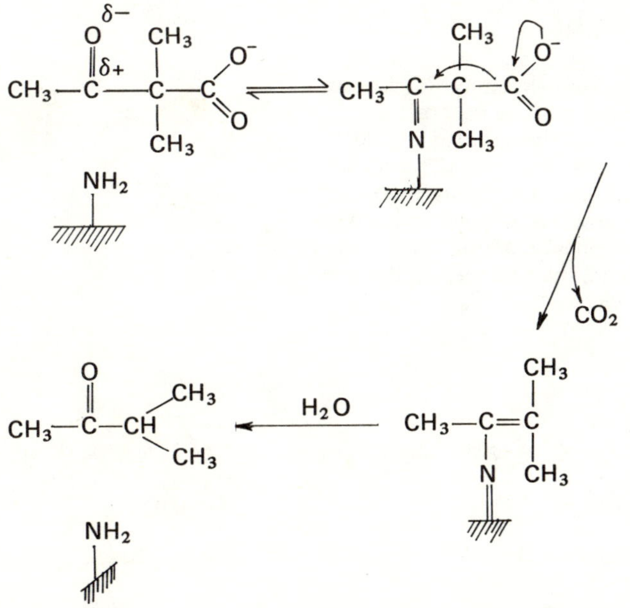

The positively charged $-NH_3^+$ group may be important for substrate binding in some enzymes Many substrates may carry a negative charge at the pH of the reaction, e.g. organic phosphates, and attraction by $-NH_3^+$ may be very helpful in bringing the substrate into the best position for bond cleavage. Alternatively, a negatively charged transient intermediate in the reaction sequence may be stabilized by $-NH_3^+$. The stabilization of an intermediate by a charged group has already been discussed in connection with the mechanism for lysozyme.

(iii) Thiol

This group can act as a general acid, but in addition $-S^-$ is a powerful nucleophile.

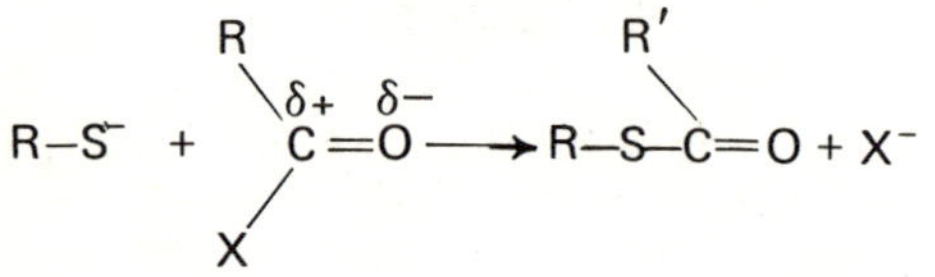

Evidence for the involvement of $-SH$ groups usually comes from inhibition experiments with iodoacetate or p-chloromercuribenzoate (PCMB) (p. 53) but whether thiols participate directly or are important for the maintenance of tertiary structure is difficult to determine from inhibition

PCMB

experiments. Apart from enzyme thiol groups others may be present that are part of the molecule of an essential cofactor such as co-enzyme A and dihydrolipoic acid (Chapter 5).

There is good evidence for the participation of an enzyme $-SH$ group in the glyceraldehyde phosphate dehydrogenase reaction. In the absence of phosphate, the first step of the reaction is as follows:

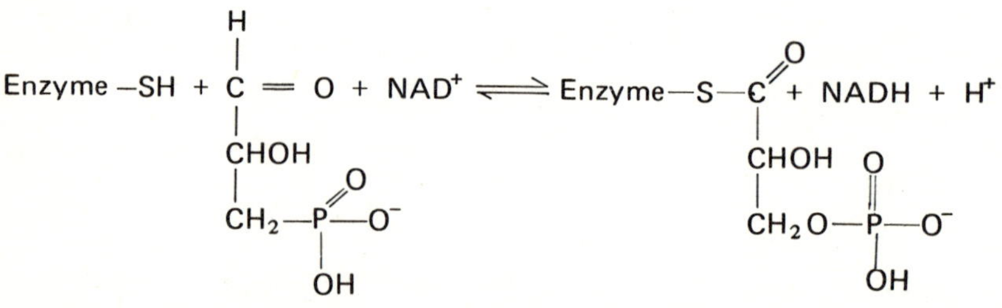

S-acylated cysteine may be isolated from hydrolysates of the denatured enzyme.

Papain is a proteolytic enzyme in which an acylated cysteine is formed as an intermediate.

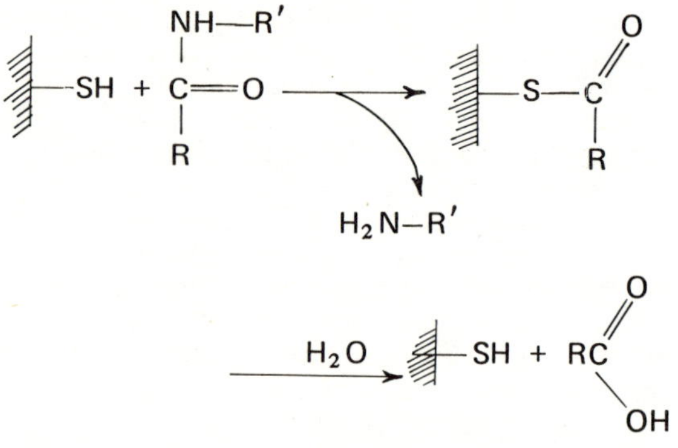

The enzyme is irreversibly inhibited by a tosyl derivative and digestion of the inhibited enzyme followed by separation of the modified amino acid residue led to the identification of a particular cysteine which is presumably in the active site.

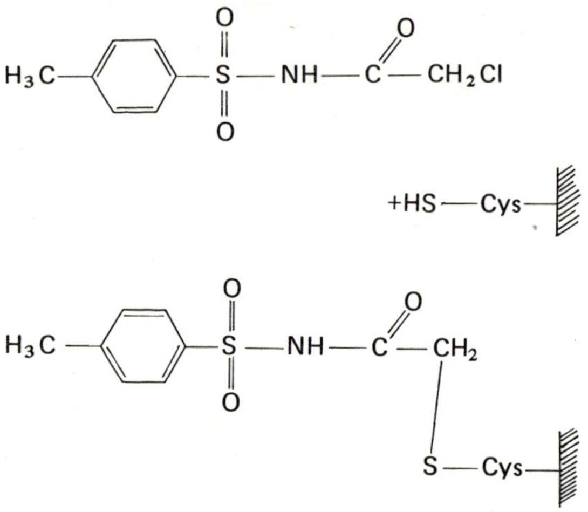

Thiol groups have been shown to be important in the multi-enzyme complex that brings about the synthesis of long chain fatty acids. Elongation of fatty acids occurs by the sequential addition of malonyl residues. The fatty acid and malonyl residues are covalently linked by thioester bonds to reactive thiols of pantothenyl derivatives.

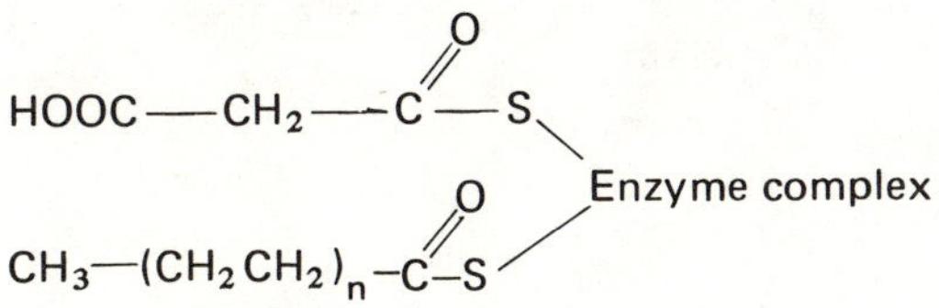

(iv) Imidazole

The properties of imidazole make histidine residues eminently suitable for a direct role in catalysis. The imidazole ring is weakly basic and can accept a proton. The pK_a value for the

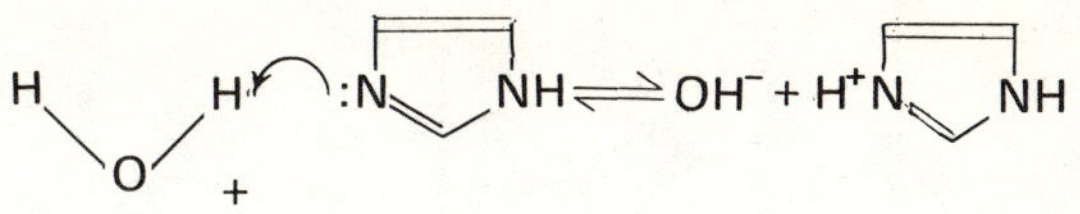

The reversal, i.e. donation of the proton to OH^- occurs very fast ($k = 2 \times 10^{10} M^{-1} sec^{-1}$) so that the overall rate for the base catalysed reaction is limited to $10^3 sec^{-1}$ by the slower step. The turnover numbers of hydrolytic enzymes range from 10 to $10^5 sec^{-1}$ and since from studies of catalysis it is known that neighbouring groups can increase the rate of proton transfer to imidazole by a factor of 10^2, it is apparent that base catalysis by imidazole can account for the observed rates of these enzymic hydrolyses. Further, it is possible to deduce that the rate limiting step in enzymic hydrolytic reactions involving a proton transfer in the mechanism, is the proton transfer step itself. For those enzymes

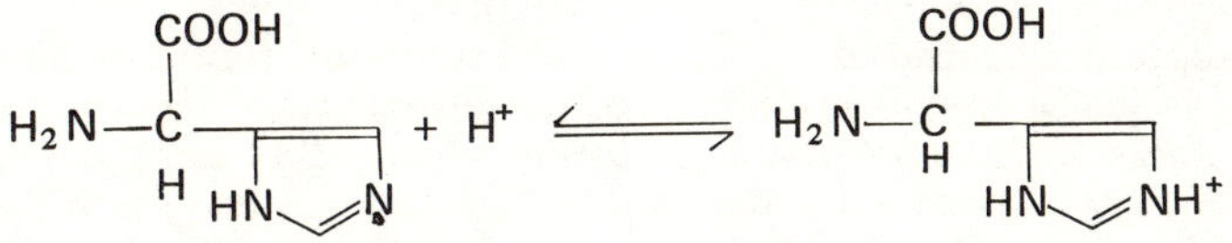

group is 6·7—7·1. Thus at pH 7 about half of the histidine imidazoles in an enzyme will be in the conjugate base form and the other half will be protonated. For a substance to be a good catalyst, i.e. to be effective in very small concentration, it is important that the return to its original state is easy so that it may recycle in the reaction scheme. A good base catalyst should therefore accept a proton readily but also give it up again so that more substrate can be attacked. A strong base accepts protons very easily at pH 7 but holds them tightly and the release process is rather slow. Similarly, a strong acid donates protons rapidly at pH 7 but reversal is slow. A good acid base catalyst at pH 7 requires the conjugate acid and base forms to be of equal strength, i.e. the pK_a values need to be close to pH 7. Imidazole fulfils this requirement particularly well.

The rates of proton transfer to and from water and a number of conjugate pairs, including imidazole, have been measured. The rate of the first step in a base catalysed hydrolysis reaction involving imidazole, i.e. abstraction of a proton from water, is $10^3 sec^{-1}$.

where the turnover number is much less than $10^3 sec^{-1}$ then some other step must be rate limiting.

In model systems general base catalysis by imidazole appears to be less common than nucleophilic catalysis. Imidazole catalyses the hydrolysis of p-nitrophenyl acetate with the formation of acetylimidazole as an intermediate providing proof of a nucleophilic mechanism (*see* top of p. 90). The intermediate possesses a characteristic absorption spectrum and can be identified spectrophotometrically. Enzymologists find the reaction interesting because many esterases hydrolyse p-nitrophenyl acetate and imidazole groups have been implicated in catalysis by many of these enzymes.

From studies of the pH-activity dependence of pancreactic ribonuclease participation of two histidine residues in catalysis was proposed. Chemical modification with iodoacetate, brings about independent alkylation of histidine 12 and histidine 119 with loss of activity suggesting that these residues are important active site constituents. The three dimensional structure of ribonuclease determined by X-ray crystallography shows that these histidine residues together with lysine 41 are grouped together in a cleft in the molecule. The effects of chemical modification of lysine 41 have indicated its involvement also in the catalytic process.

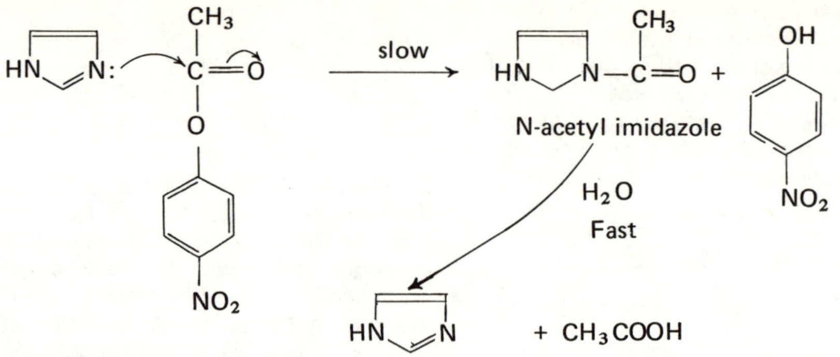

$2',3'$ Nucleotide cyclic phosphates are substrates for ribonuclease and cyclic phosphates are formed during the digestion of RNA by the enzyme. This finding suggests a concerted mechanism of catalysis in which one histidine acts as a general acid catalyst and the other as a general base catalyst. The basic histidine (I) catalyses an attack on the phosphate group by the O at the 2' position of the ribose ring. Simultaneously the acidic histidine (II) catalyses the breakage of the bond linking the phosphate group to the 5' position of the ribose of the next nucleotide in the polynucleotide chain.

The completion of the step leaves the histidines with reversed roles, i.e. I is now in the acid form and II has taken up the conjugate form. The next step in the proposed mechanism envisages a further concerted acid-base action resulting in the hydrolysis of the $2'-O-P$ bond and the regeneration of the histidines in their original form (p. 91).

The rate — pH profile of the enzyme is commensurate with a concerted mechanism. Lysine 41 probably aids the binding of the substrate by interacting with the negatively charged phosphate group.

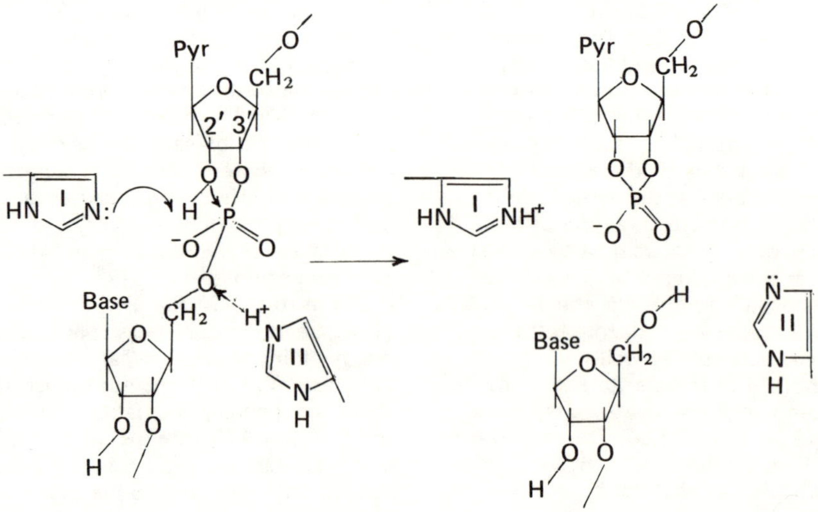

Ribonuclease is specific for pyrimidine nucleotides, i.e. cytidine and uridine. Crystallographic studies of enzyme-cytidine-3'-phosphate complexes have given an indication of how substrates are bound to the enzyme and why pyrimidine substrates are favoured. The mononucleotide fits into a groove on the enzyme molecule and several hydrogen bonds are probably formed between the pyrimidine ring and residues threonine 45 and serine 123. The hydroxyl groups of threonine and serine can act as hydrogen bond donors or acceptors and molecular models indicate that uridine can also bind to these residues provided that rotation of the important hydroxyl groups occurs. Molecular models also show that although purine nucleotides can be fitted into the active site of ribonuclease, the position of the ribose-phosphate moiety is displaced relative to the catalytic histidines and thus hydrolysis would not be expected to occur.

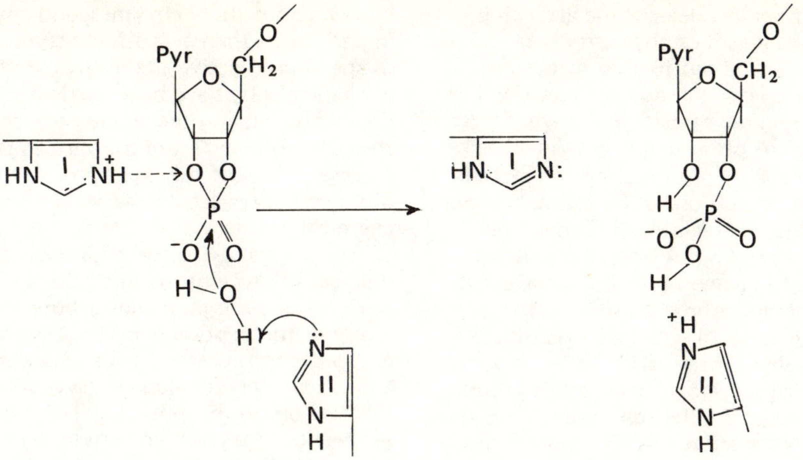

(v) Other functional groups

Other groups that may be important in certain enzymes include the hydroxyl group of serine and threonine, guanidino of arginine and phenolic of tyrosine. Serine hydroxyls have been identified at the active site of a number of hydrolases, a finding that is surprising to chemists because of the relatively low reactivity of the group. Interactions with neighbouring groups may increase the reactivity of the hydroxyl so

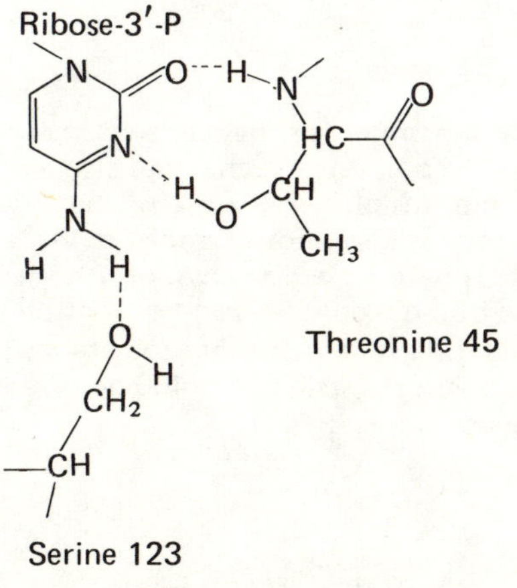

Binding of Cytidine-3'-Phosphate to Ribonuclease

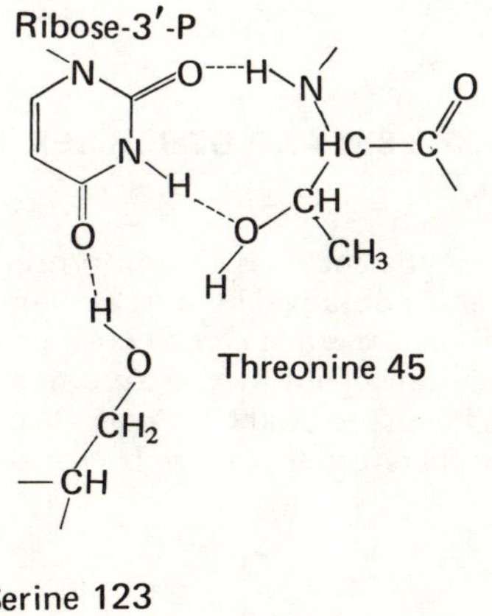

Binding of Uridine-3'-Phosphate to Ribonuclease

making it an effective nucleophilic attacking centre. The mechanism of chymotrypsin illustrates this feature but further discussion of this enzyme is left to a separate section. The side chain of arginine is positively charged over the whole range of pH at which enzymes are stable and is probably not involved in catalysis directly. It may serve a useful function, however, in stabilizing charged reaction intermediates and in substrate binding. A phenolic group of tyrosine seems to be an important constituent of the active site in carboxypeptidase A. Nitration of tyrosine 248 abolishes peptidase activity and molecular models built from X-ray crystallographic data show that the OH group lies close to the bond to be split when a molecule of substrate fills the active site. The same tyrosine residue is involved in the "induced fit" that occurs in this enzyme (p. 76). The role of the tyrosine in catalysis is believed to be that of a proton donor to the free —NH liberated by the breakage of the peptide bond.

known about these enzymes, and chymotrypsin in particular, than any other enzyme. Studies of specificity, active site investigations and X-ray crystallography have been carried out with chymotrypsin and the piecing together of all the data obtained from the various techniques represents one of the high spots in the molecular biological approach to the solution of bio-chemical problems.

Trypsin catalyses the hydrolysis of peptide linkages in proteins in which the basic amino acids lysine or arginine contribute the carboxyl group of the peptide, whereas chymotrypsin is specific for peptide bonds where the aromatic amino acids phenylalanine, tryptophan or tyrosine are in this position. The specificity properties of trypsin and chymotrypsin make the enzymes very useful reagents for determining the amino acid sequences of proteins and poly-peptides. However, the enzymes are very un-specific with regard to the nature of the bond that is hydrolysed. In proteins, which are the

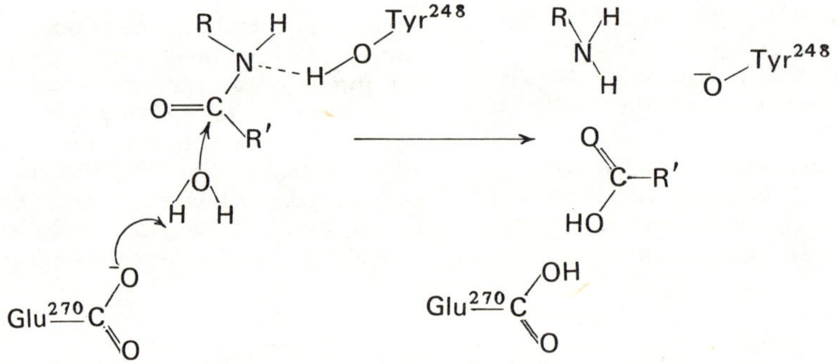

CHYMOTRYPSIN AND OTHER SERINE HYDROLASES

The proteolytic enzymes chymotrypsin and trypsin can be obtained in large amounts from beef pancreas and can be purified to a crystalline state. Thus they are useful enzymes for a biochemist to study in depth. It is not surprising therefore that more is probably

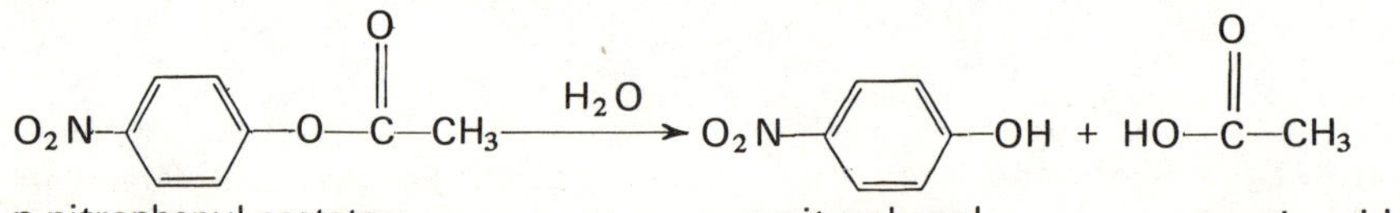

natural substrates for these enzymes, peptide bonds are split, but the enzymes also act upon esters and this property is often made use of in laboratory studies. For example, p-nitrophenyl acetate is hydrolysed by chymotrypsin and the release of p-nitrophenol can be monitored continuously in a spectrophotometer as this product absorbs strongly at 400 nm in alkaline solution:

p-nitrophenyl acetate → p-nitrophenol + acetic acid

Trypsin and chymotrypsin are synthesized in the pancreas in the catalytically inactive forms (zymogens) trypsinogen and chymotrypsinogen. In all mammalian species so far studied, including man, trypsin and chymotrypsin have very similar amino acid sequences indicating that they are almost certainly related and that the genes controlling their structures probably evolved from a single ancestral gene that coded for a single proteolytic enzyme. Activation of trypsinogen and chymotrypsinogen is achieved by hydrolysis of a particular peptide bond in the proteins close to the N-terminus generating a free isoleucine residue (Fig. 6.7). The biosynthesis of enzymes begins at the N-terminus (Chapter 3) so it can be seen that at no stage in the biosynthesis of these pancreatic enzymes is active enzyme produced.

The biological importance of producing the enzymes in an inactive form becomes apparent if consideration is given to what might happen to cellular proteins in the presence of active trypsin and chymotrypsin.

When it was found that anticholinesterases, e.g. DFP (p. 52), also reacted with trypsin and chymotrypsin it seemed reasonable to assume a common, general mechanism of action for esterases and proteolytic enzymes. Nucleophilic attack on the substrate by a group on the enzyme to give an acyl intermediate seemed possible:

$$-X: \rightarrow \quad \overset{\delta+}{C} \overset{\delta-}{=} O \quad \longrightarrow \quad -X-C=O + R'OH$$

acyl intermediate

Decomposition of the intermediate by the nucleophilic attack of a water molecule completes the proposed scheme:

$$-X-C=O + H_2O \longrightarrow -X: + \quad \overset{R}{\underset{HO}{C=O}}$$

Studies of pH-rate profiles, affinity labelling with TPCK (p. 79) and selective destruction of histidine residues led to the identification of histidine 57 as an important active site constituent. Serine 195 is phosphorylated in the reaction with DFP and it becomes acetylated during enzymic hydrolysis of p-nitrophenyl acetate. Studies with charged and neutral competitive inhibitors indicated that a $-COO^-$ group of aspartic acid was important in catalysis and modification of the N-terminal isoleucine showed that the free amino group was implicated in reaction also.

Two different schemes were suggested to account for the central roles of serine 195 and histidine 57. One scheme assumed that N-acyl

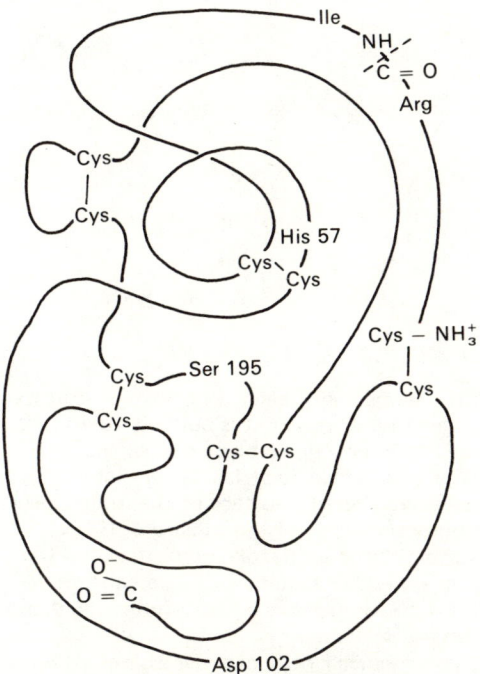

Fig 6.7. Primary structure of chymotrypsinogen showing the isoleucine residue at which cleavage of the zymogen occurs. Further peptide-bond cleavage takes place at a few specific points to form the three-chain structure of α-chymotrypsin. The chains are held together by the disulphide linkages of cystine residues. The important active site residues, His 57, Asp 102 and Ser 195 are widely separated in the primary structure.

imidazole was formed as an intermediate while the other assumed that a serine-acyl intermediate was formed. In support of the former was the known instance of an imidazole intermediate during the non-enzymic hydrolysis of p-nitro-phenyl acetate by imidazole (p. 89). The demonstration of DFP-modified serine in hydro-lysates of the enzyme lent support to the second scheme but the relatively low reactivity of the OH group of serine caused enzymologists to wonder whether such an intermediate could be formed during the hydrolysis of a normal sub-strate. A model for the active centre in which the —OH of serine was assumed to be H-bonded to a nitrogen of the imidazole ring was proposed. It was suggested that the O—H bond in the serine hydroxyl group would become polarized making the O atom more nucleophilic, i.e., more reactive than expected.

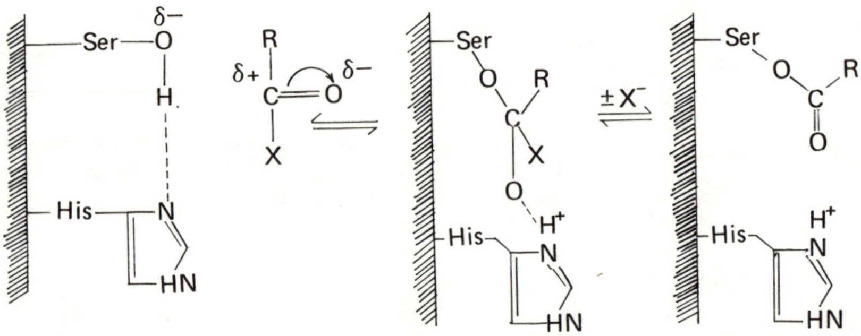

stabilized by aspartic 102. Above pH 7 the active centre is likely to be:

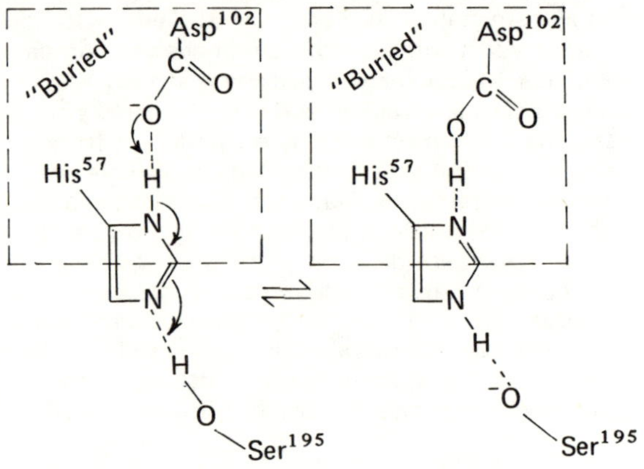

X-ray diffraction studies on crystalline chymotrypsin showed that one of the N-atoms of the imidazole ring of histidine 57 is separated from an O-atom of aspartate 102 by a distance of 2·8 Å. The other N-atom of the same imidazole is about 3·0 Å from the O-atom of serine 195. These distances suggest that hydrogen bonding occurs between the pairs of atoms mentioned. In addition, models of the tertiary structure of the enzyme show that aspartic 102 is shielded from the solvent by neighbouring amino acid side chains and access of a proton to —COO⁻ of aspartic is impossible unless there is a change in tertiary structure. Protonation of a group of pK_a 6-7 inactivates the enzyme and the group is usually identified as imidazole. One of the N-atoms of the imidazole ring of histidine 57 is "buried" and is unlikely to be the site of protonation leading to inactivation. Thus the "buried" N must be protonated already and

With a sequence of electron transfers indicated by the curved arrows, it is possible to visualize how a negative charge on buried aspartic 102 can be channelled to serine 195 which is positioned near the surface of the molecule within easy access of the substrate. If both N-atoms of the imidazole are protonated then the charge relay system is blocked and the inhibition of enzymic activity following proton-ation is explicable.

Conveyance of the negative charge to serine makes it nucleophilic and a likely mechanism of reaction for chymotrypsin is shown on page 95.

Trypsin has a similar arrangement of reactive amino acids and is assumed to act in the same way as chymotrypsin. A comparison of other amino acid residues close to the active site of the two enzymes suggests an explanation of the differences in specificity. In chymotrypsin the aromatic portion of the substrate probably lies

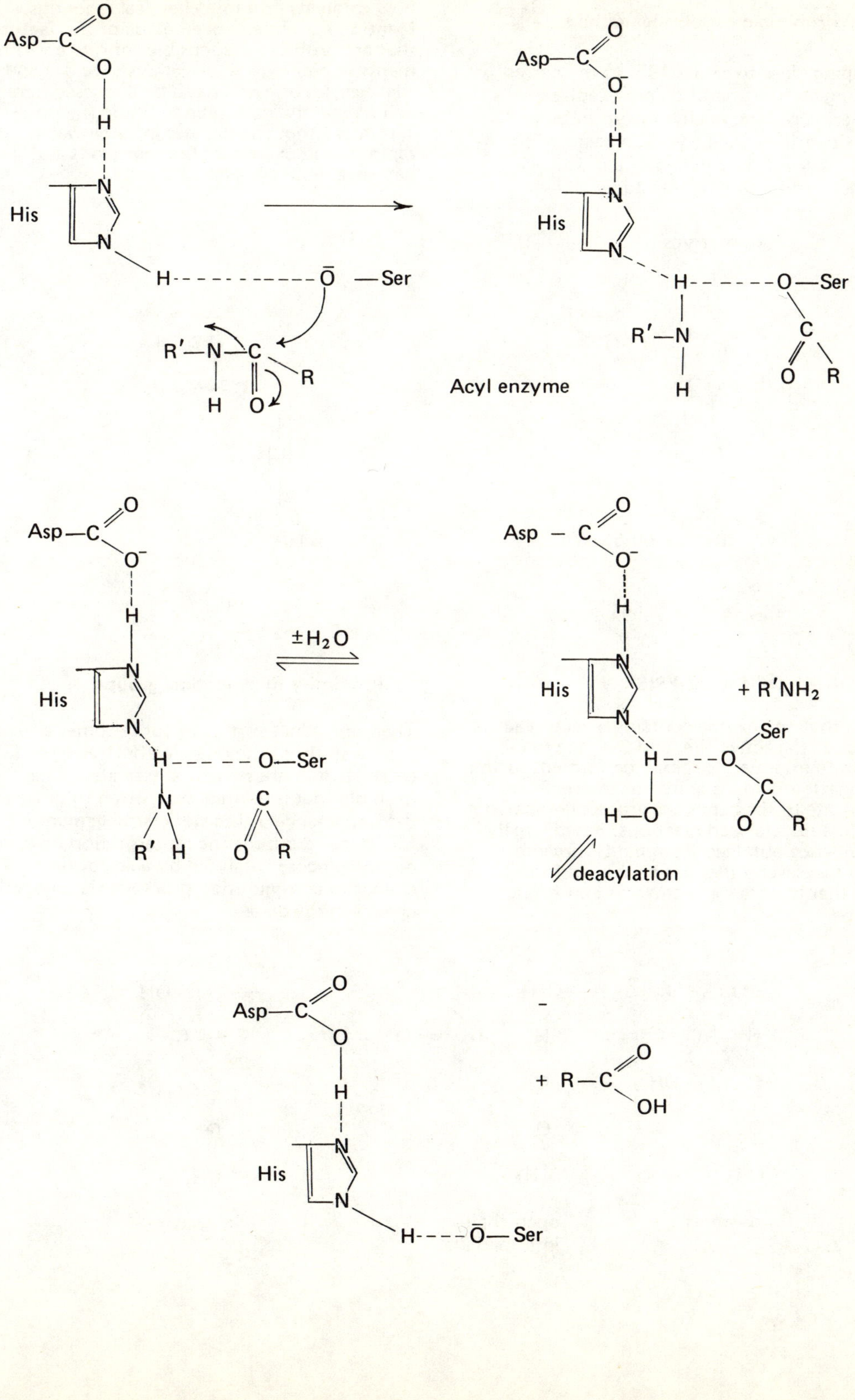

Asp—C
Asp—C ... O
O
H
His
N
N
H--------------Ō—Ser
R'—N—C
H   O   R

→

Asp—C
O
O⁻
H
N
His
N
H-----O—Ser
R'—N
C
H   O   R

Acyl enzyme

Asp—C
O
O⁻
H
N
His
N
H----O—Ser
N
R' H   O   R

⇌ ±H₂O

Asp — C
O
O⁻
H
N
His
N
H----O—Ser
H—O   C
O   R

+ R'NH₂

deacylation

Asp—C
O
O
H
N
His
N
H----Ō—Ser

+ R—C
O
OH

in a hydrophobic cleft bringing the $-\overset{\displaystyle O}{\underset{\displaystyle \|}{C}}-N$

-grouping close to serine 195. In the corresponding cleft region of trypsin, replacement of a serine residue in chymotrypsin by an aspartic residue gives trypsin a negative charge which may aid the binding of a positively charged lysyl or arginyl residue.

how catalysts promote chemical reactions. Mention has already been made of the factors that are probably responsible for rate enhancements in heterogeneous catalysis (p. 81) and the theories of catalysis to be discussed here will be mainly applicable to homogeneous systems, although enzymes in some ways represent an overlap of homogeneous and heterogeneous systems.

Chymotrypsin

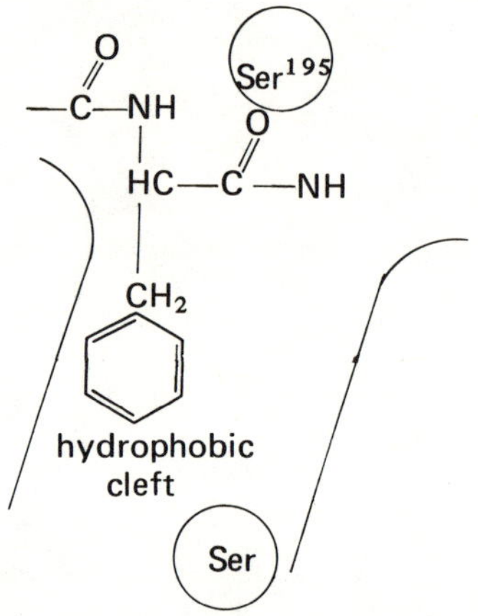

Trypsin

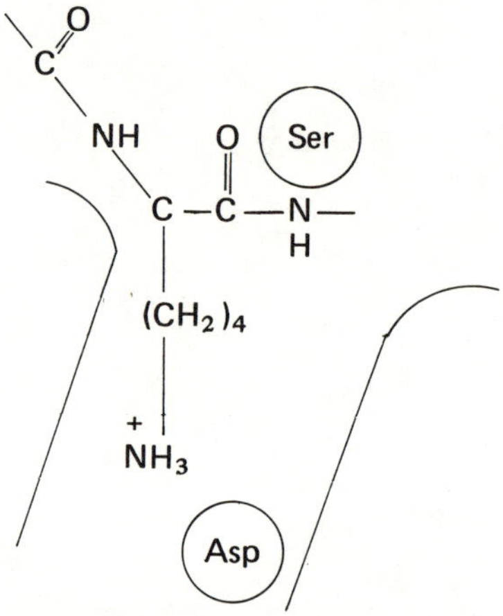

THEORIES OF CATALYSIS

Given that an enzyme contains several reactive groups at the active site, and that enzyme-substrate intermediates may be formed, further explanation is still required of the very much faster rate of the catalysed process compared with the uncatalysed reactions. Based largely on evidence obtained from model organic reactions, several theories have been formulated that go some way towards explaining

(i) Proximity of functional groups

The importance of this factor becomes apparent from a study of the relative effectiveness of catalysts that are structurally related. The example that is quoted most often to illustrate the importance of the steric arrangement of functional groups is the mutarotation of tetramethyl glucose catalysed by acids or bases. The change in configuration proceeds via an open chain intermediate.

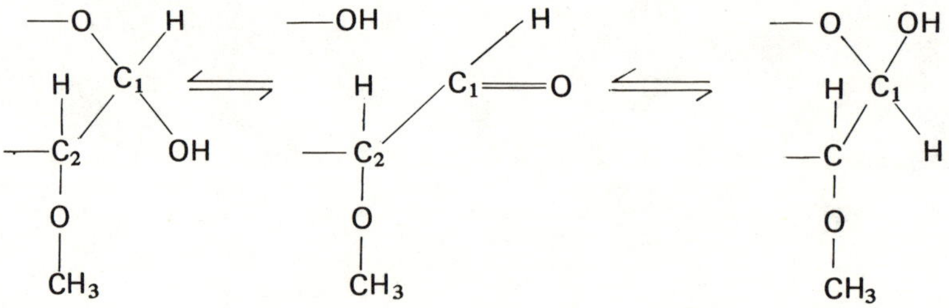

α-isomer

open chain intermediate

β-isomer

Phenol and pyridine both catalyse the reaction but a mixture of these compounds is a better catalyst than either of the components alone. 2-Hydroxypyridine is several thousand times more effective, however, than a mixture of phenol and pyridine as judged by the relative rates of mutarotation. The alignment of acid and base groups probably allows a concerted acid-base catalysis.

of all the reactants, i.e. in this case five-body collision would be necessary. The probability of this occurring is very low so the formation of AB would proceed at a rate close to zero. In the enzymic mechanism, only three-body collision is required, i.e. between enzyme and A and B because C, D and E are already orientated on the enzyme (Fig. 6.8). Also the specificity properties probably channel A and

Phenol   Pyridine   2-Hydroxypyridine

α-isomer   β-isomer

Enzymes seem to contain several active groups at the active site and the steric arrangement of these is known to be important because alteration of the protein tertiary structure, e.g. by treatment with urea, usually results in a total loss of enzymic activity. The specificity requirements of the enzyme that are largely determined by "contact" amino acids aid the positioning of the substrate for effective attack by the active site groups. Consider a reaction

$$A + B \longrightarrow AB$$

catalysed by an enzyme known to possess three functional groups denoted C, D and E all located in the active site. In free solution the reaction would occur following simultaneous collisions

Enzyme

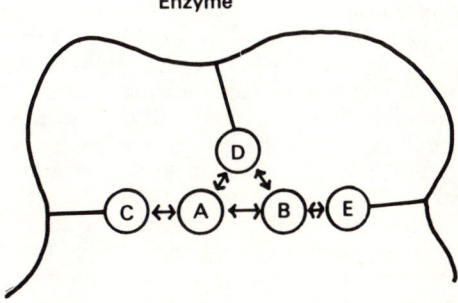

Fig 6.8. Reaction between A and B catalysed by three groups, C, D and E. The orientation of the active groups on the enzyme increases the probability of fruitful collisions between the five reactants.

B into the active site. If an induced fit occurs, then the presence of A and B also aid their own destruction by influencing the protein conformation at the active site.

An estimate of the rate enhancement that is likely to ensue from the proximity effects of reacting groups has been made and it appears that these factors alone cannot account for the rate of enzymic processes. It seems, however, that correct orientation by atoms already brought close enough to react can increase the reaction rate by several orders of magnitude. The result of orientation may at least partly be explained by assuming that the permitted angle of approach for the orbitals of bonding electrons is rather limited. It has been suggested that an enzyme is able to bring reacting molecules together and direct or "steer" electron orbitals along the most favourable path for reaction. A combination of proximity effects and electron orbital steering could be the major factors accounting for the very fast rate of enzyme-catalysed reactions, although the orbital steering theory has been criticized on the grounds that bonding orbitals are unlikely to possess the rigid directionality implied by the theory.

(ii) Strain

When the substrate becomes bound to the enzyme at the active site, it is possible that the substrate is forced into a molecular conformation that is thermodynamically less stable than the conformation it takes up in free solution. One way to interpret this strain theory is to envisage that the substrate is made to take up a structure that more closely resembles the activated intermediate in the reaction, i.e. less energy is then required to form the fully activated intermediate complex and as a consequence the reaction rate is increased. The deformation of the substrate on binding is really the converse of the induced fit hypothesis. Lysozyme appears to be an example of an enzyme which forces the substrate into a strained condition as it is bound (p. 86). It is possible that there is no connection between the deformation of the hexose ring and the breakage of the bond linking this ring to its neighbour, but since the two events are so closely juxtaposed it seems likely that the deformation has a mechanistic advantage.

Some enzyme molecules, e.g. carboxypeptidase, may be in a state of strain in the absence of substrate. The entry of substrate into the active

site may induce a conformational change that allows the enzyme to relax to a less strained condition. The free energy release associated with such a change may contribute to the energy of binding of the substrate and to the energy of activation.

(iii) Formation of intermediates

Consider a reaction represented by:

$$A + B \longrightarrow C + D$$

that proceeds via a single intermediate X characterized by a high energy of activation. In the absence of a catalyst, the rate of the process is likely to be very slow, but if an agent is introduced into the reaction medium which allows the reaction to proceed by a number of intermediate steps, each of which is separated from the next by relatively small energy barriers (Fig. 6.9) a system is provided for accelerating the overall reaction. Jumping the small energy barriers, even though there are more of them,

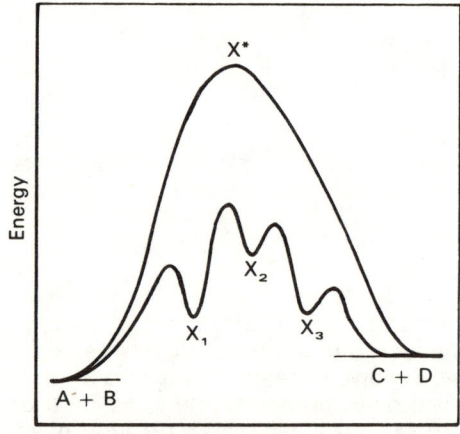

Fig 6.9. Energy diagram for a reaction proceeding via a single intermediate X* and for a catalysed reaction involving several intermediates X_1, X_2 and X_3. The intermediates in the catalysed reaction are separated from each other by relatively low energy barriers.

may be easier than jumping the single, high barrier. Thus an important factor in the increased reaction rate in the presence of a catalyst may be the promotion of a multi-intermediate route for reaction.

The reaction rate advantages derived from

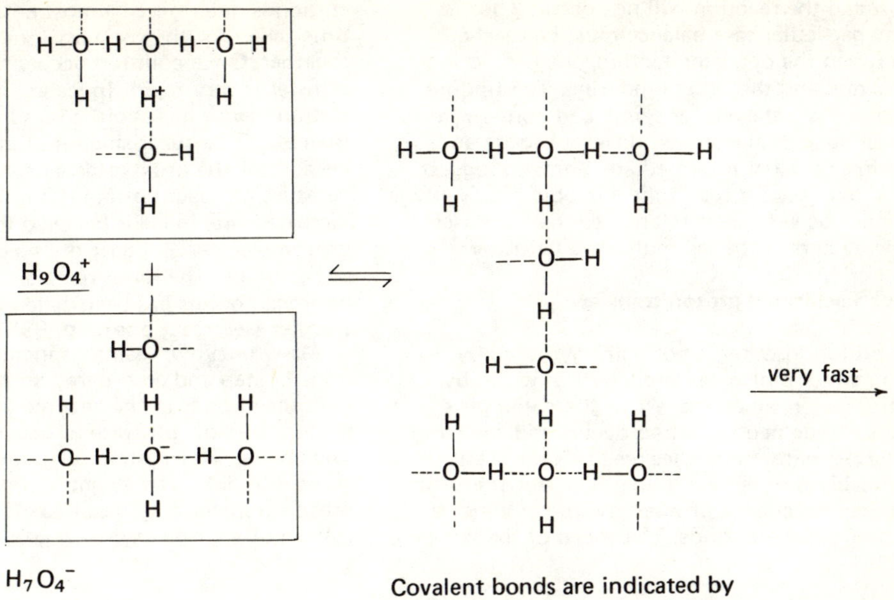

$$H-O\cdots H-O\cdots H\cdots O-H$$
$$\overset{|}{H}\quad \overset{|}{H^+}\quad \overset{|}{H}$$
$$\cdots O-H$$
$$\overset{|}{H}$$

$$\underline{H_9O_4^+}\quad +$$

$$\overset{|}{H-O}\cdots$$
$$\overset{H}{|}\qquad \overset{H}{|}\qquad \overset{H}{|}$$
$$---O-H\cdots O^-\cdots H-O---$$
$$\overset{|}{H}$$

$$H_7O_4^-$$

$$\rightleftharpoons$$

$$H-O\cdots H-O\cdots H\cdots O-H$$
$$\overset{|}{H}\qquad \overset{|}{H}\qquad \overset{|}{H}$$
$$---O-H$$
$$\overset{|}{H}$$
$$H-O---$$
$$\overset{H}{|}\qquad \overset{H}{|}\qquad \overset{H}{|}$$
$$---O-H\cdots O\cdots H-O---$$
$$\overset{|}{H}$$

very fast $\longrightarrow$

Covalent bonds are indicated by
solid lines, hydrogen bonds by
broken lines.

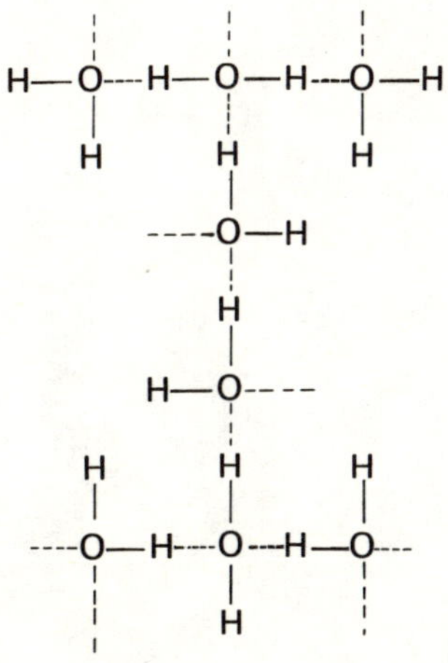

$$H-O\cdots H-O-H\cdots O-H$$
$$\overset{|}{H}\qquad \overset{|}{H}\qquad \overset{|}{H}$$
$$----O-H$$
$$\overset{|}{H}$$
$$H-O----$$
$$\overset{H}{|}\qquad \overset{H}{|}\qquad \overset{H}{|}$$
$$----O-H\cdots O-H-O---$$
$$\overset{|}{H}$$

(Note that neutralization occurs by the interchange of existing H-bonds and covalent bonds.)

smaller activation energies will be offset to some extent by the unfavourable entropy of the multi-step pathway. When a substrate binds to an enzyme a certain portion of the vibrational and translational energy of the molecule is restricted and there may be a decrease in entropy of the system. Similar constraints will be applied to substrate molecules at each step in the reaction. If the total entropy decrease is large enough to make the free energy of the overall reaction positive, then all the advantages of introducing intermediates will have been lost and the reaction will not occur. Thus in any particular case balance must be reached between the opposing factors; one-promoting reaction and the other hindering. The finding that many catalysts, enzymic and non-enzymic, react nucleophilically with the substrate and form a covalent intermediate seems to suggest that in many processes the entropic disadvantage cannot be very great relative to the decreased energy barriers of the multi-step pathway.

(iv) Facilitated proton transfer

A proton in water is normally hydrated by four water molecules and a hydroxyl ion by three water molecules. When these complex ions collide neutralization occurs and four uncharged water molecules result. Neutralization probably occurs by a flow of electrons resulting in the interchange of weak hydrogen bonds and strong covalent bonds. The speed of the neutral-ization is determined almost entirely by the time taken for the complex ions to diffuse together. Once collision occurs the proton transfer is very rapid. In ice at $-15°$ the neutral-ization reaction is about a hundred times faster than at $25°$: a surprising result at first sight because of the usual effects of increase in tem-perature on reaction rate (Chapter 4). The increased rate in ice is believed to result from proton transfer along rigidly held hydrogen bonds in ice. The rapid transfer along preformed hydrogen bonds has been called facilitated proton transfer (*see* reactions on p. 99).

Many enzymic reactions include a proton transfer step and an ordered arrangement of hydrogen bonds in the enzyme may enable the protonation of substrate to occur very quickly. The charge relay system suggested for α-chymo-trypsin (p. 94) is an example where facilitated proton transfer may occur to effect the polar-ization of a serine hydroxyl group.

7 Regulation of Enzyme Activity

When foodstuffs or metabolites pass into a cell there are several different ways in which they can be fitted into the overall economy of the cell. Conversion to storage material, breakdown for energy provision or utilization for the biosynthesis of new cellular material are the three main ways in which a substance may be dissimilated. The requirements of the cell are not static but are varying continuously to match changes in the external environment. Thus, distinct biological advantage can ensue from the ability to direct and redirect metabolites into those pathways which provide the best immediate answer to a particular challenge from the environment.

Once the steps in the main metabolic pathways were understood and well characterized, more attention could be given to the study of how metabolism is regulated and what factors proportion the flow of metabolites between different pathways. Since every metabolic step in every pathway is associated with an enzyme, the regulation of metabolism is largely a matter of controlling the activity of all or some enzymes. Preventing the flow of metabolites through one pathway, for example, by inhibiting some enzymes, could divert the flow through another route by an overspill mechanism. Recent research, much of it purely enzymological, has produced a considerable amount of information on the subject of control but we are far from a complete understanding of this important subject. It is important because life itself depends on orderliness of structure and on organized cellular activity. The loss of both cellular organization and control of normal cellular function is a marked feature of cancer cells.

CONTROL BY ENZYME CONCENTRATION

The most obvious way to alter the flux of metabolites through a pathway is to increase or decrease the molecular quantities of the constituent enzymes. However, an increase in enzyme concentration is only accompanied by increased flux when substrate is freely abundant. If there is insufficient substrate to saturate existing amounts of enzyme, production of more enzyme will not make any considerable difference to the rate of reaction.

Regulation through altered concentrations of enzymes involves the stimulation or inhibition of protein biosynthesis or a change in the rate of protein breakdown. Because a finite time is required for biosynthesis or breakdown there will be a delay between the signal that a change in enzyme is required and the establishment of the enzyme at the new concentration. Similarly, in returning to the original state there will be a time lag between the reduction of the rate of synthesis or breakdown and the attainment of the original level. The duration of the time lag will depend on the normal rates of biosynthesis and destruction of the enzyme. Obviously, if the rate of turnover and an enzyme is normally very low the lag period could be quite lengthy and span several hours or days. Very rapid adjustments of reaction rates are not easily accomplished, therefore, by changes in the concentration of enzymes. Not every challenging stress to a cell is going to demand a rapid, and transient metabolic change, however, and it is in longer term adjustments to metabolic flow that altered rates of enzyme synthesis become important.

In untreated *diabetes mellitus* there is a stimulation of gluconeogenesis in the liver (i.e. the formation of glucose from non-carbohydrate sources such as amino acids), and increased levels of glucose-6-phosphatase, fructose-1-6-diphosphatase and phosphoenolpyruvate carboxykinase can be demonstrated. These three enzymes are essential for the reversal of glycolysis for they bypass physiologically irreversible steps of the forward route. (Fig. 7.1).

Treatment with insulin increases the entry of glucose into the cells and reduces the need for gluconeogenesis resulting in a return of the liver enzymes to normal levels.

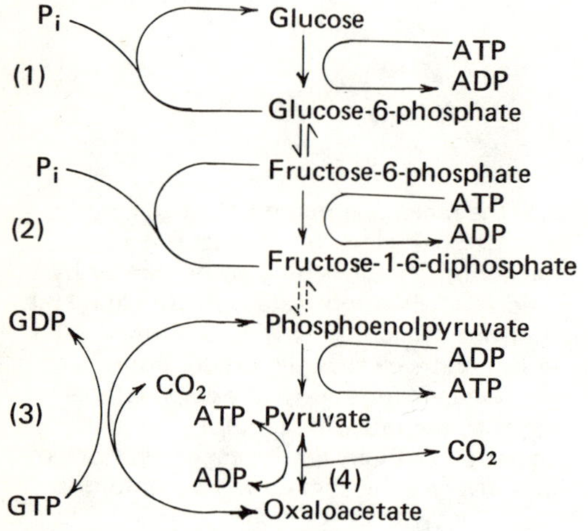

Fig. 7.1. Bypass reactions in the reversal of glycolysis: (1) Glucose-6-phosphatase; (2) Fructose-1-6-diphosphatase; (3) Phosphoenolpyruvate carboxykinase; (4) Pyruvate carboxylase.

A glutaminase enzyme in kidney seems to be very important for acid-base balance in mammals. It catalyses the breakdown of glutamine with the release of ammonia from the amide group

ingestion of mineral acid can be neutralized by ammonia formed in the glutaminase reaction and excreted in the urine as ammonium salts. The protein nitrogen pool represents a very large store of potentially available NH_4^+ cation, and ammonium production in the kidney enables anions of strong acids to be excreted as NH_4^+ salts while other important cations which are less readily available, e.g. Na^+ and K^+, can be conserved.

An induced metabolic acidosis in animals has been shown to result in an increased glutaminase activity of the kidney, suggesting that acid-base balance is at least partly regulated by alterations in the tissue level of a key enzyme. The activity of the gluconeogenic enzyme phosphoenolpyruvate carboxykinase also becomes increased, however, a few hours after a fall in pH or decrease in bicarbonate concentration. It is possible therefore that increased gluconeogenesis may be the chief factor in stimulating increased production of ammonia, since it is known that glutamate is an inhibitor of glutaminase and tissue levels of glutamate are likely to decrease during gluconeogenesis (Fig. 7.2). Thus increased glutaminase activity could be a result of the removal of a regulatory inhibitor. Whichever is the true mechanism of increased ammonia production in acidosis, it seems certain that the controlling process involves a change in the rate of synthesis of one, if not more, regulatory enzymes. Similarly the increased levels of microsomal hydroxylases in the liver are a long-term response to the toxicity of drugs such as barbiturates (Chapter 8), and an increase in liver fatty acid synthetase on refeeding after

of the substrate, and studies with isotopically labelled substrate have shown that this group is the primary source of urinary ammonia.

Acidic excretory compounds formed during metabolic breakdown of foodstuffs or after

a period of starvation, or following insulin treatment of diabetes, reflects a switch-over from the necessity of fat breakdown (to provide energy during starvation or diabetes) to the replenishment of fat stores.

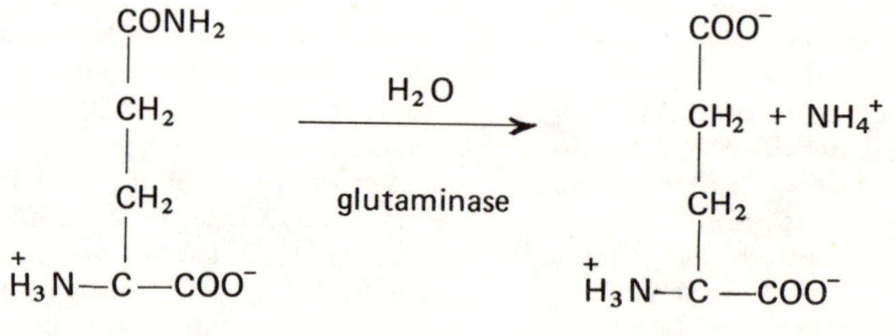

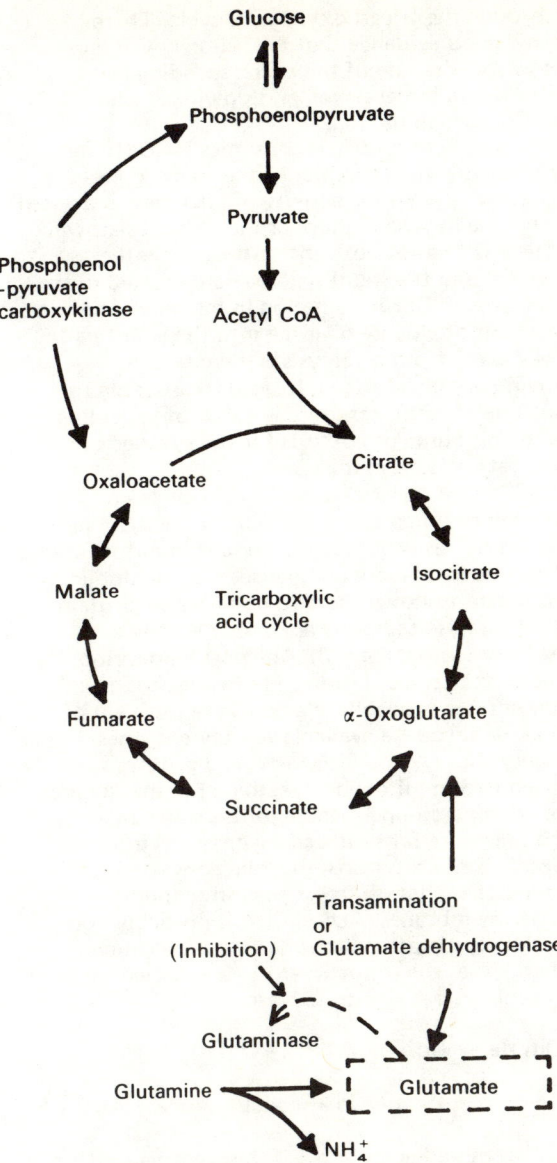

Fig 7.2. Relationship between gluconeogenesis and tissue levels of glutamate. Conversion of oxaloacetate to phosphoenol pyruvate tends to decrease the levels of intermediates of the tricarboxylic acid cycle including α-oxoglutarate. The level of glutamate falls because it is interconvertible with α-oxoglutarate, and thus the inhibition of glutaminase is relieved.

CONTROL BY SUBSTRATE CONCENTRATION AND COENZYME AVAILABILITY

In the discussion of enzyme kinetics it was shown how the substrate concentration affects the velocity of an enzyme reaction (Chapter 4). If the concentration of substrate in the cell is usually lower than the K_m value, the rate of the reaction will be very sensitive to changes in the concentration of substrate. Under these circumstances increased enzyme levels could not bring about a large acceleration of the reaction without a simultaneous increase in substrate concentration. Maintaining a balance of substrate concentration just below K_m has significant potential for metabolic control and evolutionary pressures appear to have developed this potential into important physiological mechanisms. An example of a physiological use of control by substrate is found in certain enzymes which phosphorylate glucose.

$$ATP + Glucose \longrightarrow Glucose\text{-}6\text{-}phosphate + ADP$$

The phosphorylation is irreversible and is catalysed in the liver by an enzyme specific for glucose called glucokinase. The same reaction is catalysed in brain tissue by hexokinase, which as its name implies, is less specific and can catalyse the phosphorylation of several hexose sugars. The K_m for glucose of glucokinase is approximately 2×10^{-2} M, whereas for hexokinase the K_m is approximately 10^{-5} M. The concentration of glucose in the blood passing to the liver through the portal and arterial circulations can reach nearly 10^{-2} M after absorption of a carbohydrate meal. The K_m *of* glucokinase therefore has a value such that the rate of phosphorylation is governed by the amount of glucose reaching the liver. It makes teleological sense that the organ which acts principally as a biochemical manufacturing plant and storehouse should have the capacity to deal with all glucose levels likely to reach it. By contrast, brain tissue relies very heavily on the glucose supply carried to it in the blood for its own metabolic needs. The very much lower K_m of hexokinase ensures that the enzyme is close to being saturated with glucose whatever the blood level, provided that the latter stays within normal limits. This implies that under physiological conditions the consumption of glucose by the brain proceeds at a steady rate and even when the blood glucose concentration approaches

10^{-2} M there will be little effect on the hexokinase reaction.

Within the cell, enzymes associated with a particular pathway are usually grouped together in one organelle or region of the cell (Table 7.1) so that a continuous reaction chain can be set up. The pathways are not totally independent, however, and various metabolites need to be

TABLE 7.1. DISTRIBUTION OF METABOLIC PATHWAYS WITHIN THE CELL

Soluble (Cytoplasmic) Fraction	Mitochondria	Microsomal
Glycolysis	Tricarboxylic Acid Cycle	Protein synthesis
Hexose monophosphate shunt	Fatty acid oxidation	Nucleotide biosynthesis
Fatty acid synthesis	Electron transport	Metabolism of foreign organic compounds
	Oxidative phosphorylation	

shuttled from one region of the cell to another. For example, glucose is converted to pyruvate by glycolysis in the soluble fraction of the cell, but if pyruvate is to be oxidized further through the tricarboxylic acid cycle it must be transported into the mitochondria. The imposition of restraints on the availability of substrate and coenzymes by permeability barriers, for instance, may be exploited for purposes of metabolic control.

Some pathways sharing common metabolites are located in the same organelle so that control by transport limitations could not apply. However, the opportunity can exist in these situations for controlling the rate of one pathway by that of another where the two processes share common intermediates. The enzymes of the tricarboxylic acid cycle, the electron transport system and the enzymes which catalyse the phosphorylation of ADP coupled to electron transport are located in mitochondria. The cycle generates reduced coenzymes which are then reoxidized with the generation of the energy needed to drive the phosphorylation of ADP to form ATP. The coupling of these systems means that ATP production can be limited by the rate of flux

through the tricarboxylic acid cycle. There is now good evidence that the cellular ATP level controls the rate of the cycle; speeding it up if the cellular levels fall and slowing it when ATP is accumulating.

Fatty acid synthesis consumes NADPH and one of the main supplies of the reduced coenzyme is from the activity of enzymes associated with the hexose monophosphate shunt pathway. The enzymes of both the fatty acid synthetase system and the shunt pathway are located in the cytoplasm. During lactation in mammals the amount of glucose diverted into the shunt pathway away from glycolysis is increased in mammary gland tissue. It seems reasonable to assume that this response is a way of providing extra amounts of NADPH for the synthesis of the fats present in milk.

Some of the tricarboxylic acid cycle and related enzymes are duplicated in the cytoplasm of the cell as well as being present in mitochondria. In some cases close comparisons of the duplicate forms have shown them to be different proteins that catalyse the same reaction, i.e. they are isoenzymes (Chapter 3). The parallel development of enzymes inside and outside the mitochondria is probably related to the problem of making substrate available for key enzymes. Acetyl-CoA needed for fatty acid synthesis is generated in mitochondria, the inner membranes of which are impermeable to this substance. To reach the fat synthesizing enzymes in the cytoplasm, acetyl-CoA must be converted to a metabolite that can cross the mitochondrial inner membranes. Acetyl-CoA is probably converted by citrate synthetase to citrate which then leaves the mitochondrion and is cleaved by cytoplasmic ATP-citrate lyase:

Citrate + CoA + ATP $\longrightarrow$

Acetyl-CoA + oxaloacetate + ADP + Pi

The required acetyl-CoA is thus regenerated outside the mitochondrion. Oxaloacetate formed in the lyase reaction can be converted by cytoplasmic malate dehydrogenase, to malate which in turn may be converted to pyruvate and transported back into mitochondria. Figure 7.3 illustrates these processes diagrammatically.

One of the chief catalytic differences between the two forms of malate dehydrogenase is shown by their patterns of inhibition by excess substrate. The mitochondrial isoenzyme, but not the cytoplasmic one is inhibited by oxaloacetate whereas

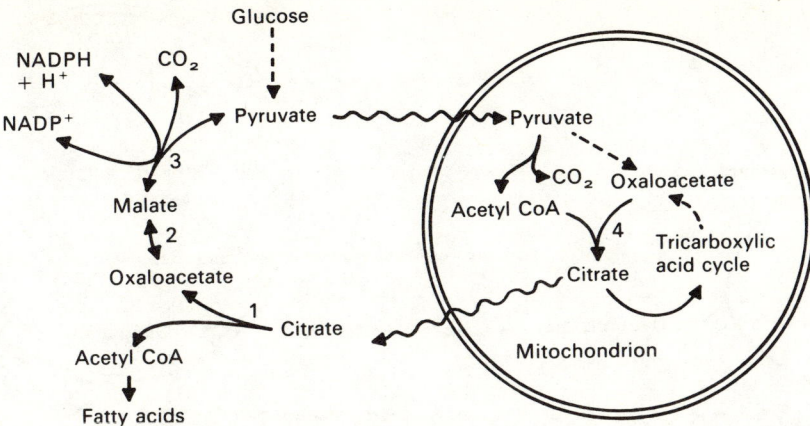

Fig 7.3. Transfer of acetyl-CoA from mitochondria to cytoplasm. 1, ATP-citrate lyase; 2, cytoplasmic malate dehydrogenase; 3, malic enzyme; 4, citrate synthetase.

the cytoplasmic isoenzyme is inhibited to a greater extent by malate than is the mitochondrial. Although the malate dehydrogenase reaction is reversible, these findings suggest that malate *oxidation* is favoured *inside* the mitochondria, but that *outside* oxaloacetate *reduction* occurs preferentially. The proposed directions of the reactions inside and outside the mitochondrion is in keeping with the role for the isoenzymes outlined above.

The two forms of malate dehydrogenase have also been suggested to be important in the transfer of reducing equivalents from the cytoplasm into mitochondria. During aerobiosis, glycolysis generates NADH which is reoxidized by the respiratory chain linked to atmospheric oxygen. The NADH formed in the cytoplasm cannot penetrate mitochondria so the reducing equivalents must be carried to the electron transport chain by an indirect route. The couple involving malate dehydrogenase is shown diagrammatically in Fig. 7.4.

The malate dehydrogenases working in opposite directions can couple NADH oxidation in the cytoplasm to NAD$^+$ reduction in the mitochondrion provided that there is no barrier to the transfer of malate and oxaloacetate. The system as it stands is too simple because unlike malate, oxaloacetate cannot readily cross the mitochondrial inner membrane. By proposing a role for the transaminase isoenzymes which, like malate dehydrogenase, are found inside

and outside mitochondria, the barrier to the movement of oxaloacetate can be overcome. Thus a suggested possible scheme for the transmission of reducing equivalents from outside to inside the mitochondrion is outlined in Fig. 7.5.

From the few examples discussed above it can be seen how interplay between metabolic cycles is essential if a substrate is to get to the appropriate enzyme at the right time. Also, if a stop or start signal is applied to the enzymes in one pathway, connected pathways are bound to be affected too. Much of the research into regulatory mechanisms is concerned with the identification of specific agents that act as regulatory signals and with how the signal is translated into increased or decreased activity at the enzyme level.

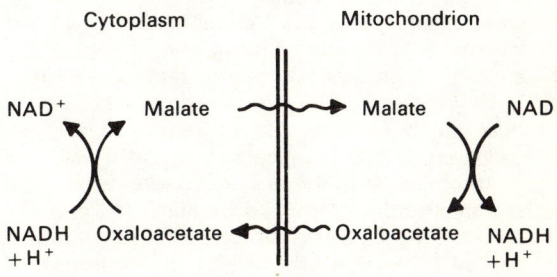

Fig 7.4. Transport of reducing equivalents into mitochondria.

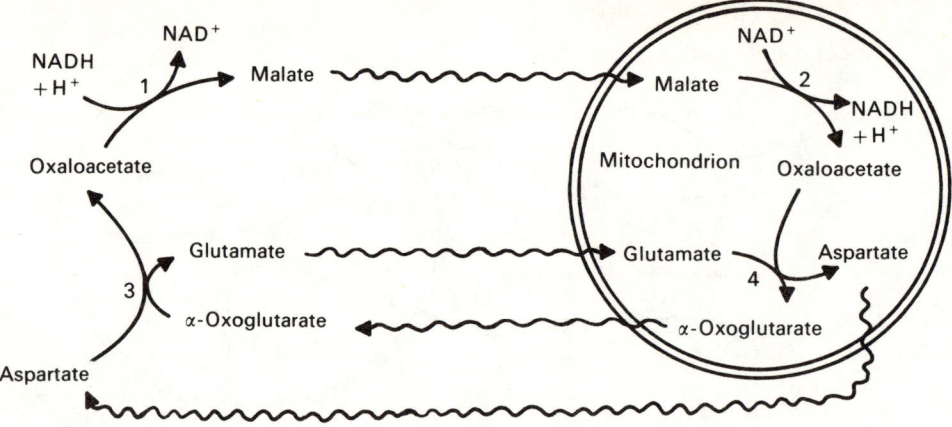

Fig 7.5. **Transport of reducing equivalents into mitochondria. 1, Cytoplasmic malate dehydrogenase; 2, mitochondrial malate dehydrogenase; 3, cytoplasmic aminotransferase; 4, mitochondrial aminotransferase.**

CONTROL BY SPECIFIC ACTIVATORS AND INHIBITORS

Most enzymic reactions *in vivo* are not isolated chemical events, but are part of an ordered sequence of reactions which may even form a closed cycle, e.g. the tricarboxylic acid cycle. Consider a linear system that converts substrate A into product D via two intermediates B and C. E_a, E_b, E_c are enzymes which act on A, B, etc.

$$\begin{array}{ccccc} E_a & & E_b & & E_c \\ A \longrightarrow & B & \longrightarrow & C & \longrightarrow D \end{array}$$

The maximum rate at which A can be converted to D is governed by the enzyme in the series with the lowest V_{max}. To make this clear consider what happens if E_b has a much lower V_{max} than its partners. The rate of formation of C (and hence D) is greatest when E_b is saturated with B. Any increase in B beyond saturation level does not speed up the rate of the reaction catalysed by E_b. The greater V_{max} of E_a will tend to bring B to the saturation level very rapidly and B may even accumulate. If the step catalysed by E_a is reversible then accumulation of B will slow down the rate of conversion of A to B by the back reaction. Once E_b is saturated an increase in concentration of any of the substrates can have only a transient effect on the rate of flux through the pathway. Thus grouping enzymes in a sequence confers certain intrinsic self-regulatory properties.

The potential for self-regulation is emphasized by considering what happens when a non-competitive inhibitor of E_b is introduced into the system. The velocity of the E_b reaction decreases with a tendency for B to accumulate. By acceleration of the back reaction the net rate of conversion of A to B is automatically adjusted to a new velocity. The rates of formation of D and C will also fall because they are dependent on the activity of E_b. Conversely, activation of E_b, i.e. increasing its V_{max} will be followed by a rapid adjustment upwards of the velocities until a new steady level is reached. Thus a change in the velocity of just one rate limiting or "pacemaker" enzyme in a sequence is sufficient to change the flux through the whole pathway because of the self-regulatory nature of coupled systems. It is interesting to note that a competitive inhibitor of E_b could only slow the pathway transiently, for the accumulation of B when E_b was inhibited would tend to overcome the effect of the competitive inhibitor.

In an unbranched sequence, location of the pacemaker enzyme at the first or second step minimizes the number of metabolites which accumulate when the activity of the pacemaker is inhibited. Placing the pacemaker last in a long chain of enzymes would be much less efficient in this respect and although regulation of flux through the pathway is still possible by alteration of the activity of the pacemaker, the diversion of one of the metabolites into an alternative route would have to compete with the continuing flow through the pathway up to the point of stoppage. The system becomes even more efficient for regulatory purposes if the activity of the pacemaker enzyme is controlled by signals originating in the pathway itself. It is commonly found that the terminal product of a pathway is an inhibitor of the pacemaker enzyme. Thus if the product is accumulating because there is more than sufficient for immediate metabolic needs, the rate of flux through the pathway will be slowed until the surplus is dissipated. This kind of control is called negative feedback.

A more complex situation arises when a pathway is branched and two different end metabolites share part of a common production line. In the example below, G and E represent two metabolites formed from a common set of precursors, A, B, C and D. Shutdown of the pathway by inhibition of E_a by excess E would

$$A \xrightarrow{E_a} B \xrightarrow{E_b} C \xrightarrow{E_c} D \begin{array}{c} \xrightarrow{E_d^1} \text{\textcircled{E}} \\ \searrow_{E_d^2} F \longrightarrow \text{\textcircled{G}} \end{array}$$

lead automatically to a decrease in the rate of production of G also. This might be appropriate in some cases, but the system loses in flexibility because the ability to convert A predominantly into G at the expense of E and vice versa is gone. Flexibility can be built into the system by making E_d^1 and E_d^2 sensitive to feedback signals. If E inhibits E_d^1 and at the same time activates E_d^2, the pathway effectively becomes a system for making G from A. Actions of G on E_d^2 and E_d^1 can bring about the preferential synthesis of E. Notice that E_d^1 and E_d^2 represent the first enzymes in linear sequences following a branch point. Thus if the branches each contain several steps wasteful accumulation of unecessary metabolites is avoided. An alternative scheme can be envisaged where an excess of both E and G is required for effective feedback inhibition of E_a. This provides a simpler answer to the problem of avoiding a deficiency of one of the end products when the other is in excess.

Many enzymes which are sensitive to feedback control have now been recognized and studies of their kinetics have shown that ordinary Michaelis-Menten theory (Chapter 4) does not

generally fit their behaviour. For regulatory enzymes when the velocity (*v*) of the catalysed reaction is plotted against substrate concentration, an s-shaped (sigmoid) curve is obtained instead of the usual rectangular hyperbola of Michaelis theory (Fig. 7.6). A sigmoid curve means that at low values of *s* the value *v* is much less than

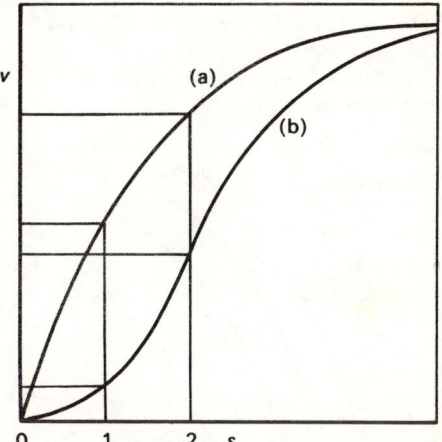

Fig 7.6. Relationship between *v* and *s* for an enzyme conforming with Michaelis-Menten kinetics (a) and for a regulatory enzyme with sigmoid kinetics (b). Note that the *percentage* increase in *v* when *s* is increased from 1 to 2 is very much greater for the regulatory enzyme.

its predicted value from the Michaelis equation. The sigmoidicity introduces a threshold for *s* and there is little increase in *v* until the threshold level, represented by the flat early part of the curve, is passed. Once beyond this point, however, *the rate of increase* of velocity is greater than predicted by the Michaelis equation. Figure 7.6 shows that when the substrate concentration is increased by 1 unit, velocity increases by 50 per cent for the non-regulatory enzyme but by 350 per cent for the regulatory one. Thus, sigmoid kinetics make the velocity very responsive to changes in *s*, and, depending on the degree of sigmoidicity, can make the enzyme function almost as an on-off switch triggered by a change in the concentration of *s*.

The shape of the velocity-substrate curve can be influenced by the presence of feedback inhibitors and activators. For many regulatory enzymes increasing the concentration of inhibitor tends to make *v* against *s* plots more sigmoid and extends the flat threshold part of the curve to higher *s* values. Considerably higher concentrations of substrate are then required before any significant rate of catalysis is reached. Activators push the curves in the other direction, i.e. to the left, and may bring the binding curve to a hyperbolic shape. Thus the velocity of the catalysed reaction for a given amount of substrate can be varied over a wide range depending on the concentrations of inhibitor, activator and substrate. Figure 7.7 shows how changing the shape of the curves can increase the velocity from 5 to 90 per cent with no alteration in substrate concentration.

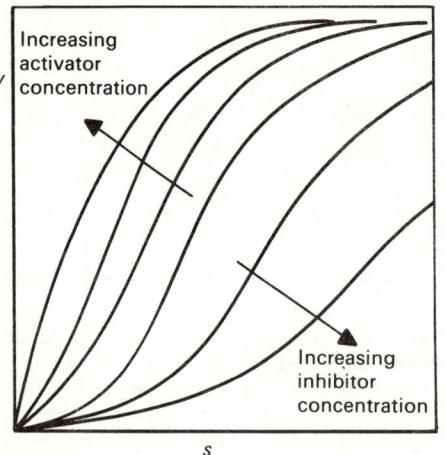

Fig 7.7. Effect of modifiers on the shape of *v* against *s* curves for a regulatory enzyme. Note that, depending on concentration of inhibitor or activator, the velocity can be varied over a wide range with no change in *s*.

Sigmoid curves are usually interpreted as evidence for the binding of more than one molecule of substrate to an enzyme and that some kind of interaction occurs between bound molecules. By interaction it is meant that the binding of one molecule of substrate facilitates the binding of the next and so on, until the enzyme is saturated and all binding sites are filled. Cooperativity is another term that is used to describe interactions between molecules of bound substrate. Investigations of reaction velocities as a function of inhibitor concentration have shown that cooperativity also exists between molecules of bound inhibitor as evidenced by sigmoid inhibition curves. A similar situation exists for activators. Activators and inhibitors are collectively referred to as effectors.

From studies of the feedback inhibition of aspartate transcarbamoylase, an enzyme important for nucleotide biosynthesis, Gerhard and Pardee in 1962 had deduced that separate sites existed on the enzyme for the binding of the substrate and effectors, i,e. the inhibition could be likened to the non-competitive type. Monod, Changeux and Jacob (1963) proposed that the possession of separate sites for binding of substrates and effectors may be a general property of regulatory enzymes. Arguing teleologically it was suggested that separate sites were to be preferred for efficient control because there need not then be close structural similarity between the substrate and effector. If two different molecules are to bind to the same site the specificity of enzymes demands certain structural similarities between them. A terminal metabolite acting as a feedback effector is unlikely to resemble the substrate of the first enzyme in the pathway. Also, attachment of both substrate and inhibitor to the same site implies a competitive type of inhibition and as we have already seen this is not very effective for altering the rate of a coupled pathway. Allosteric, meaning *other shape or position*, was a term introduced by Monod *et al*., to describe those enzymes which seem to contain binding sites for more than one kind of ligand (substrate and effectors) and for which the binding of effector brings about a change in protein conformation (or shape) thereby influencing the binding of substrate. The word has now tended to be loosely applied to all enzymes that appear to have a regulatory function and sigmoid *v* against *s* curves.

An essential feature of the second site hypothesis is that effector sites should be as specific for their ligands as the catalytic site is for substrate. Therefore, just as some chemical analogues of the substrate can bind to the catalytic site, analogues of effectors should act as alternative ligands for effector sites. It is to be expected that some analogues will act as effectors because of a good "fit" at the binding site while others will compete with the true effector and thus tend to abolish any inhibition or activation. These predictions are borne out by the behaviour of the enzyme threonine deaminase in *E.coli*. This enzyme catalyses the conversion of threonine to α-oxobutyrate on the pathway leading to the biosynthesis of L-isoleucine. Isoleucine inhibits threonine deaminase but nor-leucine is very effective in overcoming the inhibition. L-leucine is an inhibitor when added by itself and cooperates with isoleucine when both amino acids are added to the enzyme solution.

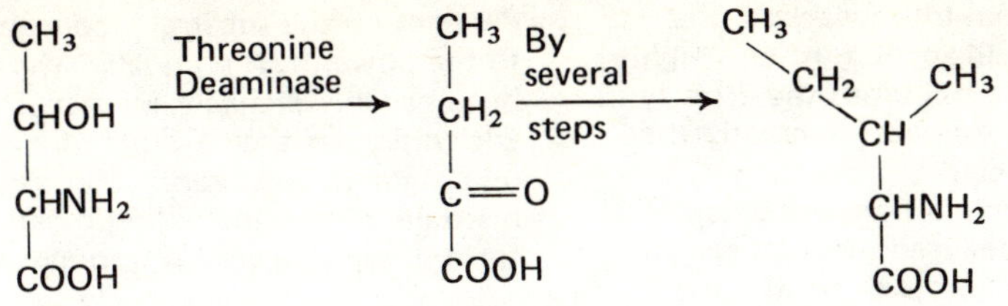

Chemical structures

$$
\begin{array}{c}
CH_3 \\
| \\
CHOH \\
| \\
CHNH_2 \\
| \\
COOH
\end{array}
\quad \xrightarrow{\text{Threonine Deaminase}} \quad
\begin{array}{c}
CH_3 \\
| \\
CH_2 \\
| \\
C{=}O \\
| \\
COOH
\end{array}
\quad \xrightarrow[\text{steps}]{\text{By several}} \quad
\begin{array}{c}
CH_3 \\
CH_2 \quad CH_3 \\
CH \\
| \\
CHNH_2 \\
| \\
COOH
\end{array}
$$

L-threonine   α-oxobutyrate   L-isoleucine

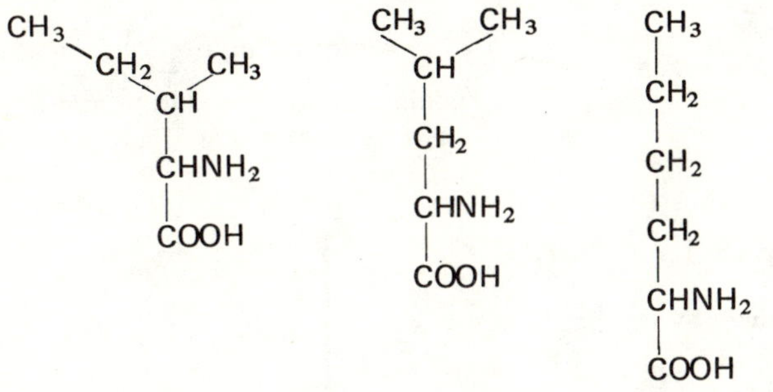

L-isoleucine   L-leucine   nor-leucine

MODELS PROPOSED FOR REGULATORY ENZYMES

The finding that a sigmoid ligand-binding curve is a general feature of regulatory enzymes has stimulated the development of theoretical models to account for the kinetic behaviour. Cooperativity automatically implies that more than one molecule of a substrate is bound and that separate but presumably identical sites exist to accommodate each molecule of substrate. If the enzyme is sensitive to an effector, which again shows cooperativity, then a second series of identical sites is required. The allosteric requirement is that the sites in one series are quite different and distinct from those in the other. Cooperativity between molecules of like ligands (i.e. substrate-substrate or inhibitor-inhibitor, etc.) is called *homotropism*. An effect on the cooperativity of one ligand by another of a different kind (i.e. substrate-inhibitor, substrate-activator, etc.) is called *heterotropism*. Several sites could be distributed along a single polypeptide chain or alternatively the enzyme could be made up of a number of identical sub-units, each carrying a substrate site, inhibitor site and activator site. All the enzymes showing cooperativity investigated so far have been shown to possess sub-unit structure and most of the theoretical models assume the existence of quaternary protein structure.

(i) Atkinson Model

One of the simplest models to account for sigmoid ligand-binding curves has been examined by D. E. Atkinson in connection with the properties of yeast isocitrate dehydrogenase.

Suppose that an enzyme E has n binding sites for a ligand S and that in the presence of S, equilibrium is rapidly attained between free enzyme and the enzyme with all n sites filled (ES_n)

$$
E + nS \underset{k_{-1}}{\overset{k_{+1}}{\rightleftharpoons}} ES_n
$$

The dissociation constant K for the dissociation of ES_n into $E + nS$ is given by

$$
K = \frac{k_{-1}}{k_{+1}} = \frac{[E][S]^n}{[ES_n]} \tag{1}
$$

For such an equilibrium to be reached, the binding of each molecule of S must be "highly cooperative", i.e. as each S binds the affinity of the remaining sites increases so greatly that forms of the enzyme carrying less than n molecules of S have only a transient existence. The velocity of the catalysed reaction can then be accounted for by the rate of breakdown of ES_n. If the appropriate rate constant is denoted by k_{+2}

$$v = k_{+2} [ES_n] \qquad (2)$$

At equilibrium the total enzyme in the system is given by

$$[E_0] = [E] + [ES_n]$$
$$\text{and} \qquad [E] = [E_0] - [ES_n] \qquad (3)$$

The maximum velocity V is given by $k_{+2}[E_0]$ so from (2) and (3)

$$k_{+2}[E] = V - v$$
$$\text{From (1)} \quad k_{+2} \frac{[ES_n]\ K}{[S]^n} = V - v$$

Substituting again for v and rearranging

$$\frac{v}{V - v} = \frac{[S]^n}{K} \qquad (4)$$

Equation (4) can be rearranged to equation (5) which is of identical form to the equation proposed by Hill to describe the

$$\frac{v}{V} = \frac{[S]^n}{K + [S]^n} \qquad (5)$$

oxygen saturation curves of haemoglobin

$$\frac{y}{100} = \frac{(x)^n}{1/K + (x)^n} \qquad (6)$$

where y is per cent saturation, (x) is the partial pressure of oxygen and n is a constant that determines the degree of sigmoidicity of the curves.

Equation (5) predicts that the velocity depends on the substrate concentration raised to the power n. If n equals 1 the equation becomes the Michaelis-Menten equation, but with n greater than 1 plots of v against s take on a sigmoid appearance. For expressing experimentally determined kinetic data a logarithmic form of equation (4) is frequently used.

$$\log \left(\frac{v}{V - v} \right) = n \log [S] - \log K \qquad (7)$$

A plot of $\log \left(\frac{v}{V-v} \right)$ against $\log [S]$ is a straight line of slope n and intercept $\frac{1}{n} \log K$ (Fig. 7.8). If n is greater than 1 cooperativity is indicated.

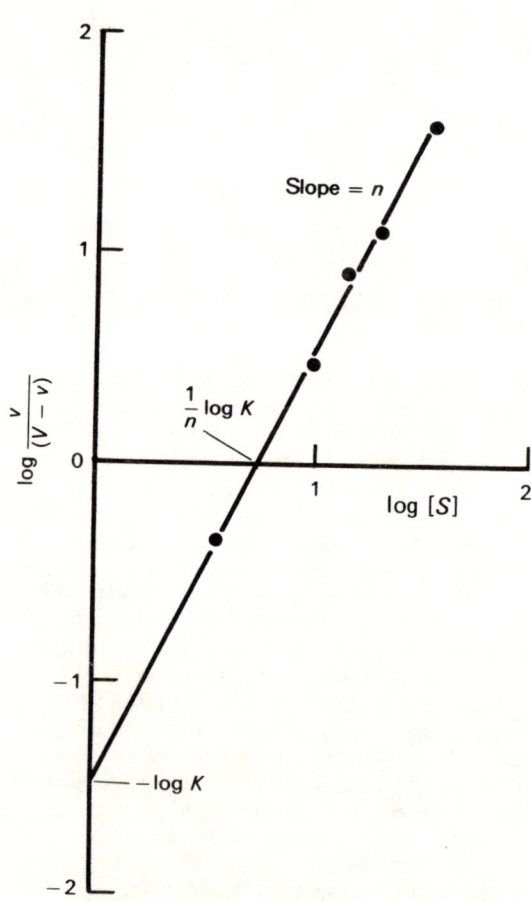

Fig 7.8. Logarithmic plot of data according to the Hill equation in order to determine the values of n and K for an enzyme exhibiting sigmoid kinetics.

For the model we have been considering, n equals the number of binding sites. It needs to be stressed, however, that this is only true for a model where the strength of the interactions is so great that the fully loaded enzyme is the only significant form that gives rise to products. For haemoglobin, to which the Hill equation was first applied, n is found to be approximately 2·5. As there are 4 sub-units in haemoglobin each with an oxygen binding site, it can be said immediately that this simple model does not describe the binding properties of haemoglobin. Hill plots are useful for testing for cooperativity but are not, by themselves, a very reliable guide to the number of binding sites on enzymes.

The activity of yeast isocitrate dehydrogenase is modified by several effectors, notably AMP which increases the affinity of the active site for isocitrate. Unlike many other regulatory enzymes the effector has no demonstrable heterotropic effect on the interactions between bound molecules of isocitrate; binding curves for isocitrate remain sigmoid as AMP is increased but the curves are displaced towards the origin. Hill plots at different concentrations of AMP have identical slopes. The addition of AMP would need to reduce the strength of the interactions between substrate molecules for the slope to be affected. For the yeast enzyme it has been calculated that the binding of each molecule of isocitrate increases the affinity at the other sites by a factor of twenty at least. Thus it seems reasonable to assume that the yeast enzyme fulfils the requirements of the Atkinson model (that E and ES_n reach equilibrium very rapidly), and no doubt other enzyme examples will be found that behave similarly.

The physical mechanism underlying the changes of affinity induced by binding of ligands can only be guessed at. The mechanism can be seen as an extension of the "induced fit" idea whereby a bound ligand springs a conformational change, the effects of which are not restricted to the region of its own site, but are transmitted through the protein to other sites also. Certainly, intact tertiary structure seems to be essential for cooperativity; treatment of allosteric enzymes by such means as gentle heating or incubation with urea and thiol-reactive reagents often returns ligand-binding curves to a hyperbolic shape without decreasing V_{max}. Sometimes abolition of the inhibitory effect of a ligand can be demonstrated with no loss whatsoever of catalytic activity on the substrate (Fig. 7.9). These experiments on "desensitization"

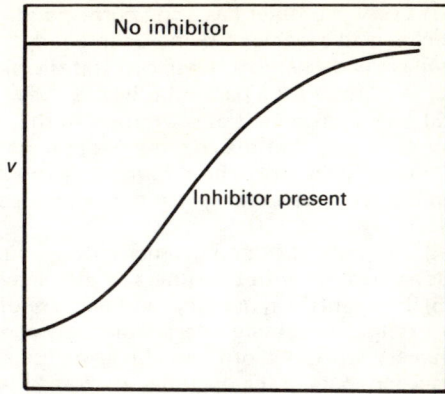

Fig 7.9. **Desensitization of an hypothetical allosteric enzyme. The enzyme is heated at 55° and at various time intervals samples are withdrawn for measurement of activity with and without inhibitor. Heat treatment abolishes the effect of the inhibitor but does not affect catalysis.**

are often used to test for cooperativity. A return to Michaelis-Menten kinetics implies that interactions between ligands are destroyed by desensitization, but the actual binding sites themselves are not damaged. An enzyme with four binding sites that are totally independent of each other is analogous to four molecules of enzyme each with only a single site and in both cases substrate binding properties are adequately described by Michaelis theory.

(ii) Monod, Wyman and Changeux Model

This model is a little less restrictive than the one just described because it does allow for heterotropic interactions between dissimilar ligands. Thus it is applicable to those enzymes where homotropic interactions are weaker than in isocitrate dehydrogenase and where enzyme molecules carrying less than the full complement of substrate molecules can give rise to products.

The Monod model is based upon a set of assumptions about the properties of allosteric enzymes which may or may not be realized in actual biological systems. The assumptions are as follows:

(1) Each allosteric enzyme is a polymer or *oligomer* of a definite number of identical sub-units called *protomers*.

(2) Every protomer has a site corresponding to each ligand capable of being bound and the protomers can exist in at least two states which differ in affinity for a particular ligand.

(3) The arrangement of protomers in the oligomer is such that the enzyme has at least one axis of symmetry. The symmetry is conserved when the protomers change from one state to another.

(4) The state of each protomer is constrained by its association with the others in the oligomer.

(5) To maintain symmetry, and because of very strong constraining interactions between the protomers, when one of them changes state all of its neighbours in the oligomer are forced to do likewise. Thus the number of different states available to an oligomer equals the number of permissible states for each protomer.

Consider an oligomer containing n protomers that can exist in two states. In the absence of any ligand the two states are assumed to be in equilibrium. Let the states be designated R_0 and T_0 and the equilibrium constant for the R–T transition be L, the *allosteric constant.*

$$R_0$$ $$T_0$$

Therefore $T_0 = LR_0$

Let K_R and K_T be the dissociation constants for a ligand S bound to protomers in the R and T state respectively. The binding of a molecule of S to each protomer is independent of the binding to neighbouring protomers so K_R and K_T describe binding to any sites in the oligomer. The equilibrium equations describing the binding of S to R and T are as follows, where the subscripts denote the number of molecules of ligand that are bound:

$$R_0 \rightleftharpoons T_0$$

$$R_0 + S \rightleftharpoons R_1 \qquad T_0 + S \rightleftharpoons T_1$$
$$R_1 + S \rightleftharpoons R_2 \qquad T_1 + S \rightleftharpoons T_2$$

$$R_{n-1}+S \rightleftharpoons R_n \qquad R_{n-1}+ S \rightleftharpoons T_n$$

Let the ratio of ligand concentration to K_R be α and the ratio of the dissociation constants be c, i.e. $\alpha = \dfrac{(S)}{K_R}$; $c = \dfrac{K_R}{K_T}$. Allowing for probability factors, it can be shown that the *fraction* of the *total number* of sites filled on the enzyme $\bar{Y}_s$ is given by equation (8)

$$\bar{Y}_S = \frac{Lc\alpha(1 + c\alpha)^{n-1} + \alpha(1 + \alpha)^{n-1}}{L(1 + c\alpha)^n + (1 + \alpha)^n} \qquad (8)$$

This rather complicated equation accounts for cooperativity when binding occurs to both states provided that K_R and K_T differ. If the postulate is made, however, that the ligand binds only to one of the states considerable simplification is possible and it is easier to see how the model accounts for cooperativity. If binding to T does not occur K_T becomes infinite and

$$c = \left(\frac{K_R}{K_T}\right)$$ becomes zero. Under these conditions equation (8) becomes

$$\bar{Y}_S = \frac{\alpha(1 + \alpha)^{n-1}}{L + (1 + \alpha)^n} \qquad (9)$$

This equation accounts for cooperativity. When L is large the protein is mainly in the T state with no affinity for S. Thus α (which is directly proportional to the concentration of S) needs to be very large before any significant increase in saturation can occur. Such a situation is characterized by a high degree of cooperativity and sigmoid plots of $\bar{Y}$ against α. Looked at in another way it can be envisaged that adding ligand to the protein, which at equilibrium is mainly in the T state, titrates it into the R state by pulling the equilibrium towards R.

T   R

Because of the symmetry requirements of the model a change in state of one protomer tends to change that of the others. Hence it is much easier for a second molecule of S to bind because more of the protein has been forced into the favourable R state. The sequence of binding to a tetrameric enzyme is represented below.

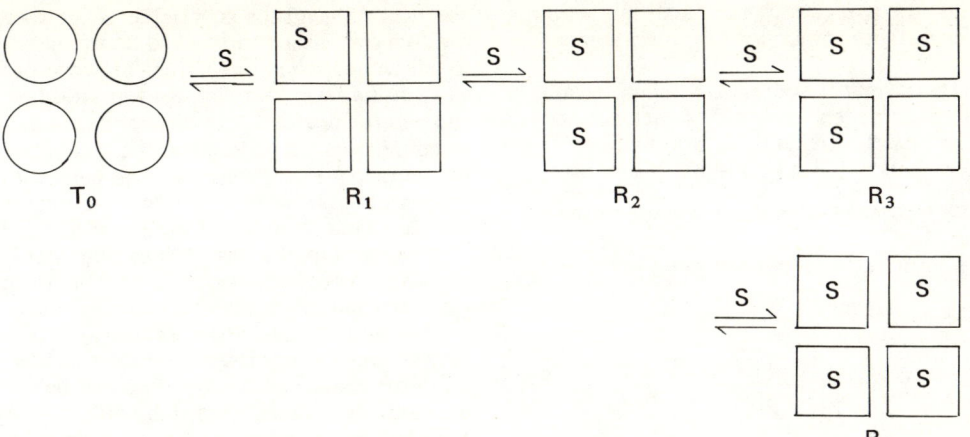

If L is zero, i.e. the protein exists only in the favoured R state, equation (9) shows that

$$Y_S = \frac{\alpha}{1+\alpha}$$

$$= \frac{S}{K_R + S} \qquad (10)$$

Setting $\overline{Y}_S$ equal to $\frac{v}{V}$, i.e. the reaction velocity as a fraction of the maximum at saturation, we have

$$\frac{v}{V} = \frac{S}{K_R + S} \qquad (11)$$

which is the Michaelis-Menten equation and verifies that cooperativity is impossible when the enzyme exists in only one form.

The effects of inhibitors and activators are explicable on the basis of the model by assuming that activator and substrate bind to the same state of the enzyme R (but at different sites), and inhibitors bind to enzyme in state T. I constrains the equilibrium in the T state but A helps S in pulling the equilibrium towards the active form and stabilizing it there. In the titration of T to R, molecules of A and S can be regarded as additive, and saturation plots as a function of (S) will be less sigmoid, i.e. cooperativity is reduced. I, on the other hand, is antagonistic towards S during the titration and high concentrations of S will be needed to

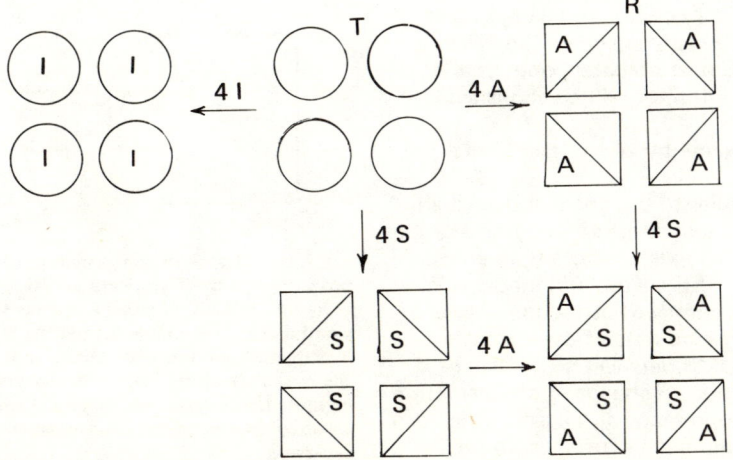

overcome the antagonism and reach saturation, i.e. cooperativity is increased in the presence of I.

This behaviour is expressed mathematically by extending the equation (9) to include more than one ligand. If K_I and K_A are the dissociation constants for bound inhibitor and activator respectively, and $\frac{I}{K_I}$ is represented by β and $\frac{A}{K_A}$ by γ, the saturation function for S becomes:

$$\overline{Y}_S = \frac{\alpha(1+\alpha)^{n-1}}{L\frac{(1+\beta)^n}{(1+\gamma)^n} + (1+\alpha)^n} \tag{12}$$

Equation (12) shows that the allosteric constant L is modified by a function that depends on the relative concentrations of the heterotropic ligands.

The nature of the change in structure when a protein changes back and forth from the R and T states is presumed to involve a redistribution of the non-covalent bonds holding the sub-units together. Because the assumption is made when defining the model, that the conformation of each protomer is dependent on interactions with its neighbours, it can be argued that a change in the interactions may allow each protomer to "relax" to a different tertiary structure. To fit the model, however, the new tertiary configurations of each protomer must be identical to conserve symmetry. A change is shown diagrammatically in Fig. 7.10.

If substrate binds to R and inhibitor to T, any damage to the protein which makes it impossible for the T state to be taken up will desensitize the allosteric enzyme. The binding of S will obey Michaelis-Menten kinetics and the effects of an inhibitor will be abolished.

(iii) Koshland, Nemethy and Filmer Model

This model is similar to the one just described in that it assumes that allosteric enzymes contain sub-units which can exist in more than one state. For the purposes of the Koshland model it is assumed that the energy of attraction between protomers in the same state of conformation is different from that between protomers in unlike conformation. A change of conformation is induced by the presence of a ligand which can then bind to a site on the protomer. In the

absence of ligand the energy barrier between the two states is considered to be too great to allow more than just a trace of protomer to exist in the ligand-binding conformation. It is allowed for protomers in different conformations to co-exist in the oligomer and conservation of symmetry is not essential. The two points made in the last sentence chiefly distinguish the Koshland from the Monod model.

The binding of a ligand to an oligomer is assumed to proceed sequentially. The protomers isomerize and bind ligand one by one until all the binding sites are filled (see diagram on p. 115). Any substance that causes the same conformational change as the substrate will be an activator, but a substance stabilizing the resting state will be an inhibitor. When a particular ◯ is adjacent to S , interaction between the two is considered to lower the energy barrier for the transition of this ◯ to S when another molecule of ligand approaches. Thus with every sequential isomerization (and binding

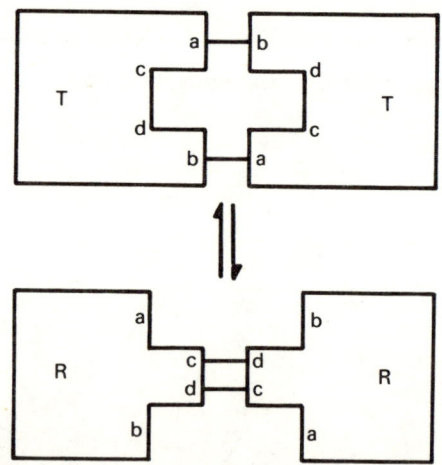

Fig 7.10. Change in conformation of protomers. Two protomers in the T-state are joined by non-covalent links through a-b pairings. A change to the R-state makes c and d groups available for pairing but separates a and b. Substrate binds preferentially to R. In haemoglobin, the change from the deoxy to the oxy configuration involves breakage of salt linkages formed with the carboxyl groups of the C-terminal amino acids of the sub-units.

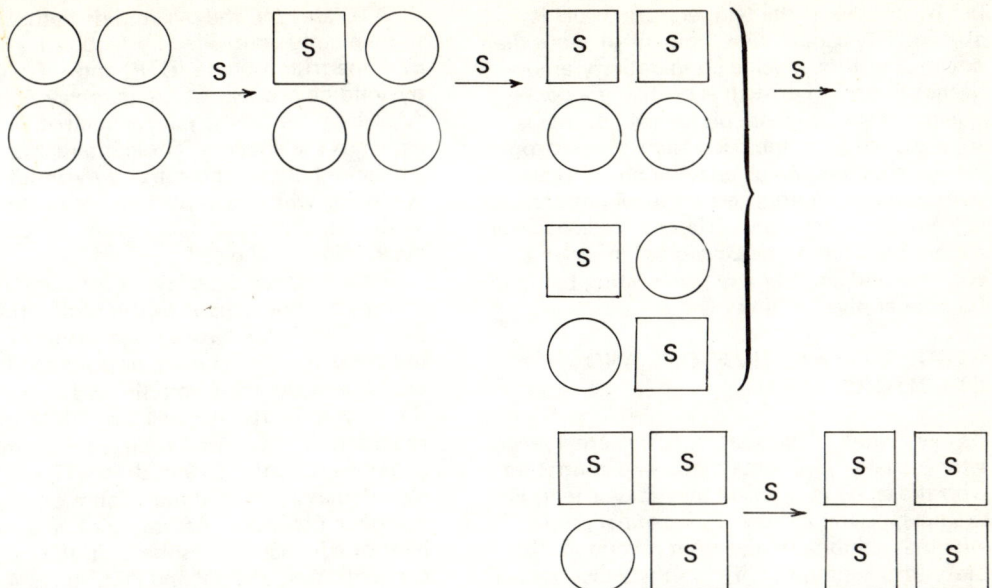

of ligand) the increasing ease of transition is manifest by a progressive increase in the affinity of each unfilled protomer for substrate. An increasing affinity with filling of sites means cooperativity.

The mathematical treatment of the model is beyond the scope of this chapter but it is worth mentioning here that the mathematics shows that Michaelis-Menten kinetics can arise even when sub-unit interactions are occurring. Thus hyperbolic v against s curves do not necessarily mean that cooperative effects are totally absent.

(iv) Other Models

Models have been proposed to account for cooperativity which show that if separate sites exist for the binding of substrate and effectors, sigmoid binding curves are possible under some conditions without the assumption of interacting sub-units. One scheme which does rely on sub-unit structure predicts cooperativity on the basis of a loss of binding ability on polymerization. A high degree of cooperativity is favoured by a predominance of the polymeric form in the absence of ligand. A ligand aids dissociation of the polymer freeing the binding sites on the protomers.

It seems unreasonable to assume that one particular model is "correct" and all the others are "wrong". For each model, examples of allosteric enzymes have been found for which there is a good fit with experimental data. In any case the different models should not be seen as rivals to explain the phenomenon of cooperativity, but as related simplifications of a general theory of interacting protein sub-units. A change in the assumed magnitude of an equilibrium constant can easily turn one model into another. It is even conceivable that an allosteric enzyme may be fitted by one model under one set of *in vitro* conditions

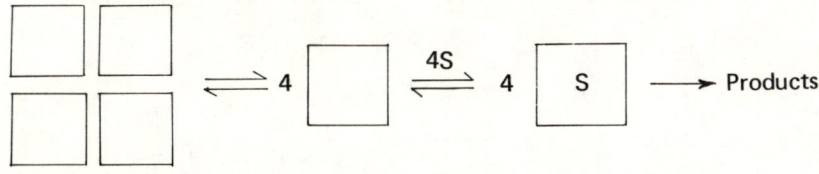

inactive polymer

but by another if the temperature or pH is altered. The parallel development of more than one means of producing cooperativity among ligands bound to protein is probably a consequence of the immense biological advantage to be gained from the possession of the property by key enzymes. As all enzyme proteins are not built upon an identical geometrical and structural pattern or master-plan, different kinds of interactions have had to be developed in order to conserve and amplify any pre-existing tendency for cooperative binding.

REGULATORY ACTIVATORS AND INHIBITORS

As mentioned at the start of this chapter, regulators usually have some metabolic connection with the enzyme reaction that they affect. For example, accumulation of a terminal metabolite may inhibit an important reaction in its biosynthetic pathway. Thus substances that act as regulators are, in general, common biochemical compounds which are substrates themselves in other reactions. A few examples of known allosteric effectors are discussed below, but descriptions of many more will be found in recent textbooks of biochemistry.

The widely studied allosteric enzyme aspartate transcarbamoylase (ATC'ase) of *E. coli* is a good example of an enzyme controlled by feedback inhibition. The enzyme catalyses the formation of carbamoyl aspartate from carbamoyl phosphate and aspartate, and the reaction is the first of a sequence for the biosynthesis of pyrimidine nucleotides. An alternative metabolic fate for carbamoyl phosphate is conversion to urea through entry into the ornithine cycle. Hence an adjustment to the flow of carbamoyl phosphate through the pyrimidine pathway is likely to affect the synthesis of urea by an overspill mechanism.

ATC'ase, and the pyrimidine pathway, seems to be mainly controlled by feedback effects of cytidine triphosphate (CTP) and ATP. CTP, a pyrimidine product of the pathway, inhibits ATC'ase whereas the purine nucleotide ATP activates the enzyme. These interactions ensure a balance between purine and pyrimidine synthesis, which is important for nucleic acid synthesis, and prevent an accumulation of pyrimidines in the cell.

Most metabolic pathways consume ATP at one or more steps with the production of ADP or AMP. Oxidative catabolism, e.g. the breakdown of glucose or glycogen by glycolysis and the tricarboxylic acid cycle, allows the ATP to be regenerated from ADP by the reoxidation of reduced coenzymes in mitochondria (Chapter 8) and thus ATP can be regarded as a terminal metabolite of such catabolic processes. Assuming a constant *total* level of adenosine phosphates in the cell, the concentration of ATP and AMP (plus ADP) will be inversely related; when ATP levels are high, AMP is depleted and vice versa. Several enzymes in the glycolytic and tricarboxylic acid cycle sequences have been shown to be very sensitive to changes in the cellular ATP/AMP ratio and energy metabolism is probably regulated by such changes. AMP activates phosphorylase *b* (see p. 118), phosphofructokinase and isocitrate dehydrogenase so that a low ATP/AMP ratio accelerates the pathways and thereby increases the rate of production of ATP. Phosphorylase *b* and phosphofructokinase are inhibited by ATP and therefore a high ATP/AMP ratio slows ATP production. Reversal of the phosphofructokinase reaction in gluconeogenesis is effected by a separate enzyme, fructose-1-6-diphosphatase. The diphosphatase is activated by ATP and inhibited by AMP so that a high ATP/AMP ratio not only inhibits glycolysis but favours the reverse pathway for

$$
\begin{array}{c}
NH_2 \\
| \\
C=O \\
| \\
OPO_2H_2
\end{array}
\;+\;
\begin{array}{c}
COO^- \\
| \\
CH_2 \\
\backslash \\
HCCOO^- \\
/ \\
NH_3^+
\end{array}
\;\longrightarrow\;
\begin{array}{c}
\qquad COO^- \\
\qquad | \\
NH_3^+ \;\; CH_2 \\
\quad \backslash \;\; | \\
O=C \;\; HCCOO^- \\
\qquad N \\
\qquad H
\end{array}
\;+\; H_3PO_4
$$

Carbamoyl phosphate   Aspartate   Carbamoyl aspartate

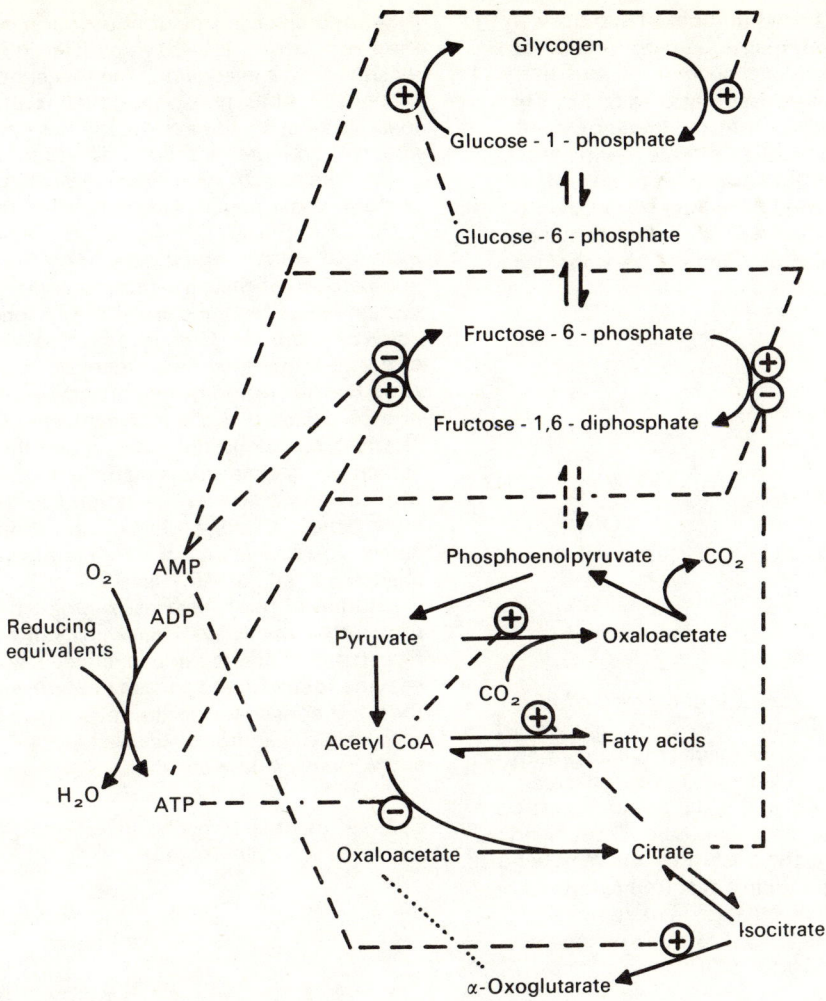

Fig 7.11. Control of energy metabolism by ATP and AMP. Certain metabolites, e.g. citrate and acetyl-CoA, also act as enzyme effectors.

glycogen synthesis. The interactions of ATP and AMP that are believed to be important in energy metabolism are shown diagrammatically in Fig. 7.11.

Some of the effects of AMP and ATP are reinforced by the accumulation, during shutdown of a pathway, of intermediates which are themselves allosteric effectors of other enzymes. This is illustrated by the activating effects of glucose-6-phosphate on glycogen synthetase and of citrate on the fatty-acid synthetase system. A reduction in the rate of glycolysis allowing levels of glucose-6-phosphate and citrate to rise will

favour glycogen synthesis and divert acetyl-CoA to fat synthesis instead of oxidizing it through the tricarboxylic acid cycle

INTERCONVERTIBLE FORMS OF ENZYMES

It is now clear that some enzymes can exist in more than one form and that the interconversion between the forms can have important regulatory significance. The different forms may be regarded as "secondary" or "derived" isoenzymes (Chapter 3) because they arise from slight chemical modifications that may be catalysed by separate enzymes.

A change in the total number of sub-units in the enzyme can accompany the modification but, more importantly, a change in the sensitivity to feedback modifiers seems also to occur. The enzymes concerned with the metabolism of glycogen in mammalian muscle are the most widely characterized examples of enzymes exhibiting this kind of regulatory behaviour.

Glycogen phosphorylase catalyses the breakdown of glycogen to glucose-1-phosphate:

$$\left[C_6H_{10}O_5\right]_n + P_i \longrightarrow \left[C_6H_{10}O_5\right]_{n-1} +$$

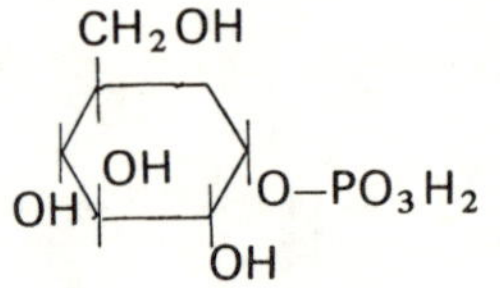

Glucose-1-phosphate

Muscle phosphorylase *a* is composed of four identical sub-units and has a molecular weight of 500 000. The other form of the enzyme, phosphorylase *b* is a dimer, has a molecular weight of 250 000 and is the predominant form of the enzyme in resting muscle. Phosphorylase *a* is converted to the *b* form by the removal of inorganic phosphate in a reaction catalysed by phosphorylase phosphatase. The removal of phosphate seems to facilitate the dissociation of the tetrameric *a* form:

$$\text{phosphorylase } a + 4H_2O \xrightarrow{\text{phosphorylase phosphatase}}$$

(tetramer)

2 phosphorylase *b* + 4P$_i$

(dimer)

The phosphatase reaction is irreversible but *b* can be converted back to *a* by phosphorylase *b* kinase, an enzyme utilizing ATP.

$$2 \text{ phosphorylase } b + 4ATP \xrightarrow{\text{phosphorylase } b \text{ kinase}}$$

phosphorylase *a* + 4ADP

AMP is an activator of phosphorylase, but the two forms of the enzyme differ in their sensitivity to the activator. At low concentrations of AMP, phosphorylase *b* is almost inactive whereas the *a* form still has considerable activity. To increase their activities to the same extent the *b* form requires an approximately tenfold higher concentration of AMP than does the *a* form. Because of the difference in the extent to which *a* and *b* depend on the activator and because of their interconvertibility, a dual system exists for the control of phosphorylase activity in muscle. Glycogen breakdown can either be stimulated by an increase in the cellular AMP concentration or by conversion of inactive phosphorylase *b* to the active *a* form. The relative physiological importance of the two activating mechanisms is difficult to assess but it seems probable that regulation could occur most rapidly through a fast chemical interconversion rather than through relatively slow changes in cellular AMP levels.

Studies of phosphorylase *b* kinase have shown that this enzyme also exists in two forms; one of them being inactive. The inactive enzyme is stimulated to some extent by Ca^{2+} ions, but conversion to the fully active form requires ATP and Mg^{2+} (as well as Ca^{2+} ions) and a kinase kinase enzyme.

$$Ca^{2+} + \text{Inactive phosphorylase } + ATP \xrightarrow[\text{kinase}]{\text{kinase}}$$
$$b \text{ kinase}$$

Active phosphorylase + ADP
b kinase

The reverse reaction is catalysed by phosphorylase kinase phosphatase.

The phosphorylase kinase kinase enzyme is activated allosterically by $3'-5'$ cyclic AMP, a compound that is formed enzymically from ATP.

$$ATP \xrightarrow{\text{adenyl cyclase}}$$

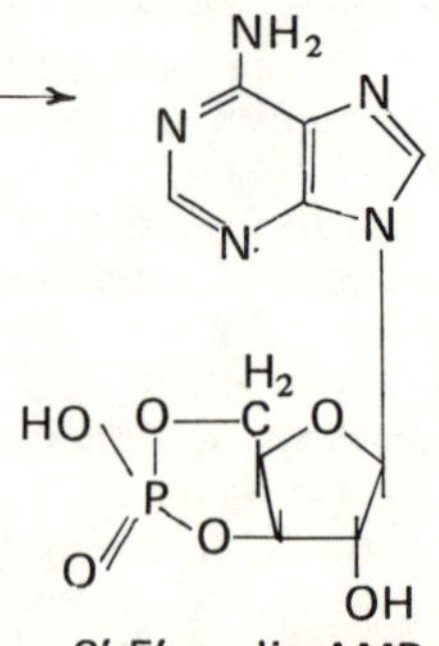

3'-5' cyclic AMP

Adrenalin and glucagon stimulate the production of cyclic-AMP by activating adenyl cyclase and can thus stimulate glycogen breakdown by setting in motion a cascade of reactions leading to the production of AMP-independent phosphorylase *a*.

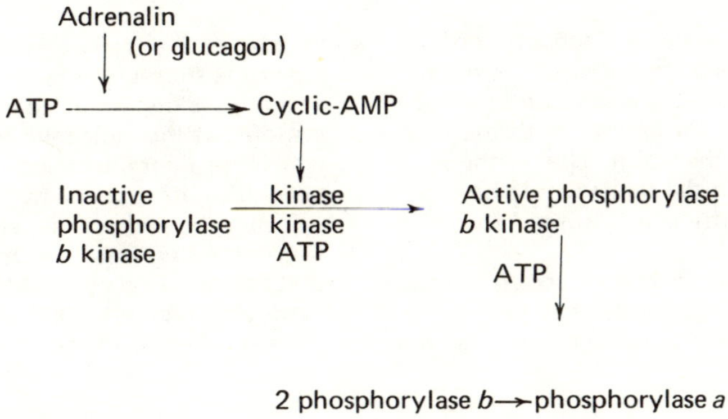

Glycogen synthesis can be stimulated by raising the level of glucose-6-phosphate in the cell or by converting more of the synthetase enzyme to the I form. Insulin changes the ratio of D to I species in muscle and may bring this about by a direct activation of the phosphatase.

$$\text{Adrenalin (or glucagon)}$$
$$\text{ATP} \longrightarrow \text{Cyclic-AMP}$$

Inactive phosphorylase *b* kinase $\xrightarrow[\text{kinase ATP}]{\text{kinase}}$ Active phosphorylase *b* kinase

ATP

$$2 \text{ phosphorylase } b \longrightarrow \text{phosphorylase } a$$

There is some evidence to suggest that insulin antagonizes the effects of glucagon in promoting cyclic-AMP synthesis.

The phosphorylase reaction is not reversible under physiological conditions; glycogen synthesis requires a separate synthetase enzyme that takes a glucose residue linked to uridine diphosphate and adds it to a growing poly-glucose chain.

Adrenalin and glucagon on the other hand stimulate the kinase activity through their effects on the level of cyclic-AMP, an activator of the kinase. Taken together, the effects of insulin on glycogen metabolizing enzymes favour synthesis while adrenalin and glucagon both promote breakdown.

A sequence of enzyme-catalysed modifications arranged in a cascade may be of general

$$\text{UDPG} + (C_6H_{10}O_5)_n \xrightarrow{\text{Glycogen synthetase}} \text{UDP} + (C_6H_{10}O_5)_{n+1}$$
(uridine diphospho-glucose)

Like phosphorylase, however, glycogen synthetase exists in two forms. One of these needs glucose-6-phosphate (D form for dependent) for activity but the other does not (I for *i*ndependent). Interconversions of D and I resemble those of phosphorylase *a* and *b* and it is possible that the same ancestral gene gave rise to both glycogen synthetase and phosphorylase. The I to D conversion of synthetase requires ATP and a kinase enzyme and the reversal is catalysed by a phosphatase (p. 120).

importance in biological systems because it enables amplification of the primary stimulus. At the start of the cascade, a very small amount of effector activates a catalyst which in turn stimulates the activation of the next enzyme in the sequence and so on. Each catalyst can activate a much greater amount of the next and so the initial signal that starts the cascade is greatly amplified. Blood clotting seems to involve an enzyme cascade. The initial stimulus, derived from the contact of blood with a foreign surface

$$\text{Glycogen synthetase} + \text{ATP} \xrightarrow[\substack{\text{kinase} \\ \text{(cyclic-AMP)}}]{Mg^{2+}} \text{Glycogen synthetase} + \text{ADP}$$
$$\text{(I form)} \qquad\qquad\qquad\qquad\qquad \text{(D form)}$$

$$\text{Glycogen synthetase} \xrightarrow{\text{phosphatase}} \text{Glycogen synthetase} + P_i$$
$$\text{(D form)} \qquad\qquad\qquad\qquad \text{(I form)}$$

or damaged tissue, causes the activation of clotting factors which are present in minute quantities relative to fibrinogen, the activation of which terminates the cascade with the formation of fibrin at the wound site. Without amplification, the initial signal would be too weak to produce sufficient fibrin to block the wound.

The effect of hormones on glycogen metabolism are interesting because they provide direct clues to the ways in which some hormones interact with enzymes. It has been known for a long time that hormone imbalance can bring about gross metabolic changes but an understanding at the molecular level of the way in which hormones influence enzyme activity has been difficult to come by. There is increasing evidence, however, that very many hormones act, in the first instance, on adenyl cyclase and that the resulting cyclic-AMP is the regulatory modifier which interacts with enzymes to produce changes in metabolic flux.

8 The Organization of Enzymes in Cells and Tissues

Almost all the enzymes of the body function entirely within the cells that produce them. The exceptions are the digestive enzymes such as pepsin, trypsin, etc., which are secreted into the digestive juices, and a relatively few enzymes which are active in the blood plasma, notably enzymes concerned in blood clotting. Apart from these instances, levels of enzyme activity are normally very low in the extracellular environment compared with those inside the cells.

The cellular basis of the organization of tissues has been recognized for 130 years as one of the universal principles of biology and several morphological details of the structure of cells became known through studies with the light microscope. During this period of investigation also, histochemical methods of locating chemical constituents (including some enzymes) in tissues and cells were pioneered. These methods have continued to develop and at the present time the sites of action of nearly 100 enzymes can be located in tissue sections prepared for the optical microscope.

ENZYME HISTOCHEMISTRY

The methods employed in the histochemical location of enzymes involve, firstly, preparation and fixation of thin sections of tissue in such a way that their cellular morphology is preserved, with minimal destruction of enzyme activity. The simplest way of achieving this is to cut sections in a cryostat from pieces of tissue which have been rapidly frozen and to use these sections without further fixation. Lyophilization of the tissue under high vacuum at low temperature also preserves its fine structure and enzyme activity often survives such treatment well. However, changes in cell structure take place on subsequent thawing or rehydration, so that some additional fixation is usually

necessary if sharp definition under the microscope is to be retained and is essential for the study of enzymes present in the cell sap. Strongly dehydrating organic solvents are also powerful enzyme denaturing agents and are therefore generally unsuitable as fixatives, although acetone has been used successfully. Dilute formalin solutions are also widely applied as fixatives in enzyme histochemistry.

When a suitable tissue section has been obtained it is flooded with a solution of the substrate of the enzyme under study. One of the products of the action of the enzyme reacts as it is released with a "trapping" reagent, so that an insoluble product, visible under the microscope, is formed. The need for insolubility is that, if the visible product is not precipitated immediately on formation, it will spread by diffusion over the whole field and any possibility of identifying the exact site of action of the enzyme will be lost.

The first enzymes to be located by these means included phosphatases and other esterases. These enzymes are usually of low specificity and will split synthetic substrates which can be selected, therefore, for the reactivity and solubility properties of the products released by the enzyme. Thus, phosphate esters of naphthols are often chosen for visualizing phosphatase activity since the naphthols react rapidly with diazonium salts (particularly at alkaline pH) to form very insoluble, intensely coloured dyes: The naphthol itself should be of restricted diffusibility to reduce blurring of the definition before the capture-reaction with the diazonium salt can take place; naphthol derivatives of lower solubility are therefore often used in place of 1- or 2-naphthol in preparing the substrate. The other product of the enzymic reaction, inorganic phosphate, can also be used to locate the site of action by carrying out the reaction in the presence of calcium ions so

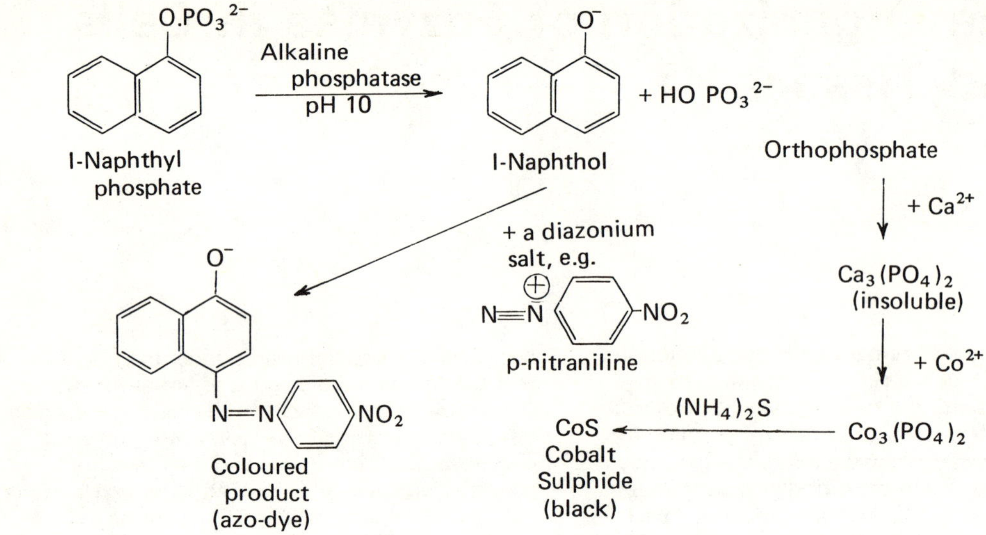

I-Naphthyl phosphate $\xrightarrow[\text{pH 10}]{\text{Alkaline phosphatase}}$ I-Naphthol $+ HO\,PO_3^{2-}$

Orthophosphate $\xrightarrow{+ Ca^{2+}}$ $Ca_3(PO_4)_2$ (insoluble) $\xrightarrow{+ Co^{2+}}$ $Co_3(PO_4)_2$

+ a diazonium salt, e.g.

p-nitraniline

Coloured product (azo-dye)

CoS Cobalt Sulphide (black) $\xleftarrow{(NH_4)_2S}$ $Co_3(PO_4)_2$

that insoluble calcium phosphate is deposited. This is rendered visible by conversion to cobalt phosphate and then to cobalt sulphide by the action of ammonium sulphide.

Another important group of enzymes for which histochemical methods of location have been devised are the dehydrogenases. In these methods a series of coupled electron-transfer reactions is arranged, beginning with the action of the dehydrogenase on its substrate and ending with the reduction of a tetrazolium salt to produce an insoluble formazan.

The oxidation-reduction potentials of the coupling agents are critically chosen so that each is able to oxidize the preceding member of the chain by accepting electrons from it and then reduce the succeeding member by donating electrons to it. Inhibitors are included in the reaction mixture to prevent side reactions taking place in which other enzymes and substrates present in the tissue section might react with the intermediates of electron transfer. Many different dehydrogenases have been located in this way by appropriate choice of substrates and reaction conditions.

The range of types of enzyme reactions

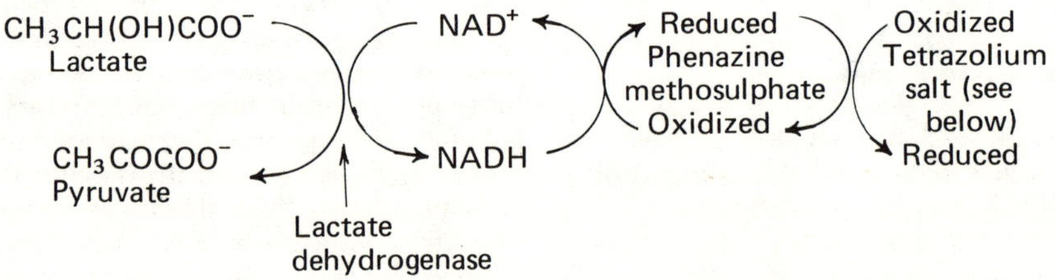

$CH_3CH(OH)COO^-$ Lactate / CH_3COCOO^- Pyruvate — Lactate dehydrogenase — NAD^+ / $NADH$ — Reduced Phenazine methosulphate Oxidized — Oxidized Tetrazolium salt (see below) Reduced

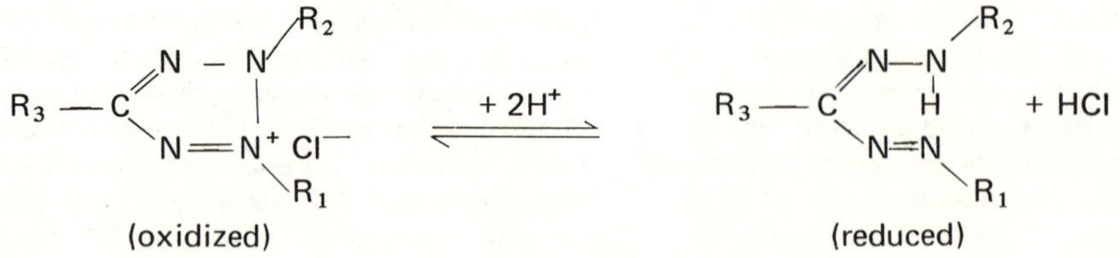

(oxidized) $+ 2H^+ \rightleftharpoons$ (reduced) $+ HCl$

$R_1, R_2 + R_3$ = aromatic residues

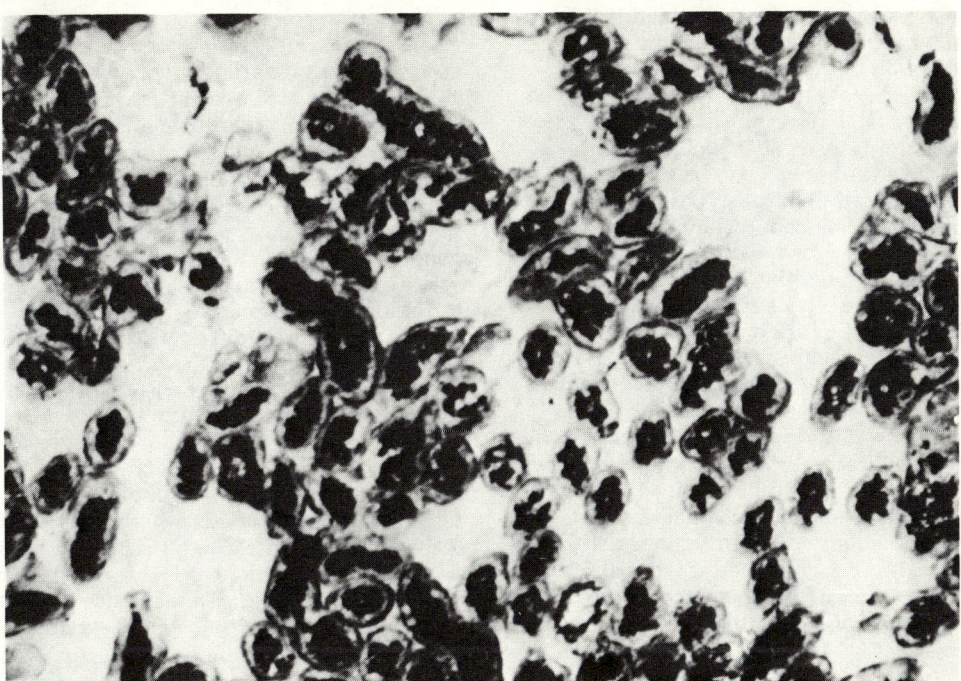

Fig 8.1. A section of rat kidney stained for alkaline phosphatase activity by Gomori's method. Inorganic phosphate liberated enzymically from the substrate (β-glycerophosphate in this case) was precipitated as calcium phosphate. This was converted to a black deposit of cobalt sulphide, which can be seen lining the cross-sectioned renal tubules.

which can be observed histochemically has been greatly extended, often by ingeniously arranged coupled reactions. Among the few enzyme classes which have not yet been made accessible to histochemical study are those enzymes which form, rather than break, covalent chemical bonds (the ligases), and enzymes which catalyse intramolecular conversions (isomerases). When zone electrophoresis began to be applied to the separation of mixtures of enzymes or isoenzymes, the reactions developed for localization of enzyme activities in tissue sections found a ready application to the problem of identifying zones of active enzyme directly on the supporting medium, e.g. paper, cellulose acetate or gel. Examples of these separations are given in Chapter 9.

The results obtained by enzyme histochemistry show that certain enzymes are present to a greater or lesser extent in some types of cell than in others. Compared with biochemical methods of analysis, in which a sample of tissue is homogenized and extracted with water or other solvent and the enzyme activity of the extract is determined, histochemistry offers the certainty that only one type of cell is being examined: the heterogenous population of cell types which together make up most tissues is sampled indiscriminately in making a tissue-extract. Against this must be set the rather involved reactions which are often necessary in enzyme histochemistry and which raise questions as to the specificity of the methods for the enzymes they are intended to detect, and also the difficulty of obtaining quantitative data by histochemical means.

Enzyme histochemistry has also revealed that, within a single cell, the distribution of a particular enzyme is not uniform. Thus, the concentration at the brush-borders of the cells lining the small intestine and renal tubules of enzymes which may have a transport function is clearly demonstrated by these methods (Fig. 8.1). Information which has resulted from the identification of sub-cellular organelles (see below) by electron microscopy and biochemical fractionation and analysis techniques can now be applied in enzyme histochemistry to identify these specific intracellular regions by light microscopy. Enzymes can thus be used as histochemical markers for changes affecting particular sub-cellular structures, e.g. the state of the lysosomes in the cells of the intestinal mucosa of patients with coeliac disease can be assessed by staining sections from biopsy specimens for lysosomal acid phosphatase.

Specific enzymes can also be used as markers in the identification of cell types (e.g. suspected tumour cells). Malignant melanomata contain high levels of the enzyme dihydroxyphenylalanine oxidase, which functions in melanin formation, and the presence of this enzyme can be recognized by a specific histochemical reaction. In muscle biopsy specimens, also, from patients with suspected muscular dystrophies, changes in enzyme levels revealed by enzyme histochemistry can assist differential diagnosis, while a further application of the technique is in the examination of foetal cells obtained during pregnancy by amniocentesis to aid pre-natal detection of inherited enzyme abnormalities (Chapter 10). It is probable that the number of such applications of enzyme histochemistry to diagnostic problems will continue to increase.

ELECTRON MICROSCOPY AND THE ULTRASTRUCTURE OF CELLS

In probing the intracellular regions the light microscope approaches its limits of resolution and the fear that artefacts introduced during fixation of tissue sections might be unrecognized as such haunted microscopists for many years. Larger objects in the cytoplasm of cells, including mitochondria (which are $1-3 \mu$ long), were discovered with the light microscope but their morphology and even their reality remained in doubt until the electron microscope was brought into use.

The limits of resolution with any microscope are set ultimately by the wavelength of the radiation which is used to form the image: points which are closer together than half a wavelength cannot be resolved. For a light microscope the limit of resolution is thus about 250 nm, but the "wavelength" of a beam of electrons is much less than this and resolution down to a few Ångstrom units is theoretically possible Practically, the resolving power of electron microscopes is already reaching levels at which the shapes of larger macromolecules can be discerned.

The first electron micrographs were surface views obtained by "shadowing", i.e. depositing thin films of metal atoms such as gold on the surface of the section. This technique was necessary because sections could not be cut thin enough to be transparent to the electron beam, but the art of cutting sections has now developed sufficiently to enable sections only 50 nm thick to be cut. Fixation of the tissue section to preserve its ultrastructure is even more critical in electron microscopy than with the optical instrument, and also introduces the danger of fixation artefacts. The most widely used fixative in electron microscopy is osmium tetroxide, but buffered solutions of aldehydes such as glutaraldehyde have also been applied successfully as fixatives in both morphological and histochemical studies. Freeze-drying and freeze-substitution (in which ice is removed from frozen tissue sections by exposure to dehydrating organic solvents such as methylcellulose or ethanol) are examples of other useful techniques. The methods of locating enzyme activity used by histochemists are now being extended to sections prepared for the electron microscope. Techniques for the demonstration of phosphatase activity in which electron-opaque lead salts are precipitated are particularly suited to electron microscopy and were among the first enzyme methods to be applied to the new instrument. While at present fewer enzymes can be studied with the electron microscope than with the optical instrument, because of the difficulties of fixation and the need to devise substrates and reaction products with appropriate properties of transparency and opacity to the electron beam, the range of enzyme methods available to electron microscopists is being extended rapidly.

The electron microscope has revealed that cells are structurally far more complex than was ever imagined when only the light microscope was available for their exploration. To the larger sub-cellular structures such as the nucleus and mitochondria must now be added smaller particles, lysosomes and peroxisomes, but perhaps the most striking feature of animal cells seen in electron micrographs is their complex internal system of membranes (the endoplasmic reticulum) and the various structures associated with them (Fig. 8.2).

In fact, in contrast with the older view of the cell as a nucleus surrounded by a fairly amorphous cytoplasm containing a few particles and bounded by a membrane, many cells seem to consist almost entirely of organized structures with a few clear areas of cytoplasm. As knowledge of intracellular anatomy has grown with the aid

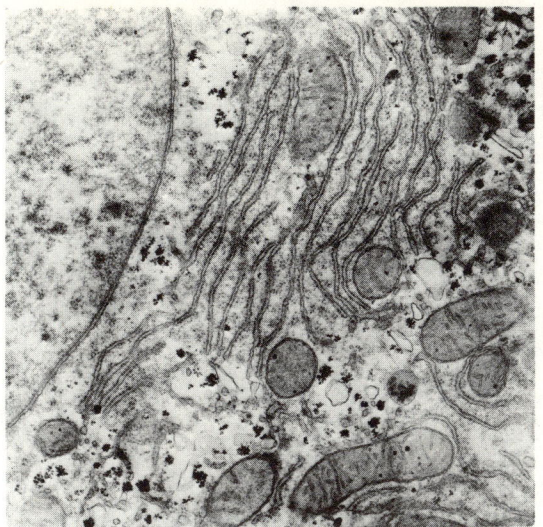

Fig 8.2. Electron micrograph of part of a rat-liver parenchymal cell. Part of the double membrane surrounding the nucleus can be seen at the left; several mitochondria with their characteristic cristae are visible in the cytoplasm, together with double membranes forming part of the endoplasmic reticulum. The dark granules in the cytoplasm are glycogen. (Final magnification X15 000.)

of the electron microscope, a parallel growth of information on the distribution of metabolic functions within the cell has resulted from biochemical studies on sub-cellular components.

SEPARATION AND ANALYSIS OF SUB-CELLULAR PARTICLES

The realization that enzymes are not uniformly distributed throughout the cell may be dated from O. Warburg's discovery, in 1913, that cellular respiration is associated with sedimentable cell particles, but the most important advances in defining the intracellular locations of enzymes have been made in the last 25 years. This has been due to the development of the technique of fractional ultracentrifugation, in which components of a cellular homogenate are successively spun down at higher and higher centrifuge speeds. These fractionation procedures yield four clearly distinguishable fractions, in proceeding from lower to higher gravitational fields: the nuclear, mitochondrial, microsomal and supernatant ("soluble") fractions. Microscopy confirms that the first two fractions are

indeed composed predominantly of the larger particles which give them their names, while the microsomal fraction derives from the endoplasmic reticulum (Fig. 8.2), together with fragments from the plasma membrane and Golgi apparatus. Further subdivision within a single fraction is also possible, e.g. the separation of lysosomes from the lighter mitochondrial fractions.

The first stage in cell fractionation (Fig. 8.3) is the preparation of a suitable homogenate of the tissue specimen. This must be done in such a way as to preserve as far as possible the structures of the cellular components and to minimize leakage of enzymes from one intracellular region to another. The tissue is therefore homogenized in a solution with osmolarity equal to that of the tissue-fluid; 0·25 M sucrose is often chosen, since this sugar is highly soluble and relatively inert. Buffers may be added to control pH, while metal ions affect the stability of subcellular particles. Additions of specific ions such as Mg^{2+} may therefore be made, or EDTA may be added to chelate heavy metals. Disruption of the cells is achieved usually by grinding in some form of mechanical mixer. A type of homogenizer which is frequently employed, due to V. R. Potter and C. A. Elvehjem, consists of a rapidly rotating pestle ("Teflon" or glass) which fits closely within a glass test-tube. As the pestle and tube are moved vertically relatively to each other, the tissue cells are subjected to shearing forces in the thin film of liquid between the pestle and the wall of the tube, which is cooled in an ice-bath to dissipate frictional heat. Cells may also be disrupted by exposure to ultrasonic vibration.

Preparative ultracentrifuges capable of sedimenting the smaller sub-cellular particles are now usually driven by electric motors (early ultracentrifuges were driven by air- or oil-turbines), and have compact, delicately balanced rotors of high-tensile steel or titanium alloys rotating in an evacuated, refrigerated chamber. The centrifugal force achieved in such a centrifuge is expressed in relation to the normal force of gravity, i.e. $n \times g$ (g = the gravitational constant, 980 cm/s^2). The gravitational field reached in the centrifuge is directly related to the square of its speed of revolution and the radius of the rotor. Therefore, to increase the field by a factor of 2 the radius must be doubled, but the speed need only be raised by a factor of $\sqrt{2}$, i.e. approximately 1·4 to produce the same effect. Since the inertia of rotors and their

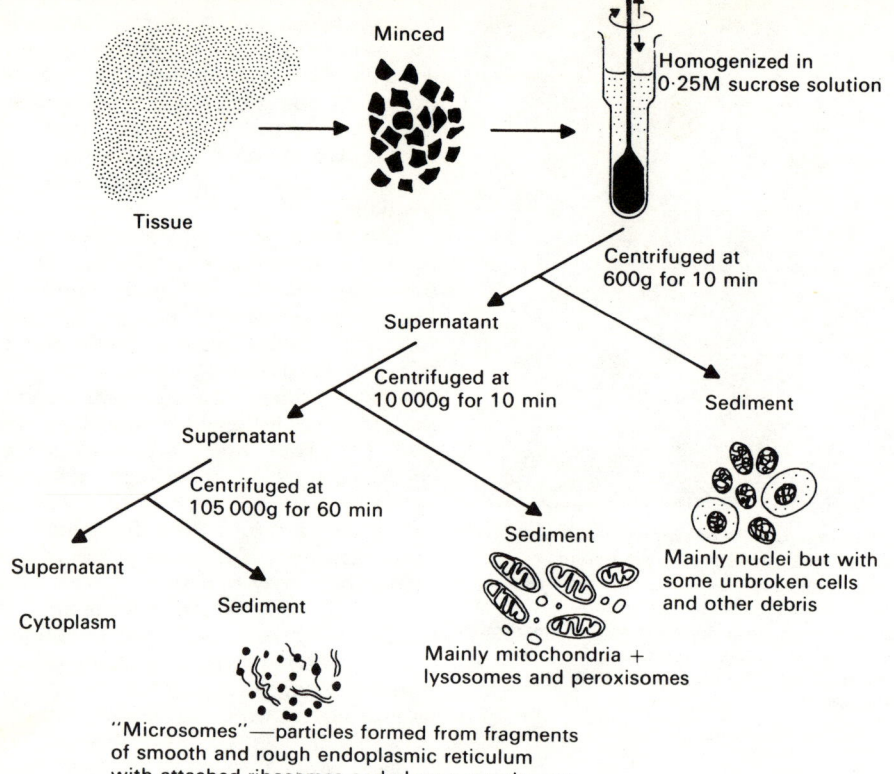

Fig 8.3. Schematic representation of sub-cellular fractionation of liver by ultracentrifugation. Details of buffer solutions and centrifugation times and speeds vary from one tissue to another, while the preparation of an homogeneous fraction (e.g. of mitochondria) requires re-suspension and further fractional centrifugation of the initial crude fractions.

tendency to disintegration increases with their size, the highest gravitational fields are achieved in practice in small rotors spinning at high speeds, although the amount of material that can be processed in a single operation is small.

For spherical particles sedimenting in a gravitational field, the time required for them to reach the bottom of the tube is inversely proportional to the difference between the densities of the particles and the fluid in which they are suspended and the viscosity of the medium, and also inversely proportional to the square of the radii of the particles (i.e. smaller particles sediment more slowly than larger ones of the same density). These principles allow the approximate conditions necessary to effect a separation of sub-cellular particles by centri-

fugation to be calculated, but the exact conditions required remain a matter for experiment and conditions established for one type of tissue cannot necessarily be applied without modification to other samples. Furthermore, particles which are not spherical do not conform to the relationships calculated for spheres.

When the fractionation protocol (gravitational fields, times of centrifuging, etc.) has been decided upon, the procedure consists of successive centrifugations after each of which the sediment is removed and supernatant liquid is recentrifuged. The particles of decreasing size which are thus separated are tested to ensure that fractionation has been correct, either by examining them under the light- or electron-microscope, or by estimating

TABLE 8.1 THE INTRACELLULAR LOCATION OF ENZYMES

Soluble (Cytoplasmic fraction)	Mitochondria	Lysosomes	Microsomal fraction	Plasma membrane	Nucleus
Glycolytic enzymes	Pyruvate dehydrogenase complex	Acid phosphatase	Glucose-6-phosphatase	Alkaline phosphatase	RNA polymerase
Hexose monophosphate shunt enzymes	Citrate synthetase	β-Glucuronidase	Ribosomal enzymes of protein synthesis	Adenosine triphosphatase	NAD synthetase
Amino acid activating enzymes	Isocitrate dehydrogenases (NAD and NADP specific)	α-Glucosidase β-Glucosidase	Steroid hydroxylating enzymes	5′-Nucleotidase Adenyl cyclase	
Acetyl-CoA carboxylase Fatty acid synthetase complex	Malate dehydrogenase and other enzymes of tricarboxylic acid cycle	Cathepsin Acid ribonuclease Acid deoxyribonuclease			
Aspartate aminotransferase Malate dehydrogenase	Acyl-CoA dehydrogenase and other enzymes of fatty acid oxidation cycle	α-Galactosidase β-Galactosidase			
Isocitrate dehydrogenase (NADP-specific)	Glutamate dehydrogenase	Lysozyme Hyaluronidase			
*Glycogen phosphorylase	Aspartate aminotransferase	Arylsulphatases Collagenase			
*Glycogen synthetase					

* These enzymes are bound to glycogen in the cell.

the activity in them of enzymes which are known, from previous studies, to be confined essentially to one region of the cell. Succinate dehydrogenase is often regarded as a mitochondrial marker and glucose-6-phosphatase as characteristic of the microsomal fraction, while glucose-6-phosphate dehydrogenase or lactate dehydrogenase are recovered in the soluble fraction.

Biochemical analysis of the separated fractions show that other enzymes also appear to be associated almost exclusively with one type of particle or another, or are recovered mainly in the soluble fraction, while some are more dispersed (Table 8.1). From the location of groups of metabolically related enzymes and from the activities of the isolated cell fractions it is clear that there is an association between structure and biochemical function inside the cell. In other words, the reactions which in orderly sequence make up a particular metabolic pathway are coordinated not merely by the substrate specificities of the enzymes concerned, one acting on the product of its predecessor's reaction, but also by the spatial relationships of the enzymes themselves.

Mitochondria contain all the enzymes of the Krebs tricarboxylic acid cycle together with those involved in fatty acid oxidation and the oxidative phosphorylation process. Enzymes with similar catalytic activities to those responsible for the reactions of the Krebs cycle do occur in other parts of the cell, but the rate at which isolated mitochondria oxidize pyruvate indicates that these particles account for most of this activity. Oxidation of fatty acids and oxidative phosphorylation are exclusively mitochondrial activities. It is interesting that, where the same reaction is catalysed by enzymes occurring both in mitochondria and in other parts of the cell, the enzymes involved may not be identical. An enzyme which catalyses transfer of an amino group from aspartic acid to a-keto-glutaric acid (glutamic oxaloacetic transaminase, or, more correctly, aspartate aminotransferase) is found in both the mitochondrial and soluble fractions, but the two enzymes are dissimilar in their response to excess substrate and in electrophoretic mobility, while genetic studies have shown that the two forms are determined by separate genes. There are analogous differences between mitochondrial and cytoplasmic malate dehydrogenases. Even more distinct catalytically are the isocitrate dehydrogenase enzymes, one of which has a specific requirement for NAD as its co-factor and the other for NADP. Both enzymes are present in mitochondria but only the NADP-linked form occurs in the cytoplasm. (Isoenzymes of the NADP-dependent form also exist.) These differences in properties between functionally similar intra- and extramitochondrial enzymes probably represent adaptations to the different patterns of metabolism existing in the several regions of the cell: examples of the way in which transport of certain metabolites across mitochondrial membranes is facilitated by the existence of intra- and extramitochondrial isoenzymes are given in Chapter 7. Alternatively, they may be manifestations of a difference in evolutionary origins of the enzymes concerned, since mitochondria contain some DNA and synthesize some of their own protein, although this does not apparently include mitochondrial enzymes.

The electron microscope reveals the double membrane structure of the mitochondrion, with the inner membrane projecting into the inner space to form the characteristic cristae. The enzymes of electron transport and oxidative phosphorylation are attached to the membranes, and the close spatial relationships of enzymes which are related functionally can be deduced from experiments in which mitochondria are themselves fragmented into smaller particles which retain the ability to carry out part of the functions of the whole organelle. Submitochondrial particles have been obtained by grinding or sonication, treatment with detergents, and by the action of enzymes such as

phospholipase. They retain in varying degrees the ability to effect electron-transport, so that the sequence of reactions involved in this process can be deduced, but the ability to link oxidation to the generation of ATP is lost.

The morphology of mitochondria undergoes marked changes depending on the conditions of isolation and the presence or absence of a variety of agents in the medium. These changes include swelling and contraction, and presumably reflect changes in the metabolic activity and permeability of the mitochondrial membrane. Since some of the agents concerned are of physiological occurrence (e.g. thyroxine), or produce metabolic changes such as uncoupling of oxidative phosphorylation (e.g. dinitrophenol), their effects on mitochondrial shape have been extensively studied in the hope of disclosing details of mitochondrial function and its physiological control; however, many details of mitochondrial activity remain obscure, particularly the mechanism of oxidative phosphorylation.

A quite different type of particle was identified in 1955 among the lighter mitochondrial fractions, by C. de Duve and his colleagues. These particles contain a group of hydrolytic enzymes with pH optima in the region of pH 5: ribonuclease and deoxyribonuclease, esterases (phosphatases and sulphatase), proteases (cathepsins and collagenase) and glycosidases. Because of the hydrolytic nature of the enzymes, the particles were named *lysosomes.* Sequestration of the enzymes within the lysosomal membrane presumably prevents them from destroying the structure and metabolites of the rest of the cell, as they do when the cell dies (autolysis) — lysosomes have been vividly described as the suicide bag of the cell. In life, the lysosomal enzymes are probably active in intracellular digestion, the process taking place in digestive vacuoles formed in the cytoplasm by ingestion of exogenous material (heterophagy) or by inclusion of a small portion of the cytoplasm (autophagy).

Considerable medical interest has focused on lysosomes and lysosomal enzymes because of their apparent involvement in inflammatory processes and injury, tissue resorption and some inherited metabolic diseases. Cell damage may result from lysosome rupture caused by an attack on the lysosomal membrane by lipid-soluble compounds (e.g. some drugs, steroids or vitamins) or by the toxins of ingested bacteria. Anti-inflammatory agents such as cortisone may act by stabilizing the membrane. Tissues surrounding cells in which lysosomes are disintegrating may also suffer damage or alteration by lysosomal enzymes; Vitamin A or parathyroid hormone stimulate bone resorption by cartilage cells in tissue culture by promoting release of lysosomal enzymes. Overloading of lysosomes so that they come to occupy almost the whole volume of the cell occurs in several conditions in which there is an inherited deficiency of a lysosomal enzyme. These include a number of lipid storage diseases and the glycogen storage disease described by Pompe, in which muscular weakness and cardiac enlargement are accompanied by excessive deposition of glycogen in the heart, muscles and liver: the deficient enzyme is the lysosomal α-1,4-glucosidase acid maltase, which splits off glucose residues from the outer branches of the glycogen molecule.

The *microsomal fraction* does not represent a single type of sub-cellular particle but is made up of the debris of the intracellular membrane system. This endoplasmic reticulum is seen in electron micrographs to consist of a double membrane layer forming narrow channels or tubules which may communicate on one side with the outer boundary of the cell and on the other with the membrane surrounding the nucleus. Two types of membrane are distinguishable, smooth and rough, the latter being studded on the outer surface with dense particles. When the microsomal pellet is treated with detergent the membrane, which contains lipid material, is dissolved away and the particles (ribosomes), composed of nucleic acid and protein, are left.

Compared with the well-studied metabolic sequences which are characteristic of mitochondria, the biochemistry of the microsomal fraction has until recently been less well characterized, but the range of enzymes associated with the endoplasmic reticulum and the complexity of the reactions that they catalyse is now becoming clear. The endoplasmic reticulum appears to be the site of protein-synthesizing activity in the cell; certainly the ribosomes are the structures at which the RNA message is translated and amino acids are linked to form proteins (Chapter 3), and the number of granules

attached to the endoplasmic membranes correlates with the activity of the cell in protein synthesis. Other synthetic activities of the microsomal components include the incorporation of acetate into cholesterol, the biosynthesis of mucopolysaccharides, phospholipids and triglycerides and, in liver cells in particular, the formation of modified derivatives of metabolites and drugs. This last type of reaction is an especially interesting one from a clinical aspect since it is the means by which both naturally arising by-products of metabolism (e.g. bilirubin, hormones) and many foreign compounds (such as drugs) are disposed of. Although several types of reactions are involved, and although the compounds metabolized in this way are of a wide range of chemical structures, the effect in all cases is to increase the solubility in water of the substance undergoing modification (i.e. its polarity) while lowering its solubility in lipids: this reduces the tendency of the compound to become dissolved in the body lipids, with possible harmful consequences such as follow the deposition of bilirubin in the developing nervous system (the kernicterus of neonatal jaundice), and increases the efficiency of excretion in the urine or bile. Thus, the result in these cases is to reduce the potential toxicity of the substance. Two main reactions are involved, one or both of which may operate in a particular case. These are the introduction of hydroxyl groups into the toxic compound to increase its polarity, and the conjugation of the compound with a highly polar substance such as glucuronic acid, a sulphate radical, glycine, etc.

Microsomal hydroxylation reactions involve several factors and result in the introduction into the substrate (the potential toxin) of an oxygen atom which has been shown, by isotopic labelling, to be derived from molecular oxygen and not from water. The system involves a flavoprotein (FP), NADP, an iron-containing protein in which the iron is not combined as haem, and a cytochrome (cytochrome P-450), which differs in properties from cytochromes found elsewhere in the cell.

The sequence of events is probably:

The exact nature and composition of the cytochrome P-450 complex is not known since the whole system is extremely labile and difficult to extract and fractionate. The alteration of the substrate may not end with the introduction of a hydroxyl group into the molecule. Hydroxylation may be a prelude to removal of a group masking a potential hydroxyl group, as in phenacetin:

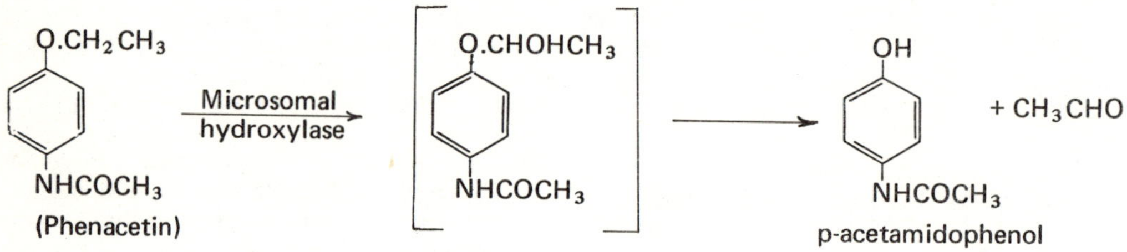

is part of the process by which metabolically active molecules are produced and it is in this way that progesterone is converted to 21-hydroxyprogesterone, for example. Hydroxylase systems which involve cytochrome P-450 are also present in mitochondria in these tissues and are responsible for introducing hydroxyl groups into other positions in the steroid nucleus.

Amino groups can also be removed from some compounds and oxygen substituted for sulphur atoms in others by the hydroxylation mechanism.

Not all cytochrome P-450-dependent hydroxylations result in metabolic inactivation, however. In the steroid-hormone producing tissues, adrenals and gonads, microsomal hydroxylation

Conjugation with glucuronic acid is probably the most important reaction of its type in man, and is brought about by microsomal glucuronyl transferases. The glucuronic acid is in an "activated" form, combined with the coenzyme uridine diphosphate (UDP) as UDP-glucuronic acid (UDPGA):

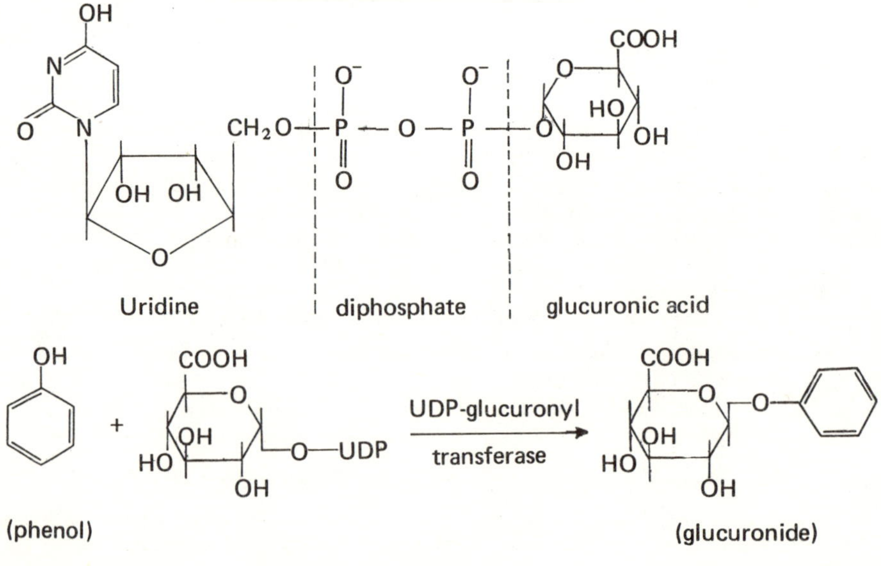

UDP-glucuronic acid is synthesized by enzymes in the soluble fraction of the cell, starting from glucose-1-phosphate:

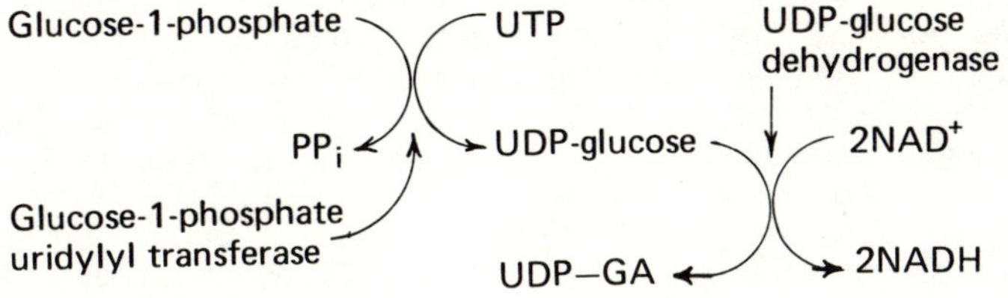

phenicol), and also of bilirubin which may be formed in excessive amounts; for example, in rhesus incompatibility. In both the newborn

$$Glucose\text{-}1\text{-}phosphate \quad UTP \qquad \qquad \text{UDP-glucose dehydrogenase}$$

$$PP_i \quad UDP\text{-glucose} \quad 2NAD^+$$

$$Glucose\text{-}1\text{-}phosphate\ uridylyl\ transferase$$

$$UDP\text{--}GA \qquad 2NADH$$

Glucuronides can also be formed by addition of glucuronic acid to nitrogen atoms (e.g. in sulphanilamide) and sulphur atoms, as in the anti-addiction drug, Antabuse (disulfiram).

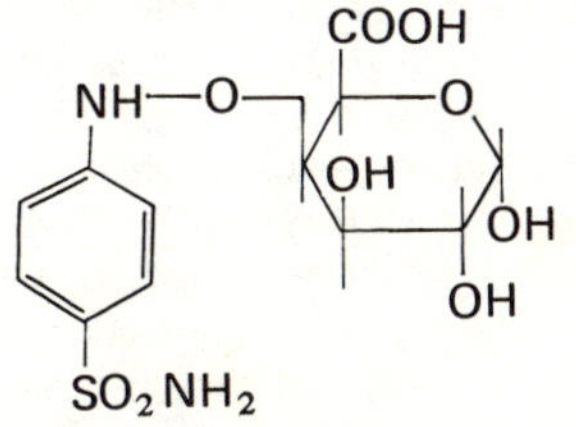

and the adult exposure to certain drugs stimulates production of the enzymes by the liver and, for this reason, treatment with barbiturates has been tried in an attempt to increase the

(Sulphanilamide glucuronide)

Antabuse

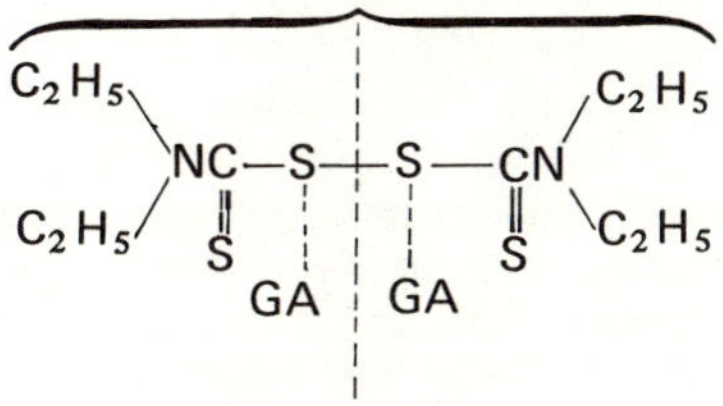

Glucuronide formation

Bilirubin is conjugated to form a diglucuronide by liver microsomes and it is in this form that bilirubin is excreted in the bile, or, if the flow of bile into the intestine is impaired (obstructive jaundice), in the urine. The effect of conjugation on the solubility of bilirubin in water is seen in the contrast between haemolytic, or pre-hepatic, jaundice, in which the excessive rate of bilirubin formation outruns its transport to and conjugation in the liver, and hepatic and post-hepatic jaundice, in which it is the disposal of conjugated bilirubin in the bile which is limiting. In the former the bilirubin is insoluble and cannot enter the urine, being kept in solution in the plasma by association with albumin, while in the latter conditions the conjugated, soluble bilirubin is readily excreted in the urine.

The level of microsomal hydroxylating and conjugating enzymes is low in foetal and neonatal life and the newborn is therefore more vulnerable than the adult to the potentially harmful effects of many drugs (e.g. chloram-

conjugation of bilirubin in the neonatal period. Induction of the enzyme systems is accompanied first by an increase in the amount of rough (i.e. ribosome-studded) endoplasmic reticulum in the cells, corresponding presumably to increased enzyme synthesis, then by an increase in smooth endoplasmic reticulum (or loss of ribosomes by rough membranes) and these areas seem to be the sites of hydroxylation and conjugation. In chronic liver disease the ability of the liver to detoxicate drugs and metabolites is impaired: this may be due to loss of enzyme activity or to the accumulation of inhibitors of the process. The products of the synthetic activity of the endoplasmic reticulum may be transported from their sites of formation to those at which they act or are stored through the channels which, under the electron microscope, are seen to be formed between the reticular membranes and which appear to communicate with the cell surface.

The metabolic activities of the *nucleus* are less well defined than those of other parts of the cell, although about forty different enzymes have been shown to be present in nuclei. The prime function of the nucleus is the replication of DNA, and the transcription of the genetic information which it carries, into lengths of messenger RNA, to be translated into proteins elsewhere in the cell. Nuclei are able to generate energy in the form of ATP by glycolysis, and this is presumably the source of energy for replication and transcription of the nucleic acids. One of the most interesting features of nuclear biochemistry is that this organelle appears to be the exclusive site of NAD synthesis, due to the presence of a key enzyme, NAD pyrophosphorylase, in the nucleus. This observation has given rise to speculation that the nucleus is thus able to regulate cellular metabolism by means other than the control of protein synthesis.

The presence of a selectively permeable membrane (*the plasma membrane*) surrounding cells is detectable in the electron microscope and can be inferred from the uptake of various substances by cells and from differences in composition between intra- and extracellular fluids. The rapid entry of lipid-soluble substances into cells suggested that the membrane must be composed largely of lipid or lipoprotein. From the physical and chemical properties of isolated cell membranes (e.g. red-cell "ghosts"), J. F. Danielli and H. Davson, in 1935, proposed a double layer structure for the cell membrane consisting of two layers of lipid with protein molecules on their outer surfaces. Under the electron microscope the cytoplasmic membrane is seen as two approximately parallel, closely spaced dense layers, each about 2-3 nm thick with a gap of about the same width between them. The double layer seems to be a universal structure common to all membranes and is one which is repeated in the membranes around the nucleus and the mitochondria, as well as in the endoplasmic reticulum. More recent evidence suggests that protein is concentrated into restricted areas of the membrane where it may penetrate through between the inner and outer surfaces, rather than being spread out in uniform layers over the surfaces. The plasma membrane is convoluted into microvilli at the free surfaces of certain cells, notably renal-tubular cells and those of the intestinal mucosa, forming a brush-border.

The cell membrane presents a simple barrier to diffusion to some substances, which move down the concentration gradient between the two sides of the membrane. The rate of this type of transport is influenced by the lipid solubility of the substance. Certain substances, e.g. water, pass through the membrane more quickly than would be expected for simple diffusion, and their passage is therefore presumed to be facilitated by some system in the membrane. Ions may be actively transported against the concentration gradient; thus, high K^+ and low Na^+ concentrations are maintained inside the cell relative to the extracellular fluid. This is brought about by a Na^+ and K^+-activated ATP-ase present in the membrane. Other enzymes located in the membrane are presumably responsible for other transport functions. A number of digestive processes also take place in the membranes of epithelial cells lining the intestine. These include the actions of the group of disaccharidases which hydrolyse the two-unit sugars resulting from the digestion by amylase of starch or glycogen, or disaccharides ingested directly, into their component monomers. Among these enzymes are maltase, lactase and invertase (which hydrolyses sucrose). A group of inherited disorders is now recognized which result from a deficiency of one or more of these enzymes.

The metabolic activities of the cell membrane require the expenditure of energy and thus they cease in the presence of inhibitors of glycolysis or oxidative phosphorylation, i.e. inhibitors which interfere with the regeneration of ATP. Energy is also required for the maintenance of semipermeability and in this case is presumably used in keeping the membrane in good repair, replacing parts that are lost by the absorption of membrane-enclosed droplets (pinocytosis) and regulating the degree of porosity. The plasma mebrane is also the site of contact between cells, and the associations so formed between like cells are the basis of the aggregation of cells to form organs.

ORGAN-SPECIFICITY IN THE DISTRIBUTION OF ENZYMES

The basic patterns of metabolism are repeated throughout the living world: the synthesis of proteins with specific activities as enzymes, generation of energy by coupled sequences of reactions involving the oxidation of carbohydrate and fat and resulting in the generation of ATP, are fundamental to the existence of living matter. However, the evolution of living matter into its present great diversity of forms has been accompanied by both quantitative and qualitative modification of this basic pattern,

with the superposition of more elaborate reaction sequences and also the loss of others in
more advanced organisms. The process of
photosynthesis in green plants and the development of muscular contraction as the basis of
locomotion are outstanding examples of biochemical divergence between the plant and
animal kingdoms. Within the animal organism
biochemical variation has accompanied differentiation into organs and tissues, so that an organ-
specific pattern of metabolism has been further
superimposed on the basic system. These differences in biochemical activity between the
several organs and tissues of the body range
from quantitative variations in the activity of
a particular metabolic pathway from one organ
to another, to the occurrence in certain tissues
of reactions which are restricted essentially to
cells of those specific types. Thus, the tissues
of the human body differ in the relative rates
at which they oxidize glucose in the presence of
oxygen, a process of which all are capable; brain
grey-matter, renal cortex and skeletal muscle
all have higher rates of glycolysis than white
matter, renal medulla and smooth muscle. The
biosynthesis of steroid hormones, however, is
more restricted, being essentially confined to
cells of the adrenal cortex, gonads and placenta.
The structural distinctions existing between
the organs can themselves also be regarded as
organ-specific metabolic differences, since
they reflect the activities of enzymes adapted
to produce these forms.

The division of biochemical labour into organs
and tissues with specialized functions clearly
has evolutionary advantages for the organism;
for example the development of a central
nervous system and sense organs and facilities
for digestion and movement make the creature
less at the mercy of its environment, but with
specialization goes loss of flexibility. Although
all cells are endowed with the potential ability,
in the form of the DNA which they receive
from the germ-cells, to make a complete range
of enzymes able to catalyse all the reactions
which make up the complete metabolic pattern
of the whole organism, this ability is selectively
lost or suppressed during the process of differentiation and development so that the differentiated
tissues are not able to assume the biochemical
functions of others damaged by injury or disease,
or can do so to a limited extent only. Tissues
can often increase the level of their normal
enzymic activities to meet an increased metabolic
demand as does the liver microsomal hydroxy-

lating system, for example, in response to
challenge by drugs, but this adaptability does
not extend to the acquisition of new functions,
If, for example, the ability of the cells of the
adrenal cortex to make the steroid hormones
normally secreted by them is lost as a result of
disease, other tissues are unable to respond to
the high levels of adrenocorticotrophic hormone
present in the blood by producing hormones to
replace those that are deficient, and the resulting
Addison's disease is fatal if not treated with
exogenous hormones.

An important situation in which an organ
may begin to lose its biochemical identity and
its characteristic pattern of enzymes is when
malignant change has taken place. Cancer cells
show a tendency, whatever their tissue of origin,
to progress towards a uniform type of metabolism in which anaerobic glycolysis is an active
metabolic pathway and in which specialized
metabolic functions typical of the cells from
which the tumour has sprung are progressively
lost. The prostate gland, for example, contains
very high levels of a non-specific phosphatase
active at acid pH and when cancer of the
prostate develops, less of this enzyme is produced by the malignant cells. Similarly, specific
functions of the liver such as urea formation
are reduced in hepatomata. The biochemical
changes accompanying the malignant transformation resemble a reversion towards the
metabolic pattern of more primitive cells, and
in some tumours proteins and enzymes can be
detected which are closely similar to, or even
identical with, counterparts which have been
lost or suppressed during differentiation. An
α-foetoprotein is present among the serum
proteins of the human foetus and this protein
declines in amount and becomes undetectable
after birth. In the presence of some malignant
tumours (e.g. primary hepatomata), however, a
protein appears which is apparently identical
with the foetal protein and which may represent
a de-repression of the gene which codes for
this particular molecule. A similar explanation
may account for the production by a small
proportion of cancers of an alkaline phosphatase
which resembles placental phosphatase (itself
determined by the foetus) in a number of ways.

From measurements of the activity of selected
enzymes in extracts of healthy tissues it is
possible to form a picture of their distribution
in which it can be seen that some enzymes occur
widely, but at different levels, in many tissues
while others are highly active in extracts from

one or a few sources only and are virtually absent elsewhere in the body. It is not easy to arrive at completely definite and reproducible estimates for the activities of every enzyme in all tissues, since methods of preparing the tissue extract and bringing the enzymes into solution which are suitable for one tissue may not be equally effective for another, and similar considerations apply to the selection of assay conditions. It is difficult to establish whether failure to detect the presence of an enzyme in the organ under study is a true reflection of its complete absence, or is due to a lack of sensitivity in the methods used. The datum to which enzyme levels should be referred when comparing extracts of different tissues (e.g. nucleic acid content, protein content, wet or dry weight of tissue, or estimated number of cells) also presents difficulty. An impression of the range of activity of some widely distributed enzymes in a number of tissues can be gained from Fig. 8.4, which

also includes some enzymes which are virtually organ-specific in their occurrence.

THE DISTRIBUTION OF ISOENZYMES

The rates at which specific reactions proceed in different tissues may reflect not only quantitative differences in the levels of activity of the enzymes which catalyse them, but also the occurrence of enzymes which are functionally similar, in that they catalyse the same reactions, but which differ in other respects. The ways in which these different forms, or isoenzymes, may arise at the molecular level were discussed in Chapter 3, and the existence of isoenzymic variation between the various sub-cellular compartments has been referred to already in this chapter.

Although examples of organ-specific variations in the properties of particular enzymes of wide occurrence can be traced back for many years, interest in the existence and distribution of isoenzymes as a general biological phenomenon dates from the discovery in the late 1950s that lactate dehydrogenase prepared from a single tissue (e.g. heart muscle) is not a single, homogeneous protein but can be resolved into two or more fractions with distinct electrophoretic mobilities. Five lactate dehydrogenase isoenzymes (LD_1, LD_2, LD_3, LD_4 and LD_5) can be identified in most human and other vertebrate tissues, and these arise by association into tetramers of two types of polypeptide chains of different amino acid composition. Each type of polypeptide is the product of a distinct structural gene locus and the relative activities of these loci in different types of cells contribute to the observed differences in proportions of the five isoenzymes. E. S. Vesell and colleagues have shown that, in the rat, liver and skeletal muscle respectively synthesize LD_5 about 4 times and twice as fast as does heart muscle. However, degradation rates also differ between the tissues, being 10 and 20 times less rapid for LD_5 in liver and skeletal muscle than in heart. These different rates result in a half-life for LD_5 in heart which is only a tenth or less of that of this isoenzyme in the other two tissues. Three principal patterns can be distinguished in human tissues (Fig. 8.5). In the first pattern, represented by heart muscle, kidney and red blood corpuscles, the two most anodic fractions on electrophoresis are predominant (LD_1 and LD_2), together accounting for 70-90 per cent of the total lactate dehydrogenase activity. Skeletal muscle and liver exhibit the second

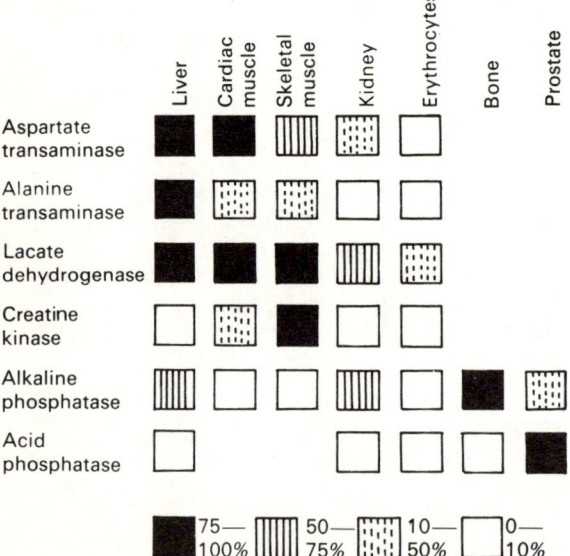

Fig. 8.4. Diagram of the relative activities of some diagnostically useful enzymes in human tissues. Activities are expressed as a percentage of the activity in the tissue with the highest concentration of that enzyme, and are approximations only since isoenzymic and methodological differences affect the results obtained with various tissues. However, it can be seen that some enzymes (e.g. lactate dehydrogenase and the transaminases) occur widely, whereas some enzymes or isoenzymes (e.g. acid phosphatase) are concentrated predominantly in one tissue.

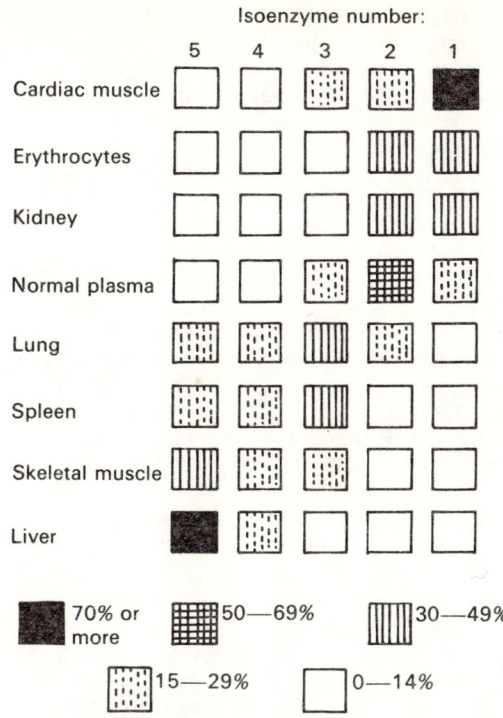

Fig 8.5. Diagram of the relative proportions of the five major isoenzymes of lactate dehydrogenase in some human tissues. The approximate percentage of the total activity contributed by each isoenzyme is represented by the different intensities of shading.

are Michaelis curves of different shapes. The reduction of pyruvate to lactate by LD_1 is inhibited by excess substrate to a much greater extent than is the case with LD_5 (Fig. 8.6), and this has prompted the suggestion that tissues

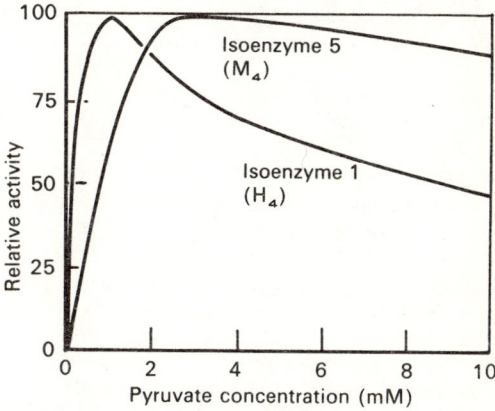

Fig 8.6. Typical Michaelis curves for isoenzymes 1 and 5 of lactate dehydrogenase, showing the greater inhibition of isoenzyme 1 (the predominant isoenzyme of cardiac muscle) at high concentrations of pyruvate, compared with isoenzyme 5, the major form present in skeletal muscle.

pattern, in which the electrophoretically most cathodal fraction (LD_5) represents 50-80 per cent of the total. Other tissues in general show a pattern in which the intermediate isoenzymes are present in the greatest amounts. A sixth isoenzyme of lactate dehydrogenase, LD_x, is found in testis of man and many other species and accounts for 80-90 per cent of the lactate dehydrogenase activity of human sperm. The properties of LD_x are different from those of the other LD isoenzymes and it appears to be a tetramer of yet another type of polypeptide sub-unit determined by a third structural gene locus. However, the LD_x sub-unit is sufficiently like the sub-units of the more common lactate dehydrogenase isoenzymes to be able to form functional hybrid enzyme molecules with them in experiments carried out *in vitro*.

Among the diversity of properties which arises from the underlying structural differences between the lactate dehydrogenase isoenzymes

with LD_5 as the predominant isoenzyme have a greater capacity for anaerobic glycolysis, in that accumulation of pyruvate does not inhibit its reduction to lactate and the coupled reoxidation of $NADH_2$: this occurs in rapid activity of skeletal muscle with the accumulation of an oxygen debt which is subsequently repaid in the oxidation of lactate. Tissues in which accumulation of pyruvate does not normally occur, e.g. cardiac muscle, with its excellent perfusion and high density of mitochondria, are much less dependent on anaerobic metabolism and therefore do not need a form of lactate dehydrogenase adapted to work at high pyruvate levels. This view of the differential functions of lactate dehydrogenase variants is supported by comparisons of the distribution of the isoenzymes among the breast muscles of birds. Those with sustained flight habits and hence a more even muscular activity have a preponderance of LD_1, compared with those of intermittent flight which have mainly LD_5. However, the theory has not gone unchallenged: it has been questioned on the grounds that inhibitory levels of pyruvate are not reached in animal tissues, and it has also been pointed out that the different responses to pyruvate of LD_1 and LD_5 are less marked under physiological conditions of pH and temperature. The comparison of velocity-substrate concentration curves may be too simple an approach and it has been suggested that the key to the difference in function between the isoenzymic forms may lie in the comparative readiness with which LD_1 forms catalytically inactive ternary complexes with oxidized NAD and pyruvate, since there is an excess of NAD in resting skeletal muscle.

Changes in the distribution of LDH isoenzymes during development are also evident in many tissues and species. In the human embryo the intermediate isoenzymes are predominant but the normal specific tissue patterns are present soon after birth. The appearance of the characteristic distributions is presumably associated with the emergence of aerobic or anaerobic patterns of metabolism, though whether causally or not is not known. The dedifferentiation associated with malignant change is also accompanied by a change in the distribution of LDH isoenzymes with a relative increase in isoenzyme 5. This change correlates well with the tendency for an increased fermentative metabolism in neoplasms but, again, it is not known whether the re-distribution is a cause or an effect of malignancy. An abnormality of isoenzyme pattern is also seen in patients with Duchenne-type progressive muscular dystrophy, in whose muscles isoenzymes 1 and 2 remain the principal ones instead of the change to predominantly isoenzymes 4 and 5 taking place as in normal individuals. There seem to be other enzyme abnormalities also in this disease, but dystrophic muscle has an impaired capacity for anaerobic metabolism as might be expected from its isoenzyme pattern and its production of lactate after exercise is less than normal.

Changes in isoenzyme pattern which in several respects are similar to that of lactate dehydrogenase are also shown by aldolase and creatine kinase. Aldolase occurs widely in vertebrate tissues and three main variants of the enzyme have been recognized. The first of these, aldolase A, occurs in all tissues and is particularly abundant in muscle; it cleaves fructose -1,6-diphosphate to triose phosphates during glycolysis. Aldolase B, found mainly in liver and also present in kidney, acts on fructose diphosphate at a lower rate than aldolase A but readily splits fructose-1-phosphate into dehydroxyacetone phosphate and glyceraldehyde.

The third isoenzyme, aldolase C, has been found in brain and some other tissues. As is the case with lactate dehydrogenase, the aldolases are made up of sub-units and it is probable that association of different types of aldolase sub-unit, each type being produced under independent genetic control, gives rise to the A, B and C forms. This interpretation is supported by the occurrence in tissues in which two forms of aldolase exist together of hybrid forms intermediate in properties between the parent forms, the hybrids presumably consisting of heteropolymers of more than one type of sub-unit. The intermediate forms can be reproduced by hybridization experiments *in vitro*, and their number (e.g. A and C aldolase produce three additional isoenzymes forming a five-membered set) suggests that aldolases, like lactate dehydrogenases, are tetramers.

Embryonic liver contains mainly aldolase A, but during foetal life the amount of the B isoenzyme increases until at birth the latter is the predominant form, and a corresponding increase in the amount of aldolase C probably occurs in brain. Malignant change may be accompanied by a reversion to the foetal distribution. The catalytic properties of aldolase A, the muscle enzyme, suggest that the presence of this isoenzyme is an adaptation favouring the cleavage of fructose diphosphate and thus the generation of energy by glycolysis, an important process in muscle. The presence of aldolase B in liver reflects the tendency to fructose diphosphate synthesis (a stage in gluconeogenesis) and thus accounts for the ability of this organ to metabolize fructose by way of fructose-1-phosphate. The different specificities of the aldolase variants are apparently related to changes in structure close to the carboxyl ends of the polypeptide chains, since treatment of the enzymes with carboxypeptidase, which removes amino acid residues from the C-terminal positions, reduces the differences between them.

Three forms of the enzyme creatine kinase, which catalyses the reaction

$$\text{Creatine} + \text{ATP} \rightleftharpoons \text{Creatine phosphate} + \text{ADP}$$

are found in human and other vertebrate tissues. The isoenzymes have a specific tissue distribution: skeletal muscle contains the electrophoretically most cathodal variant, brain the most anodal. The intermediate form is found in cardiac muscle in small amounts together with much more of the cathodal, muscle-type isoenzyme. These distributions result from the interaction of two different creatine kinase sub-units which combine to make up the active enzyme dimers. In muscle M-type sub-units are produced almost exclusively and in brain the B-type sub-units predominate. The intermediate MB dimers result from synthesis of both sub-units, although in heart the M-type is greatly in excess. Peptide mapping has shown that the M and B sub-units have different amino acid constitutions but although there are small differences in the catalytic properties of the MM and BB enzymes no tissue-specific catalytic roles have yet been identified for them. The pattern of creatine kinase variation is further complicated by the observation that in some species the enzyme can form what are believed to be conformational isoenzymes ("conformers") by the folding of a single type of polypeptide chain into different stable configurations.

As with LDH and aldolase, the adult patterns of distribution of creatine kinase isoenzymes are different from those of foetal tissue. Foetal muscle contains the brain (BB) isoenzyme; later, the intermediate (MB) form and the MM form can be detected, but maturation is accompanied by a loss of the BB and MB types. The foetal distribution may reappear in muscular dystrophy, as in the case of lactate dehydrogenase iso-enzymes.

Another human enzyme which exhibits distinctive tissue differences in isoenzyme composition and also developmental changes is alcohol dehydrogenase (ADH). At least three different structural gene loci appear to be involved and the expression of these loci varies from tissue to tissue and at different stages of development.

The most complex ADH isoenzyme patterns are seen in the liver and this tissue also shows the most striking developmental changes. In liver specimens obtained early in foetal life one single main ADH isoenzyme ($\alpha\alpha$) can be detected (Fig. 8.7) and this is believed to be made up of pairs of α polypeptides determined by a locus called the ADH_1 locus. During foetal development, however, a second locus (ADH_2) synthesizing β polypeptide sub-units becomes progressively more active in the liver and additional isoenzymes appear in increasing quantities. Thus at the time of birth three main isoenzymes can be detected, $\alpha\alpha$ made up of α sub-units determined by the ADH_1 locus, $\beta\beta$ made up of β sub-units determined by ADH_2

and an intermediate hybrid $\alpha\beta$ isoenzyme. Later, ADH_2 activity increases further in the adult liver and exceeds ADH_1 activity, while the situation becomes further complicated by the expression of the third locus (ADH_3) which synthesises γ polypeptides in the liver. These latter manifest not only as a $\gamma\gamma$ isoenzyme but also as hybrid $\alpha\gamma$ and $\beta\gamma$ isoenzymes so at least six different isoenzymic forms of ADH can be found to occur in adult human liver.

In some other tissues the ADH isoenzyme patterns appear to remain relatively constant throughout foetal and into adult life. For instance, in lung the β polypeptides determined by ADH_2 predominate in both foetal and adult tissue specimens and no ADH_1 or ADH_3 activity can be detected. In the gastro-intestinal tract the ADH isoenzyme pattern also remains much

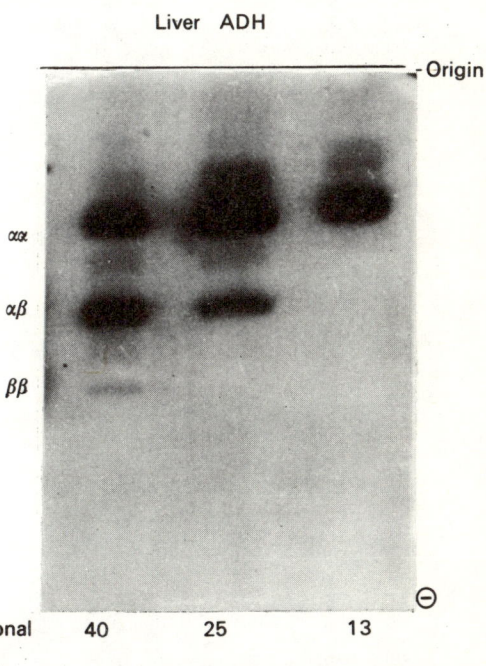

Liver ADH

Origin

$\alpha\alpha$

$\alpha\beta$

$\beta\beta$

$\ominus$

Gestational age (weeks)   40   25   13

Fig 8.7. Photograph of an electrophoretic separation on starch gel of the alcohol dehydrogenase isoenzymes in samples of human foetal liver of different gestational ages. The gradual increase in activity of the ADH_2 (β chain-containing) isoenzyme and the apparent decline of activity of the ADH_1 (α-containing) isoenzyme with increasing maturity is shown. (From M. Smith, D. A. Hopkinson and H. Harris (1971) *Ann. Hum. Genet.* **34, 251, by courtesy of the authors and Cambridge University Press.)**

the same in foetuses and adults, though in this tissue it is the γ polypeptide determined by ADH_3 which predominates. Kidney, on the other hand, exhibits mainly ADH_3 activity in foetal life but in adults most of the activity is due to ADH_2. The functional significance of these tissue differences and the developmental changes in isoenzyme composition is not yet known and will presumably have to await detailed studies on the catalytic properties of the different isoenzymes.

From the examples which have already been studied it can be concluded that isoenzymic variation contributes to differences in the metabolic patterns of tissues and organs. The existence of isoenzymes is probably a generalized phenomenon and many more examples can be expected to come to light. As well as exhibiting different substrate specificities and catalytic constants, there is already some evidence that isoenzymes may respond differently to small molecules which regulate enzyme activity (e.g. allosteric regulators), so that the existence of isoenzymes may provide another opportunity for differential control of metabolism. Variations in molecular structure between isoenzymes may represent adaptations to enable them to be incorporated into different sub-cellular locations.

9 Diagnostic Enzymology

Although the interlocked chemical reactions which constitute the living organism and the activities of the enzymes which catalyse them take place almost entirely within the cells of the body, it is possible to infer from analysis of the extracellular body fluids (particularly the blood plasma or serum) the nature of changes occurring inside the cells of different organs or tissues. The importance of this approach in the diagnosis and management of human disease can be gauged from the observation that for each stay of 10 days by a single patient in a modern hospital an average of five chemical analyses will probably be required; together with out-patient attendances, the biochemistry department of a hospital of 700 or 800 beds would expect to perform each year more than 300 000 analyses for diagnostic purposes. Amongst these analyses the estimation of the levels of particular enzymes is assuming an increasingly important place, and this activity is usually referred to as "clinical" or "diagnostic" enzymology.

FACTORS INFLUENCING ENZYME LEVELS IN EXTRACELLULAR FLUIDS

The normally low enzyme levels in the plasma and tissue fluids are the result of a balance between several factors: the rates at which enzymes are synthesized within and escape from the cells are balanced by the rates at which they are removed from the extracellular fluid by inactivation and degradation or excretion (Fig. 9.1). The presence in the extracellular fluid of substances which modify the action of enzymes, either to activate or inhibit them, also influences the level of activity. (In this respect, however, it must be remembered that measurement of enzyme activity in, for example, a specimen of blood plasma, usually involves a considerable dilution of the sample, so that the apparent

effect of these factors may be considerably reduced compared with their influence *in vivo*.) Changes in extracellular enzyme levels in disease may result from alterations affecting any one, or sometimes more than one, of these factors. However, the most important and useful from a diagnostic point of view are those changes which result from alterations in the rates at which specific enzymes are produced, either as

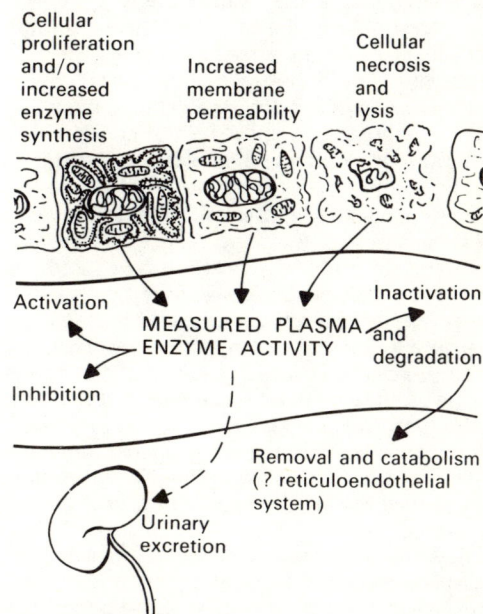

Fig 9.1. Factors which affect enzyme activities in plasma.

a result of increased enzyme synthesis by individual cells or an increase in the number of cells producing the enzyme, and the increases in enzyme levels which follow the leakage of enzymes from injured or dying cells.

ENZYME LEVELS AS AN INDEX OF TISSUE DAMAGE

The use of enzyme estimations in serum in the detection of acute or chronic damage to cells is now one of the main applications of diagnostic enzymology. Its basis lies in the high concentration gradients existing for most enzymes between the intra- and extracellular environments and the presence of an actively maintained semi-permeable barrier surrounding the cell (the plasma membrane) through which enzyme proteins cannot normally pass. Leakage of enzymes into the extracellular fluid and thence into the blood provides an extremely sensitive index of deterioration of the plasma membrane since enzymes can be detected by their catalytic activity when present in quantities which are too low to be detectable by any other means, such as estimations of protein nitrogen.

Many factors can increase the permeability of cell membranes and so cause leakage of intracellular enzymes. These include the restriction of oxygen supply to the cell, depletion of circulating glucose, and the presence of many drugs and chemical compounds which block glycolysis, the citric acid cycle, or oxidative phosphorylation, thus emphasizing the role of energy production in maintaining an effective barrier. In addition, infective agents, bacterial toxins or viruses, may act directly on the membrane, while genetically determined deficiencies in membrane structure may possibly be a factor in muscular dystrophy. Leakiness may be reversible or may progress to an irreversible stage. Reversible leakage of enzymes (e.g. aldolase) from muscle cells occurs as a result of strenuous exercise and is presumed to be due to a temporary slight anoxia, perhaps aided by an increase in osmotic pressure in the cells as large molecules are degraded to small ones, followed by uptake of water, cellular swelling and stretching of the membrane.

Enzyme leakage can be studied under experimental conditions in tissue slices or in isolated, perfused organ preparations. These experiments have demonstrated several factors which govern the rate of release of enzymes and, in particular, the role of anoxia as a causative agent. Metabolism under aerobic conditions enables perfused liver preparations to retain cytoplasmic enzymes (e.g. lactate dehydrogenase) more effectively than anaerobic metabolism. Enzyme release is increased during the anaerobic phase, but the

loss can be arrested during reversion to aerobic metabolism (Fig. 9.2). During anaerobic metabolism the cell is dependent on glycolysis, glucose units derived from stored glycogen or supplied in the circulation being broken down to lactate with formation of two molecules of ATP per glucose unit (starting from glucose) or

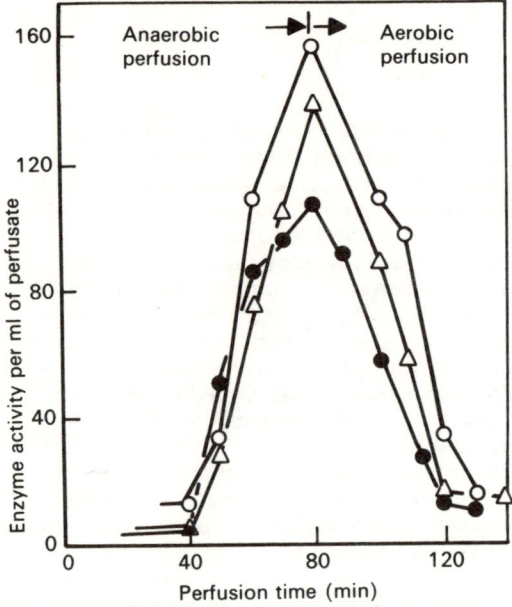

Fig 9.2. **Enzyme release from isolated, perfused rat liver. The concentration of enzymes in the perfusing fluid rises steadily under anaerobic conditions, but when aerobic conditions are restored enzyme concentrations fall. ○, alanine transaminase; ●, aspartate transaminase; △, aldolase. (Data from an experiment by Dr A. K. Walli.)**

three molecules of ATP (starting from glycogen). The importance of this mode of energy generation during conditions of oxygen deprivation is shown by increased lactate production and depletion of glycogen stores. In the prevention of enzyme release the dependence on glycolysis is shown by the greater loss of enzyme from the livers of rats which have been starved prior to death (thus depleting glycogen stores), or when inhibitors of glycolysis are added to the perfusion medium. Anaerobic glycolysis to lactate is a much less efficient means of energy production than glycolysis to pyruvate with subsequent oxidation of pyruvate through the Krebs cycle and the electron-transport chain, with its coupled oxidative phosphorylation mechanism; in the latter process the yield of ATP is more than 30 molecules per glucose unit. (However, although *necrosis* is generally accepted as a cause of leakage of intracellular enzymes, some investigators do not wholly accept the role of impaired energy metabolism in leakage of enzymes from *reversibly* damaged cells.)

When enzyme leakage has begun, three factors affect the rate at which individual enzymes appear in the perfusing fluid. The first of these is the concentration gradient across the cell membrane, which provides the driving force for enzyme leakage. The steepness of the gradient varies for different enzymes and also for different types of cell. For liver cells the concentration of lactate dehydrogenase inside the cell is about 3000-fold greater than the external concentration, but only about 200-fold greater in red blood cells. Intra- to extracellular ratios of activities for liver cells may be as high as 50 000 to 1 for sorbitol dehydrogenase, 20 000 to 1 for alcohol dehydrogenase and 10 000 to 1 for aspartate transaminase and alanine transaminase. Enzymes having higher concentration gradients in general leak out of the cells more quickly than those with lower intra- to extracellular ratios.

The second factor is the size of the enzyme molecules themselves, i.e. their molecular weights. Passage of an enzyme out of a damaged cell involves diffusion through the intracellular fluid, then through the fluid-filled gaps in the membrane and into the extracellular fluid. Smaller molecules diffuse at more rapid rates than larger ones while, in addition, deterioration of the plasma membrane may be associated with a progressive increase in pore size, so that smaller molecules escape at an earlier stage of damage. When rates of release of different enzymes are measured in, for example, a perfused liver preparation, they are found to be correlated inversely with molecular weight.

The third factor is the intracellular location of the enzymes. As would be expected, the enzymes which leak most easily from the tissues are those which in cell-fractionation experiments are recovered in the soluble fraction, i.e. the cytoplasmic enzymes. This is very clearly seen when the different forms of aspartate transaminase, the cytoplasmic and mitochondrial isoenzymes (p. 105), are studied. The cytoplasmic isoenzyme is released into the perfusing fluid at a rate which correlates well with that of

other soluble enzymes of a similar molecular weight; the mitochondrial isoenzyme, on the other hand, is released much more slowly, although both forms are of similar sizes. Release of mitochondrial enzymes is more likely to occur when deterioration in the condition of the plasma membrane progresses beyond the point of reversibility so that the cell dies and the intracellular organelles are broken down.

One of the most common causes of tissue damage is anoxia resulting from a local failure of perfusion. This happens in coronary thrombosis when formation of a clot in a branch of a coronary artery shuts off the circulation of blood to a section of the heart muscle. As a result the cells in the affected area begin to deteriorate and may die and break up, the muscle thus becoming infarcted. Myocardial cells have relatively small glycogen stores with which to support anaerobic metabolism and rely for oxidizable substrates on the supply of lactate or glucose from other tissues; thus they cannot long meet their energy requirements when perfusion by blood is interrupted. Many different enzymes are consequently released from the damaged cells and appear in the circulation, including transaminases, several dehydrogenases (e.g. lactate and malate dehydrogenases), enzymes of the glycolytic pathway (e.g. aldolase, phosphoglucomutase, phosphohexose isomerase), as well as enzymes particularly common in muscle such as creatine kinase. (The reactions catalysed by those of these enzymes that have been found most useful in the detection of tissue damage are listed in Table 9.1: their relative merits for this purpose depend on the extent to which the level in the blood rises following cellular damage and the duration of this rise, and the ease and accuracy with which small changes in enzyme level can be detected and measured.)

The direct relationship between the extent of tissue damage and the amount of enzyme released has been shown conclusively in experiments in which the coronary arteries of dogs were ligated so that the depletion of enzyme in damaged tissues could be correlated with the rise in circulating activity. Myocardial infarctions of differing extents have also been induced by injecting plastic spheres of graded sizes into the coronary arteries and, again, a balance between enzyme depletion and release was observed. Numerous studies of patients with myocardial infarction proved by positive electrocardiographic changes or *post-mortem* examination

TABLE 9.1 SOME ENZYMES WHICH ARE RELEASED INTO THE PLASMA AS A RESULT OF TISSUE DAMAGE

Enzyme	Reaction catalysed	Most useful in damage to:
Aspartate aminotransferase (formerly glutamic-oxaloacetic transaminase)	L-aspartate $\rightleftharpoons$ L-glutamate α-oxoglutarate oxaloacetate	heart, liver
Alanine aminotransferase (formerly glutamic-pyruvic transaminase)	L-alanine L-glutamate + $\rightleftharpoons$ + α-oxoglutarate pyruvate	liver
Lactate dehydrogenase	pyruvate + NADH $\rightleftharpoons$ lactate + NAD^+	heart, liver
Isocitrate dehydrogenase	L_S-isocitrate $\rightleftharpoons$ α-oxoglutarate + CO_2 + NADP + NADPH	liver
Glutamate dehydrogenase	α-oxoglutarate $\rightleftharpoons$ L-glutamate + NADH + NH_4^+ + NAD^+ + H_2O	liver mitochondria
Creatine kinase	creatine + ATP $\rightleftharpoons$ creatine phosphate + ADP	cardiac and skeletal muscle
Aldolase	fructose-1,6-diphosphate $\rightleftharpoons$ dihydroxyace-tone phosphate +D-glyceraldehyde-3-phosphate	skeletal muscle

have shown a high degree of association of raised levels of several enzymes in the blood with the occurrence of infarction; for aspartate transaminase, for example, raised levels have been reported in more than 97 per cent of cases in most series. A correlation as strong as this gives good grounds for concluding, in an individual case, that if enzyme levels remain normal throughout the first few days following the suspected infarction no significant tissue damage has taken place. Conversely, the small increases in serum enzyme activities which may be observed in cases of angina pectoris or coronary insufficiency probably are derived from small foci of cellular damage, perhaps of a reversible nature. Since other tissues besides cardiac muscle, notably skeletal muscle and liver, release enzymes as a result of anoxic or traumatic damage, possible contributions from these sources to the serum enzyme activity must be kept in mind when investigating suspected myocardial infarction.

Measurement of serum enzyme activities, typically those of creatine kinase, lactate dehydrogenase and aspartate transaminase, is thus a powerful adjunct to the electrocardiogram in the diagnosis and management of heart disease and one which is particularly valuable when interpretation of the electrocardiogram is obscured by previous infarctions or by administration of digitalis.

The rise of enzyme activity in the plasma following myocardial infarction is a transient phenomenon: the levels rise more or less rapidly to a peak value, then decline more slowly to normal. The rate at which the peak value is reached varies from one enzyme to another. This occurs at about 24-36 h after the moment of occurrence of infarction (as judged by symptoms such as pain in the chest) for creatine kinase and aspartate transaminase, and somewhat later, at about 48-72 h, for lactate dehydrogenase (Fig. 9.3). The sensitivity of the enzymes as tests of tissue damage also varies — for an infarction of a given size, a larger increase in creatine kinase activity is usually observed than of aspartate transaminase which in turn reaches higher levels than lactate dehydrogenase.

The factors which affect the rate of release

of different enzymes from injured cells and which thus influence the gradient of the rising sides of the peaks in Fig. 9.3 have already been discussed. However, the overall shapes of the curves and particularly the falling slopes are

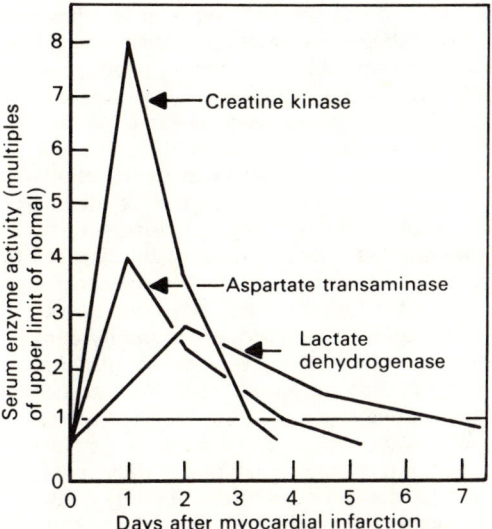

Fig 9.3. Average time-course of enzyme changes following acute myocardial infarction.

also affected by the rates at which the released enzymes are inactivated or removed from the circulation. This process remains considerably more mysterious than the mechanisms by which the enzymes enter the circulation from the cells. A few enzymes have molecules which are sufficiently small to pass through the healthy renal glomerulus and so are excreted in the urine: an example is amylase (mol. wt approximately 45 000). However, most of the intracellular enzymes that are released from damaged tissues are of molecular weights greater than 65 000, so that excretion into the urine is not an important route for their elimination. Bile is rich in a number of enzymes and this may be the way in which some enzymes are disposed of, after being taken up from the blood by the liver. It may be, however, that the presence of these enzymes in the bile arises from some function which they exercise in its elaboration and excretion. Thus, except in a few instances, it does not appear that excretion in urine or bile of active enzymes can account for the decline of excess enzyme activity in the blood, nor for

the steady removal which must be necessary to maintain the normally low levels of most enzymes.

Measurement of the rate of disappearance of purified enzymes injected into animals provides an experimental approach to the problem. Studies of this nature have shown that the injection of a quantity of enzyme is first followed by a steep fall in activity in the plasma corresponding to a rapid distribution of the enzyme throughout the whole of the extra-cellular fluid. This process may take 2 h and is succeeded by a slower fall in activity; during this phase the rate of loss of activity varies from one enzyme to another, alanine trans-aminase being reduced in activity less rapidly than aspartate transaminase, for example, while the mitochondrial form of the latter enzyme is inactivated faster than the cyto-plasmic isoenzyme. The form of the curves for the disappearance of injected enzyme activity with time is similar to that seen with labelled plasma proteins, but if the rates of disappearance of the proteins and enzymes (expressed as their half-lives) are compared those of the plasma proteins are much longer, ranging from 8 to 30 days for different proteins in contrast with the 2 to 4 days of the enzymes.

It seems certain that enzymes are not absorbed from the blood into any organ in an active form and there accumulated: probably inactivation precedes catabolism of the enzyme protein and may take place, or at least begin, in the plasma before uptake and breakdown by tissue cells, and this may account for the short half-lives of enzymes. Some reports have suggested that an "ageing" process may be detectable by an increase in the Michaelis constant of serum enzymes with time following their release from cells, but this does not seem to be an invariable occurrence. A number of experiments have attempted to identify the tissues ultimately responsible for removing enzymes from plasma and their catabolism. In some of these it has been shown that zymosan, an insoluble poly-saccharide from yeast which damages the reticulo-endothelial system and inhibits its function during the first few hours after its administration with a subsequent proliferation of reticulo-endothelial cells, causes a reduction in the rates at which injected aspartate trans-aminase and lactate dehydrogenase are cleared from the circulation. Although these results appear to implicate the cells of the reticulo-endothelial system in enzyme removal this conclusion is still rather tentative.

Lesions which lead to the loss of enzymes from cardiac muscle are of an acute nature and following them the temporal relationships of enzyme release and clearance can be most clearly seen. Leakage of enzymes from liver parenchymal cells can result from both acute and chronic injury, however. The most severe injury to liver cells results from the ingestion of poisons such as carbon tetrachloride or other organic solvents or the toxins of fungi and the increases in serum enzyme activities which follow are consequently very great, values of several hundred times normal for the transaminases and lactate dehydrogenase having been recorded. Experimental administration of graded doses of these poisons to animals has demonstrated that, as in myocardial infarction, the degree to which serum enzyme levels are raised is directly proportional to the amount of tissue destroyed. A further similarity between this type of liver injury and myocardial infarction is that both are of relatively short duration; therefore, differential rates of clearance of the various enzymes from the circulation do not distort their relative activities in the serum in the period immediately following the attack so that the enzyme pattern of the plasma resembles that of a homogenate of the tissue from which the enzymes are released. Carbon tetrachloride is metabolized to a free radical which cleaves the fatty acids of the plasma membrane by peroxidation; lysosomes, mitochondria and nuclei are progressively attacked, leading to complete destruction of large numbers of parenchymal cells. Consequently, not only the enzymes of the cytoplasm leak out but also those which are normally sequestered within the mitochondria, further strengthening the parallel between the serum enzyme pattern and the distribution of activities in a tissue homogenate.

Acute infective hepatitis is a disease in which serum enzyme changes have a very great diagnostic and management application. The activities of transaminases or of lactate dehydrogenase remain normal during incubation of the disease, but begin to rise sharply in the prodromal stage before the appearance of other signs such as jaundice. The rise in enzyme levels also occurs in the considerable proportion of acute hepatitis cases which do not go on to develop jaundice. It is particularly valuable in identifying the disease while it is in its early stages of rather vague and non-specific symptoms, thus permitting prompt isolation of suspected patients to limit the spread of hepatitis through closed communities such as schools and hospitals, or amongst those particularly exposed to a risk of infection such as patients attending chronic renal dialysis units.

The transaminases (particularly alanine transaminase) respond perhaps most promptly and sensitively to this type of liver injury: it has been estimated that involvement of as few as one liver cell out of every 750 is enough to produce a detectable rise in serum alanine transaminase activity. Transaminase levels reach 10 to 50 times the normal during the first 2 or 3 weeks and slowly decline to normal as the illness subsides. In infective hepatitis the histological appearance of the liver is of a diffuse, inflammatory process affecting most of the tissue with isolated foci of necrosis mainly in the central regions of the lobules, rather than one of massive cell necrosis, and recovery is usually complete with little scarring. Changes in mitochondrial morphology suggest that an interference with oxidative energy-productions also takes place. This type of lesion would be expected to result in leakage of enzymes, mainly from the cytoplasmic compartment of many cells through damaged cell membranes. When the enzyme content of the whole liver cell is considered, the activity of aspartate transaminase is greater than that of alanine transaminase (under the conditions usually chosen for measurement of these two activities). However, about one-third of the total aspartate transaminase is located in the mitochondria, so that in the cytoplasm the relative activities are reversed. The serum level of alanine transaminase typically exceeds that of aspartate transaminase in acute infective hepatitis, thus apparently reproducing the distribution pattern of cytoplasmic enzymes. However, the mitochondrial isoenzyme of aspartate transaminase and other mitochondrial enzymes such as glutamate dehydrogenase are detectable in increased amounts in the serum in the early stages of acute hepatitis, indicating that mitochondrial damage and cellular necrosis do take place. The relatively higher activity of alanine transaminase compared with aspartate transaminase is probably also partly maintained during the recovery phase by the more rapid clearance of the latter enzyme from the circulation.

The effect of treatment of acute hepatitis with adrenal corticosteroids (e.g. prednisolone) is usually a reduction in serum enzyme levels. Reduction of the dose or cessation of steroid administration may be followed by a return to pre-treatment enzyme levels so that it appears that the action of steroids is not curative, in the sense of removing the primary cause of cellular injury, but is palliative, lessening symptoms such as enzyme leakage. The site of action of steroids is probably at the cell membrane, reducing the inflammatory process and membrane permeability. Corticosteroids have been shown to have a stabilizing effect on lysosomes *in vitro* and if this takes place in the living cell also it would tend to reduce the extent of autolytic processes.

Chronic liver disease is also accompanied by raised serum enzyme levels which are attributed to the leakage of enzymes from damaged parenchymal cells, but the elevations of activity are generally much less marked than in acute hepatitis. The distribution of enzymes in the serum in chronic hepatitis (i.e. when the process is largely an inflammatory one) is similar to that in acute hepatitis with alanine transaminase often exceeding aspartate transaminase. In cirrhosis, however, there is a tendency for the ratio of the two activities to be reversed and this may reflect the necrotic episodes, in which groups of liver cells are lost and replaced by fibrous tissue, which mark the course of the disease. While these generalizations are helpful in understanding the nature of the pathological changes which underlie enzyme release from the chronically damaged liver, they are less useful in the day-to-day management of patients. Serum enzyme levels fluctuate considerably in parallel with the activity of chronic disease from nearly normal levels to those that are markedly raised and the levels of different enzymes may cross and re-cross each other. At low levels, also, numerical values of enzyme ratios are less reliable. A further complication is that in the chronically damaged tissue enzyme synthesis, and therefore patterns of relative enzyme activity, may not be maintained at normal levels: indeed, some authors believe that increased enzyme synthesis may also contribute to the raised serum enzyme levels which accompany acute, essentially inflammatory conditions such as infective hepatitis, although the major contribution of leakage of pre-existing enzyme is not generally disputed. Nevertheless, regular measurement of the serum activity of one or more enzymes (often alanine transaminase) is valuable in following the progress or remission of chronic liver disease.

Enzymes may be released from parenchymal cells as a result of secondary changes in the liver induced by primary disease elsewhere. Obstruction of the bile ducts outside the liver by stone or by a cancer of the head of the pancreas may cause a slight or moderate rise in serum transaminases originating from liver cells damaged by inflammation, ascending infection or mechanical disruption by the pressure of the retained bile. Restriction of bile flow in the bile canaliculi *within* the liver frequently accompanies acute infective hepatitis and may also result from administration of drugs of the chlorpromazine (phenothiazine) type. In intrahepatic bile stasis the release of intracellular enzymes (e.g. transaminases) is primarily due to the direct action of the toxic agent on liver cells, rather than to any secondary effects of the kind which operate in extrahepatic obstruction, and therefore serum transaminases can range between moderately or markedly raised levels depending on the number of damaged or necrotic cells at any given moment. Thus, although the enzymes which indicate cellular damage (i.e. the transaminases and dehydrogenases) are generally more elevated in intrahepatic obstruction compared with posthepatic obstructive jaundice, measurements of these enzymes alone cannot distinguish the two types of biliary stasis with certainty. However, certain other enzymes which appear to enter the blood by a different mechanism show some quantitative differences in behaviour in intra- and extrahepatic obstructive jaundice and may help in distinguishing the two sites of obstruction. These enzymes are discussed in a later section.

When the liver is invaded by secondary cancerous deposits the host cells are progressively destroyed by the growth of the invading cells, with an accompanying loss of enzymes into the plasma. The tumour tissue itself may also release intracellular enzymes and it is probable that some of the elevated serum lactate dehydrogenase activity that is often observed in patients with widely spread cancers arises directly from the malignant cells themselves. Fatty infiltration of the liver (e.g. in diabetes or obesity) also causes release of liver enzymes to an extent proportional to the extent and severity of the invasion.

While destructive lesions of the liver and heart are the clinical conditions which account for the greatest proportion of the enzyme tests carried out for diagnostic purposes, any form of mechanical, chemical or infective injury to cell membranes of any tissue will result in leakage of the intracellular enzymes: infarction of the kidney, for example, is followed by raised serum transaminases. In the case of the kidney the urine provides a second route by which enzymes escaping from damaged cells may be removed and one which seems to be more easily followed by enzymes originating in the cells of the renal tubules. In a number of acute or chronic renal diseases which are associated with tubule damage (e.g. acute tubular necrosis following surgical or traumatic shock, infection or poisoning) lactate dehydrogenase and alkaline phosphatase (an enzyme abundantly present in tubule cells) are detected in increased amounts in the urine without any marked elevation in serum enzyme levels.

Skeletal muscle collectively represents a very large organ rich in enzymes, and damage to the muscles by accident or surgery releases intracellular enzymes into the circulation. Extreme muscular exertion may also result in a rise in some serum enzyme levels as already mentioned. The most spectacular increases are observed, however, in the hereditary progressive muscular dystrophies, particularly the Duchenne type. The enzymes most affected are creatine kinase and aldolase. The leakage of enzymes from the muscles is at its greatest at the beginning of the active phase of the disease, before muscle wasting is very apparent, and the serum levels subsequently fall as the wasted muscle is replaced by connective tissue and fat. Curiously, the rapid muscular atrophy which may follow motor-nerve injury or disease is not accompanied by a marked change in serum enzyme activities and normal or only slightly raised levels are observed in these conditions. No explanation has yet been found for this difference between primary and neurogenic muscular disease. The membranes of the muscle cells show an increased permeability in muscular dystrophy and this is the immediate cause of the enzyme leakage, but the origin of this membrane defect is not known. A genetically determined deficiency in membrane structure may be a feature of the disease but this cannot be the only abnormality as the isoenzyme pattern of dystrophic muscle fails to show the normal changes associated with maturation, as was mentioned in the previous chapter.

The appearance of high levels of pancreatic amylase in the blood in acute pancreatitis was the basis of one of the earliest applications of enzyme tests in the detection of damage to a specific organ, and one that is still useful. The acinar cells of the pancreas become permeable to the enzyme, probably as a result of inflammation together with the effects of increased pressure resulting from occlusion of the pancreatic duct. The latter hypothesis is supported by the obser-

vation that morphine, which contracts the sphincter of Oddi, causes an increase in serum amylase activity. The enzyme then reaches the circulation either directly via the pancreatic vein or indirectly by lymphatic absorption from the peritoneal cavity. The activity of amylase in the blood begins to rise rapidly within a few hours of the onset of an acute attack of pancreatitis, reaching levels of from 5 to 10 times the normal value at about 24 h: values more than 2½ times normal are almost diagnostic of acute pancreatitis. The level then falls almost as steeply, regaining normal values in 2 to 3 days, either because the cells again become impermeable to the enzyme or because, in more severe cases, synthesis of amylase by the pancreas declines. Amylase is one of the few enzymes with a molecular weight low enough to allow it to filter into the urine and the release of enzyme is therefore detectable by measurements on urine collected during and immediately following a suspected attack, provided that renal function is normal. The slightly later occurrence of the peak of activity in the urine compared with the blood may sometimes be useful if investigation has been delayed.

Other pancreatic enzymes, e.g. lipase, behave similarly to amylase in acute pancreatitis and may even remain elevated for a rather longer period, but the methods for their estimation are less sensitive and reliable. Measurement of serum amylase is thus a valuable procedure in the investigation of acute abdominal pain, though it must be remembered that other abdominal disorders such as perforating peptic ulcer or intestinal obstruction may cause rises in serum amylase by a secondary involvement of the pancreas but these increases are usually moderate or slight. Salivary amylase may also enter the blood in mumps or bacterial parotitis but these conditions are unlikely to present a differential diagnostic problem.

Enzyme changes in chronic pancreatitis or carcinoma of the pancreas are much less reliable. However, stimulation of secretion with pancreozymin and secretin may produce an abnormal rise in serum amylase or lipase when chronic disease is present in which the pancreatic duct is occluded by inflammation or malignant growth provided that the organ is still able to produce these enzymes in adequate quantities. A more reliable assessment of chronic pancreatic disease is obtained by measurement of enzyme levels in the pancreatic secretion sampled by passage of a duodenal tube.

ENZYME LEVELS AS A REFLECTION OF THE NUMBER AND ACTIVITY OF ENZYME-PRODUCING CELLS

In the examples considered so far changes in enzyme activity in the serum have been related to destructive or damaging lesions of the enzyme-containing cells. However, a number of enzymes are able to gain access to the blood without apparent structural damage to the cells from which they originate, either because they are secreted into and function in the blood, or because the cells in which they are formed are so placed that transfer of enzyme to extracellular fluid and thence to blood is particularly easy. In these cases, therefore, changes in serum enzyme levels provide an index of the activity of the cells in enzyme synthesis or of the number of cells in which enzyme production is taking place.

One of the functions of liver parenchymal cells is the synthesis and secretion of a group of specific proteins, of which serum albumin is quantitatively the most important but which also includes the enzymes prothrombin, cholinesterase and lipoprotein lipase, as well as the copper-containing protein caeruloplasmin which has oxidase activity *in vitro*. The levels of these proteins in plasma is reduced in liver diseases in which the synthetic capacity of the liver is impaired. This is particularly marked in chronic disease such as cirrhosis. Serum cholinesterase has been reported to respond more sensitively to short-term fluctuations in the disease than albumin and to provide a better guide to the success of treatment such as portocaval anastomosis. The physiological function and substrate of this enzyme, which has a somewhat different substrate specificity pattern from the "true" cholinesterase of nervous tissue, are not known but it hydrolyses and inactivates suxamethonium (succinyldicholine) drugs which are used as muscle relaxants in anaesthesia. The apnoea which results from relaxation of the respiratory muscles by these drugs is thus likely to be prolonged in patients who have a lowered serum cholinesterase level due to liver disease. Some individuals (about 1 in 2000 of the population) have been shown to have a genetically determined abnormality in which an atypical form of serum cholinesterase is present. The mutant enzyme differs from the normal form in its response to certain inhibitors and has a reduced catalytic activity. This condition also results in prolonged succinyldicholine apnoea, if the abnormality has not previously been recognized and the use of

these drugs avoided. Deficiency of prothrombin synthesis may occur in chronic liver disease and results in a lengthened bleeding time, increasing operation risks in these patients.

An increase in the number and activity of cells which produce the enzyme alkaline phosphatase accounts for the changes in serum alkaline phosphatase activity that accompany certain types of bone disease. Alkaline phosphatases (for there are several variants) are present in many tissues and *in vitro* are able to remove an orthophosphate group from a wide range of phosphate esters at pH optima in the range from pH 8 to above pH 10, depending on the nature of the substrate and its concentration. The specificity of these enzymes is extremely low and almost any phosphate ester is attacked by them; even inorganic pyrophosphate is split into two orthophosphate radicals, and the orthophosphate groups are removed one at a time from polyphosphates such as ATP. Many of the substrates of alkaline phosphatase do not occur normally in living matter and the physiological function of the enzyme remains unknown. However, alkaline phosphatase is found throughout the animal kingdom wherever calcification processes occur and is therefore generally believed to play some part in calcification: in man, the condition of hypophosphatasia in which serum and some tissue phosphatase levels are abnormally low is characterized by defective bone mineralization. The cells of bone which produce the enzyme are the bone-forming cells, the osteoblasts, and from them the enzyme is released into the extracellular fluid and thus into the plasma where it accounts for part of the normal plasma alkaline phosphatase activity of the adult. The normal plasma level in children is rather higher than in adults and correlates well with the extent of bone growth as judged by other criteria such as X-ray examination. The plasma alkaline phosphatase falls to normal levels as growth slows down.

Serum alkaline phosphatase activity is raised in all types of bone disease in which there is increased bone formation. The number of osteoblasts is increased in these conditions and it is probable that individual cells are also more active in production of the enzyme than in the resting state. The level to which serum phosphatase is rasied is proportional to the extent of skeletal involvement. Highest activities (up to about 20 times the upper limit of normal) are reached in Paget's disease (osteitis deformans), a slowly spreading decalcification of the bones in which serum calcium and inorganic phosphate levels remain normal, and the serum enzyme level may remain steady for a period of years after which a sudden rise may presage the development of a bone-forming tumour (osteogenic sarcoma). Rather lower values are observed in osteomalacia and rickets, whether the cause is Vitamin D deficiency, intestinal malabsorption or a renal tubule defect, and again the serum phosphatase level is correlated with the degree to which the bones are involved. Where deficiency of Vitamin D is the cause, successful treatment is accompanied by a fall in serum phosphatase.

In hyperparathyroidism, also, slight to moderate increases in serum alkaline phosphatase occur, paralleling the degree of skeletal involvement, and serum alkaline phosphatase estimations are useful in assessing this aspect of the disease. However, since many cases of hyperparathyroidism do not show bone changes and hence have a normal alkaline phosphatase, a persistently raised serum calcium is a more consistent biochemical sign in this disease. The occurrence of a raised alkaline phosphatase in hyperparathyroid disease with skeletal lesions seems at first sight to contradict the hypothesis that changes in the activity of this enzyme are associated with altered osteoblastic activity, since in this condition the bone lesions are predominantly of a destructive, or osteolytic, nature. However, much abortive new bone formation, with enhanced osteoblastic activity, apparently takes place.

Malignant disease involving the bones also raises the serum alkaline phosphatase activity if its effect is to stimulate the osteoblasts. Thus, diseases such as multiple myelomatosis, which cause thinning of the bones without stimulating bone formation, are not accompanied by raised serum alkaline phosphatase levels, but secondary deposits from other cancers (e.g. of breast or prostate) can be of an osteolytic or osteoblastic nature and, if the latter, increase serum alkaline phosphatase proportionally to their extent and activity.

Another phosphatase with a substrate specificity almost as low as that of alkaline phosphatase is acid phosphatase, so called because its pH optimum is at about pH 5, in contrast to the high pH optima observed with alkaline phosphatase. Acid phosphatase, also, is an enzyme of wide distribution and one of which tissue-specific variants exist; it is present normally in serum to the extent of about one-twentieth of the alkaline phosphatase activity and probably a considerable part of this normal activity originates from blood platelets. Most tissues contain one or more acid phosphatases but by far the richest source is the prostate gland, which has about 400 times as much acid phosphatase as any other tissue, and it is this that gives the enzyme its value in the diagnosis of malignant disease of the prostate. The prostatic acid phosphatase is secreted into the seminal fluid and a little escapes into the urine, but only a small amount passes into the blood. Inflammation of the prostate or mechanical irritation (by rectal examination or catheterization) produces a slight rise in plasma acid phosphatase activity, but in spite of the huge concentration gradient for the enzyme between the prostatic cells and the extracellular fluid relatively little enzyme escapes. A similar effect is observed in cancer of the prostate as long as the disease is confined to the gland; thus, only about a quarter of all patients with prostatic cancer without metastases show an increase in the total serum acid phosphatase. The appearance of secondary deposits elsewhere in the body is, however, accompanied by a marked rise in serum acid phosphatase in more than 80 per cent of patients. The increase is due to the greater number of cells producing acid phosphatase and the easier access of the enzyme to the extracellular fluid and the blood, since the metastases are not surrounded by a capsule as is the prostate itself. The malignant primary and secondary cancer cells produce rather less acid phosphatase than a normal prostatic cell (this is an example of the de-differentiation accompanying malignant change which was mentioned in the previous chapter), but the acid phosphatase appears to be of normal prostatic type.

Metastases of prostatic cancers often lodge in the bones and stimulate surrounding osteoblasts and osteoclasts. Alkaline phosphatase activity in the serum is increased as a result of the increased osteoblastic activity, as in other types of bone disease, while osteoclasts produce an acid phosphatase of slightly different characteristics from the prostatic enzyme. Acid phosphatase arising presumably from osteoclasts can also be detected in the serum in other types of bone disease in which these cells are stimulated, e.g. in hyperparathyroidism involving the skeleton. When prostatic cancer is treated with the synthetic oestrogen analogue, stilboestrol, and the metastases regress, acid phosphatase levels fall in parallel with the decreased number of malignant cells and the reduced activity of osteolytic lesions but alkaline phosphatase may at first rise because of accelerated bone healing before eventually declining (Fig. 9.4).

A few conditions have been noted in which small amounts of acid phosphatase enter the circulation apparently from sources other than the prostate or the bones: one of these is an inherited abnormality of lipid metabolism, Gaucher's disease of the spleen, and in this instance the enzyme may be derived from the lysosomes of the spleen cells.

Changes in serum alkaline phosphatase activity are also characteristic of diseases of the

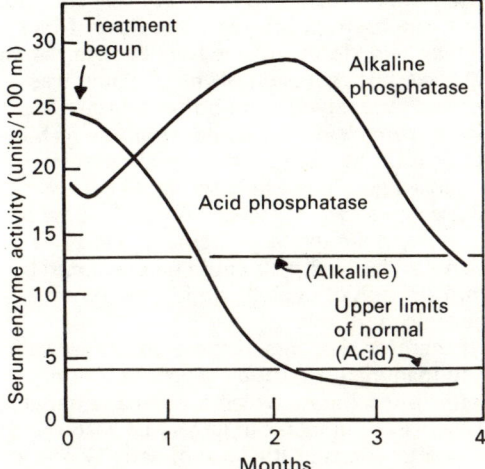

Fig 9.4. **Typical changes in serum acid and alkaline phosphatase activities in a case of metastatic cancer of the prostate treated with stilboestrol.**

liver and bile ducts, particularly those conditions in which the flow of bile is impaired. While bone cells have been accepted almost without question as the source of the raised serum alkaline phosphatase in bone disease, however, the origin of the additional enzyme in hepatobiliary disease has been the subject of much controversy. The rival theories differ in the roles which they assign to the liver in the chemical pathology of alkaline phosphatase. The retention theory views the liver as an excretory organ as far as the enzyme is concerned, removing from the blood alkaline phosphatase from other organs (principally bone) and excreting it into the bile; when bile flow is impeded, either by intra- or extrahepatic obstruction, the phosphatase of bile is regurgitated into the circulation. The hepatogenic theory, on the other hand, sees the liver itself as the enzyme producer with loss of enzyme from damaged liver cells or, according to more recent views,

increased phosphatase production contributing to the serum alkaline phosphatase.

The main basis of the retention theory is the observation that an experimental bile fistula in animals produces a phosphatase-rich flow of bile, and that, provided there is no infection, the serum alkaline phosphatase remains normal. Experimental biliary obstruction produces a rise in serum alkaline phosphatase activity. Removal of other organs in the presence of biliary obstruction has been attempted, to indicate which tissues are the ultimate sources of the bile phosphatase, but without success; thus, removal of the intestine (a phosphatase rich organ) from dogs with ligated bile ducts does not prevent the increase in serum alkaline phosphatase, and the retention theory therefore favours bone as the tissue which contributes most enzyme to that retained in biliary obstruction.

The retention theory does not specifically exclude the liver as a contributor to the retained enzyme but, in this theory, the contribution from the liver is generally assumed to be smaller than those of other tissues. The proponents of the hepatogenic theory, on the other hand, regard the liver as the major, or indeed the sole, source of the abnormal enzyme level in biliary obstruction. Again, the first relevant observations were made by experimental surgery: removal of the liver is followed by a much smaller rise in serum alkaline phosphatase than that which follows obstruction of the bile ducts in which the liver is left intact. The difference is such as to suggest that a large part of the serum enzyme increment originates in the liver. (However, not all the additional phosphatase seems to be from the liver, since there is some increase in serum alkaline phosphatase after total hepatectomy.)

A more direct approach to the problem has resulted from application of the techniques of enzyme fractionation and characterization, which have contributed so much to isoenzyme research, to the study of the properties of serum alkaline phosphatase in diseases of different kinds. These studies, also, have favoured the hepatogenic theory. In general, the increased serum alkaline phosphatase in bone disease has the characteristics of the bone enzyme, as would be expected, but in liver disease the serum enzyme resembles the form which can be purified from the liver, instead of bone or intestinal phosphatase as the retention theory requires. The comparisons of serum and tissue phosphatases which have been found most useful in arriving at this conclusion are of Michaelis constants, electrophoretic mobilities (Fig. 9.5)

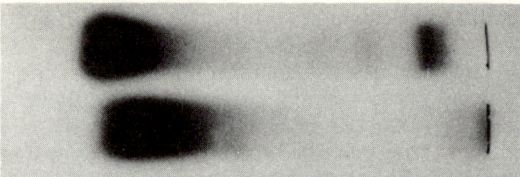

Fig 9.5. Zones of alkaline phosphatase activity separated by starch-gel electrophoresis at PH 8·6 of extracts of human liver (upper) and bone (lower). The anode is at the left and the origin towards the right.

and resistance to denaturation by heat (Fig. 9.6). Liver alkaline phosphatase probably enters the blood in obstructive jaundice both indirectly by regurgitation of bile through the sinusoidal barrier as a result of the increased intra-duct pressure

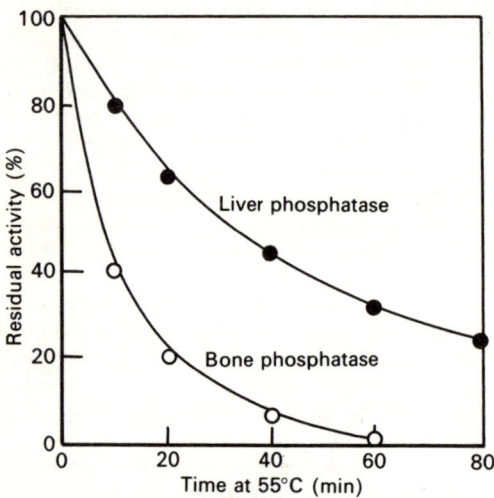

Fig 9.6. Loss of activity of human liver and bone phosphatases during heating at 55° C.

(marked alkaline phosphatase activity in the biliary canaliculi can be demonstrated by histochemical staining, while the presence of a slightly altered form of the enzyme which is present in bile can often be detected in the serum by electrophoresis), as well as directly from the liver cells themselves. This latter proportion of the serum phosphatase seems to result from an increased synthesis of the enzyme by the cells, rather than from a passive leakage from damaged tissue: alkaline phosphatase is largely a microsomal enzyme and therefore not one which would be expected to leak easily from the cells, while the

extent to which the serum alkaline phosphatase is raised in liver disease is not closely correlated with the extent of cellular destruction as assessed by other serum enzyme measurements such as transaminase levels. Recent experiments which support the assumption that the liver cells respond to biliary obstruction by making more alkaline phosphatase include the ligation of a single liver-lobe in dogs (this animal's liver, unlike human liver, has clearly distinct lobes), which produced increased serum, bile and tissue phosphatase levels, and measurements of alkaline phosphatase in the fluid perfusing isolated cat livers and in the liver tissue itself: as long as a free flow of bile was maintained the enzyme level in both remained unaltered, but occlusion of the bile duct was followed by a rapid rise in both perfusate and tissue enzyme levels. A similar response to bile-duct ligation has also been observed in rats; the consequent increase in serum alkaline phosphatase was shown by electrophoresis to be of liver type, while the large increase in enzyme activity in the liver cells could be prevented by administration of cycloheximide, an inhibitor of protein synthesis.

It therefore appears, if these results are valid for all mammals, that the changes in serum alkaline phosphatase which accompany hepato-biliary disease in man can largely be accounted for by alterations in the level of activity of the liver cells in the production of this enzyme. It is not yet known what is the function of alkaline phosphatase in liver cells but it may be concerned with the secretion of the bile into the canaliculi, so that more enzyme is needed to continue this function in the presence of increasing biliary pressure.

The long debate over the origin of serum alkaline phosphatase in liver disease has not detracted from the usefulness of this estimation in these cases, however, and it has been in continuous use as a liver function test since the association of raised levels of the enzyme with obstructive jaundice was first described, in 1930. Increased serum alkaline phosphatase activity is seen whatever the anatomical site of the obstruction to bile flow, whether this is intra-hepatic as in infective hepatitis or when the growth of primary or secondary tumours within the liver increases the pressure on the bile canaliculi, or extrahepatic as in occlusion of the common bile duct by stone or by a cancer of the head of the pancreas. The extent to which the serum phosphatase is raised is in general greater the more complete the obstruction to bile flow, and levels are on average higher (more than 2½ times the upper limit of normal) in cases of extrahepatic jaundice than in obstruction of intrahepatic origin; thus, in infective hepatitis rises in serum alkaline phosphatase are usually only slight or moderate. This provides one of the few biochemical tests which help to distinguish between intra- and extrahepatic jaundice, but it is necessary to

because of bone growth, or in women in late pregnancy when the serum alkaline phosphatase is raised by the entry of enzyme from the placenta into the maternal circulation. One such enzyme is a specific phosphatase, 5'-nucleotidase, which removes a phosphate group from the 5'-carbon atom of the ribose ring of nucleotides such as adenosine-5'-phosphate (AMP):

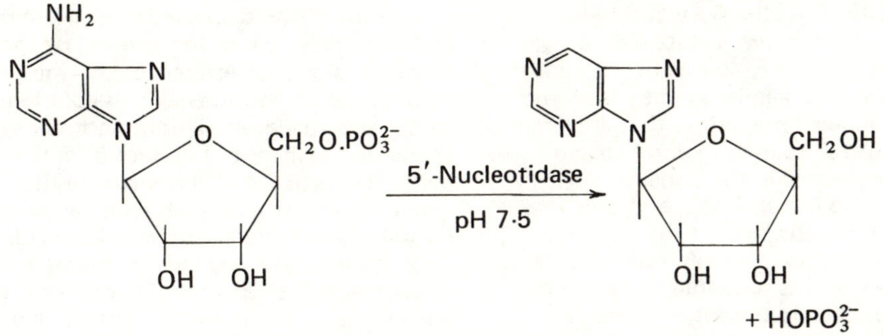

remember that individual cases may depart from the general rule — in intrahepatic cholestasis due to cancer, for example, high levels of serum alkaline phosphatase may be reached.

Although patients in whom a raised serum alkaline phosphatase level is due to liver disease are usually (though not invariably) jaundiced at some stage of their illness, the levels of serum bilirubin and phosphatase often do not rise and fall in parallel. This dissociation between two constituents which are both present in bile and which therefore might be expected to respond similarly to bile stasis has given rise to some debate. Part of the explanation probably lies in the ability of conjugated bilirubin (bilirubin glucuronide) produced by the liver to escape into the urine when the flow of bile is obstructed, while alkaline phosphatase cannot pass through the glomerulus, but more recent suggestions that the *de novo* synthesis of alkaline phosphatase by liver cells during biliary obstruction, rather than merely passive regurgitation of biliary enzyme, is the main origin of the increased serum phosphatase make the lack of complete correlation between the two constituents less surprising.

One or two other enzymes behave in a similar way to alkaline phosphatase in hepato-biliary disease and estimation of their levels is useful in deciding between disease of liver or bone as the source of a raised alkaline phosphatase when this is not obvious from other evidence; for example in children, who have a physiologically higher serum alkaline phosphatase than adults

5'-Nucleotidase, which has a pH optimum at about pH 8, is widely distributed but the normally low activity of the enzyme in the serum is increased only in hepato-biliary disease, normal levels being found in osteoblastic bone disease. In liver diseases the majority of cases which have a raised serum alkaline phosphatase also have a raised 5'-nucleotidase and vice versa. However, a small proportion of cases of liver disease are seen in which one or other enzyme level is elevated while the other remains normal, and this suggests that the two enzymes enter the circulation by different mechanisms or that they originate in different types of cells. The conditions in which a disproportionate rise in 5'-nucleotidase compared to alkaline phosphatase is likely to be found include diseases which particularly affect the cells of the bile ducts and biliary canaliculi, e.g. biliary cirrhosis, and it has been suggested that 5'-nucleotidase levels in the blood rise as a result of a proliferation of biliary duct cells from which the enzyme may be derived.

Among other examples of changes in serum enzyme levels in which the level of enzyme production may play a part may perhaps be included the greatly elevated serum lactate dehydrogenase activities which are a feature of untreated pernicious anaemia. In this condition the serum lactate dehydrogenase usually exceeds 7 times the upper limit of normal, and the level falls rapidly on treatment with Vitamin B_{12}. Congenital and acquired haemolytic anaemias show normal or only slightly raised lactate dehydrogenase activities. The origin of the greatly elevated lactate dehydrogenase levels in pernicious anaemia is not immediately obvious: the lactate dehydrogenase isoenzyme in the serum is of the same type as that predominant in red blood cells, but the rate of destruction of red cells in pernicious anaemia, though increased, is only about 3 times the normal value. It has been calculated that red cells would have to be lysed at about 15 times the normal rate to account for the high serum lactate dehydrogenase levels if this were the only source of the additional enzyme. Pernicious anaemia is a megaloblastic condition, i.e. red-cell precursors that are larger than normal but fail to mature and extrude the nucleus are present in the bone marrow. The marrow is also hyperplastic, and thus there are probably more cells present than usual. The megaloblasts contain a high level of lactate dehydrogenase and it is probable therefore that these enzyme-rich cells present in increased numbers are the source of the high serum lactate dehydrogenase activity in pernicious anaemia.

ENZYME LEVELS AS AN INDICATION OF ALTERED RATES OF INACTIVATION OR EXCRETION

The uncertainty surrounding the fate of enzymes released into the serum has already been discussed, together with the contribution that relative rates of enzyme removal may make to the time-course of changes in serum enzyme activity following acute episodes of tissue damage. These factors may also be expected to operate in the absence of an abnormal release of tissue enzymes and are presumed to be part of the normal homeostatic mechanism controlling serum enzyme levels, so that an increased activity of an enzyme in serum could result from a reduction in its rate of inactivation or removal, or an abnormally low level from an increased rate. However, there are very few instances in which serum enzyme changes can be attributed to this cause.

The ability of amylase to pass through the glomerulus into the urine has been mentioned and urinary amylase levels are correlated with serum levels in pancreatic disease. It would be expected, therefore, that in renal failure serum amylase activity would rise, even in the absence of disease of the pancreas, as a result of failure

to excrete the enzyme and slight or moderate increases have in fact been recorded. The short half-life of amylase in the circulation does not seem to be completely accounted for by excretion of the enzyme in the urine, however, so that other mechanisms of inactivation and removal probably operate also, preventing a continuing accumulation of serum amylase in renal insufficiency.

In view of the suspected role of the reticulo-endothelial system in the removal of some enzymes from the blood, interference with the function of the reticulo-endothelial cells could increase serum enzyme levels. This has not yet been shown to happen in man, but in mice infection with a particular virus (the Riley virus) results in elevated serum enzyme levels which may be attributable to reticulo-endothelial blockade.

Stimulation of anabolic or catabolic processes in general will alter rates of protein turnover and enzymes are presumably affected like other proteins. Thyroid hormones are known to shorten the half-lives of non-enzymic plasma proteins and, since enzymes may be similarly affected, the suggestion has been made that reduced levels of circulating thyroid hormones in hypothyroid patients and reduced enzyme catabolism may account for the elevated serum creatine kinase activities found in a majority of these individuals. However, it seems more probable that there is a direct effect on muscle membranes in this condition which promotes enzyme leakage. Administration of steroids increases protein turnover and consequently may also affect serum enzyme levels. Anabolic drugs have been shown to increase the production of transaminases by liver cells, resulting in a raised serum enzyme level by overflow.

Such alterations in enzyme inactivation or excretion rates as these are unlikely to confuse the interpretation of clearly raised serum enzyme levels due to disease in man, but a clearer understanding of these factors could have an important effect on diagnostic enzymology when lower enzyme levels are to be interpreted. As was mentioned earlier, the balance between the relative rates of entry of an enzyme into the circulation and of its inactivation or removal determines its level of activity; thus the normal serum level will be influenced by these rates. A knowledge of their respective magnitudes and the factors which modify them would allow the normal level under given conditions to be estimated more precisely, facilitating assessment of borderline serum enzyme values.

MODIFICATION OF ENZYME ACTIVITIES IN SERUM BY ACTIVATORS AND INHIBITORS

All enzymes are susceptible to a reduction in their activity by different types of specific or non-specific inhibitors, while their effectiveness may be increased by activators (Chapter 5). These effects presumably operate *in vivo* as well as in the assay systems in which they are studied, and many potential enzyme modifiers are present in plasma and serum, e.g. amino acids, urea, cations, anions, etc. It is difficult to assess the influence that these factors may have on plasma enzyme activities in the circulation because the conditions of enzyme assay usually used involve the dilution of the plasma or serum sample tenfold or twenty-fold or more, thus reducing the effective concentration of activators or inhibitors by a corresponding factor. At various times the suggestion has been put forward that the circulating blood contains an activator or inhibitor which accounts for the changes in serum enzyme activities encountered in particular diseases. Hypotheses of this nature can be tested by experiments in which the sample of serum suspected to contain the inhibitor or activator is mixed with a normal sample of known enzyme activity: an additive resulting activity would suggest that no modifier of activity was present, while a result below or above the calculated sum of the two activities would indicate respectively that an inhibitor or activator was present.

The existence of an activator of alkaline phosphatase has been postulated to account for the increased activity of this enzyme in serum in liver disease, but although many of the substances present in serum (e.g. amino acids) can be shown to affect the enzyme, the degree of activation or inhibition is small compared to the enzyme changes in disease and is not consistently demonstrable. This theory is therefore no longer tenable. Attempts to demonstrate the presence of a circulating alkaline phosphatase inhibitor in hypophosphatasia have failed similarly. However, the accumulation of metabolites which specifically inhibit serum lactate dehydrogenase has been shown to occur in chronic renal failure. It was noticed that the serum lactate dehydrogenase activity of uraemic patients increased after treatment by dialysis, but the levels of other enzymes such as aspartate transaminase were not affected. Fractionation and analysis of the dialysis bath fluid following treatment of these patients has shown more than one inhibitor present in

uraemic serum. One is urea, another is oxalate, and both had been recognized earlier as lactate dehydrogenase inhibitors, while dialysates probably contain others as yet unidentified.

Since many enzymes are activated by metal ions (e.g. enzymes catalysing reactions which involve phosphorylated compounds are often found to be activated by magnesium ions), deficiency of these cations might be expected to result in reduced enzyme activities. Whether relative deficiencies of metal ions play a part in regulating enzyme activities within cells is not known, but reduced serum enzyme levels do not seem to be attributable to this cause: normal serum alkaline phosphatase activities have been noted in hypomagnesaemia, for example.

Present evidence thus suggests that enzyme activation and inhibition make at most a minor contribution to measured activities in serum. However, there is some evidence that vitamin deficiency may reduce the activity of certain enzymes within cells. The role of vitamins as constituents of coenzymes and prosthetic groups was described in Chapter 5. Deficiency of vitamins of the B group has been shown to be accompanied by reduction of the activities of certain enzymes in haemolysates of red blood cells from affected patients, e.g. transaminase (prosthetic group pyridoxal phosphate, derived from pyridoxine), transketolose (thiamine pyrophosphate-dependent) and glutathione reductase, part of the FAD prosthetic group of which is derived from riboflavin. Synthesis of the apoenzymes (enzymes minus prosthetic groups) apparently continues at a normal rate since addition of the appropriate coenzyme increases the enzyme activity of the sample. Measurement of the enzyme stimulation produced by added coenzyme has been used as a test for vitamin deficiency.

DEMONSTRATION OF CONGENITAL ENZYME DEFECTS

Inherited abnormalities in enzyme synthesis which result in disease are discussed elsewhere, and in more than 100 of these inborn errors of metabolism the identity of the defective enzyme has been established. Although the diagnosis of a particular metabolic defect may be based on the demonstration of an abnormal accumulation of a metabolite, or of an unusual metabolic excretory product (e.g. the greatly increased plasma level of phenylalanine or the occurrence of the abnormal metabolite phenylpyruvic acid in urine, both of which are consequences of phenylketonuria), the definitive diagnosis is achieved by demonstration of the specific enzyme defect. This has therefore become an important branch of diagnostic enzymology, particularly in the screening of newborn infants, its importance lying in the realization that the only effective treatment in these conditions may be the prompt institution, in response to accurate diagnosis, of the appropriate diet in which the constituent which is not tolerated is reduced or withheld altogether.

Most of the enzymes concerned with these metabolic abnormalities are intracellular; few can be demonstrated by assays on blood plasma or serum, and therefore samples of tissue cells are needed. Examples of inherited enzyme deficiencies or modifications which can be detected by measurements on serum are hypophosphatasia, in which defective bone mineralization is accompanied by low serum alkaline phosphatase levels, and the presence of an abnormal form of serum cholinesterase which renders the affected individual abnormally sensitive to the effects of succinyldicholine muscle relaxants (p. 159). Many of the enzyme mutations giving rise to disease affect processes which take place in the liver, but to obtain enough tissue for enzyme studies from this organ in small babies is not often a practicable procedure. The diagnosis of the liver-enzyme defect, phenylketonuria, for example, depends for practical purposes on the demonstration of raised levels of metabolites in plasma or urine, as mentioned above. However, since enzymes of a particular function are only rarely confined to one tissue only, it is often possible to select cells from more convenient locations. In this respect, red blood corpuscles constitute a readily sampled and relatively homogeneous population of cells in which many enzyme mutations can be identified. Not all of these result in disease, but they nevertheless form a rich field of study for the enzymologist and geneticist interested in the degree of molecular variation which may be associated with a particular catalytic activity and the way in which it is inherited. A clinically important enzyme deficiency which can be confirmed by measurements of enzyme activity in red cells is galactosaemia. Affected babies cannot tolerate galactose and, if the consequences of cirrhosis, cataract and mental retardation are to be avoided by exclusion of galactose from the diet, early diagnosis is essential. The deficient enzyme is galactose-1-phosphate uridylyl transferase, a key enzyme in the conversion of galactose to glucose. This process takes place mainly, but

not exclusively, in the liver; red blood cells also normally contain the enzyme and it is absent from them in galactosaemia, so that the need to obtain liver tissue is avoided. Another technique which has recently shown promise as a means of obtaining a sample of cells for diagnosis of congenital enzyme deficiency is amniocentesis. In this, a sample of amniotic fluid is withdrawn by a needle, and the cells it contains, which are derived from the foetus, are grown in tissue-culture until enough material is obtained for reliable estimations of their complement of enzymes to be made. Failure to demonstrate a particular enzyme activity in the foetal cells then indicates that the foetus has inherited the associated deficiency disease, so that preparations for immediate post-natal treatment can be made. Some enzyme defects are not amenable to treatment and are not compatible with prolonged post-natal survival, however, and a diagnosis of this nature during pregnancy would provide a strong indication for therapeutic abortion. Diseases that have been diagnosed *in utero* by enzyme studies include a range of lipid-storage and glycogen-storage diseases and disorders of amino-acid metabolism, as well as galactosaemia.

As techniques of reliable and safe tissue biopsy develop, together with methods for quantitative and qualitative study of enzymes in very small tissue samples, it will become possible to make accurate and rapid diagnosis in more and more cases of suspected inherited metabolic defects, and it is probable that diagnostic enzymology in its more general aspect also will develop along these lines.

ISOENZYME TESTS IN DIAGNOSIS

Detection of inborn errors of metabolism is an example of isoenzyme diagnosis. These conditions may range from extreme cases in which the abnormal enzyme, or "isoenzyme", has no detectable enzyme activity at all, to those in which the presence of an atypical isoenzyme is only manifested clinically in the presence of certain drugs (e.g. the muscle relaxant, succinyl-dicholine, p. 159 or, indeed, has no known clinical consequences. In all these instances the unusual genes determining the production of the isoenzymes are present only in a proportion of the population. However, those isoenzymes such as the isoenzymes of lactate dehydrogenase or alkaline phosphatase which are the products of genes carried normally by all members of the population are also useful in diagnosis, because of their variation in properties or activities from one tissue to another (Chapter 8). Determination of the type of isoenzyme present in the serum in disease, as well as the change in total enzyme activity, can provide valuable additional information in a number of cases. The advantages to be gained from isoenzyme studies on serum or plasma are twofold: firstly, the isoenzyme type or distribution of types in the serum may be shown to resemble that of a particular tissue or organ, which is thus identified as the source of the abnormal amount of enzyme in the serum. Secondly, it may be possible to demonstrate the presence of an abnormal pattern of isoenzymes in serum even before the total activity of that particular enzyme has risen above the normal range, or after it has returned to normal: isoenzyme tests thus offer a possibility of increased sensitivity, as well as specificity, in diagnosis. Both these possibilities have been realized in clinical practice.

The use of acid phosphatase estimations in the detection of prostatic cancer is an example of isoenzyme diagnosis which has been in use for many years. In this condition early detection of the prostatic type of acid phosphatase is a valuable indication of the occurrence of metastases (p. 148). The prostatic isoenzyme differs from most other tissue acid phosphatases in that it is inhibited by the dextrorotatory form of tartaric acid. By adding this compound to the assay mixture, the test is made more specific for prostatic phosphatase and both the sensitivity and specificity of the investigations are increased. The organ-specific differences between alkaline phosphatases from different tissues have already been mentioned (p. 149) and, besides its general interest in understanding the ways in which alkaline phosphatase levels in the blood are regulated, identification of specific alkaline phosphatase isoenzymes can be useful in individual patients. For example, a raised serum alkaline phosphatase may be the only biochemical abnormality in a patient whose clinical symptoms are not very definite. Often there may be indications of malnutrition due to failure properly to absorb constituents of the diet and consequently disease of the liver or of the bones may be suspected. In such a situation, identification of the serum alkaline phosphatase as of bone or liver origin by means of the tests mentioned earlier can be diagnostically helpful. In cancer patients, also, a raised alkaline phosphatase level may indicate the spread of the tumour to bone or liver and, again, detection of the appropriate

isoenzyme may enable this process to be detected and located.

However, it is in the study of the organ-specific distribution of lactate dehydrogenase isoenzymes and its effect on isoenzyme patterns in the serum in different diseases that the use of isoenzyme tests in diagnosis has received most attention in recent years. The preponderance of the isoenzyme LD_1 in tissues which have a highly aerobic type of metabolism (such as heart muscle) and, conversely, the high levels of LD_5 in tissues such as liver or skeletal muscle which are able to undertake anaerobic metabolism, was mentioned in the previous chapter. When damage to these organs leads to the release of lactate dehydrogenase into the bloodstream, the pattern of isoenzymes in the serum takes on that of the affected tissue (Fig. 9.7). Thus, the source of an elevated lactate dehydrogenase can be identified by

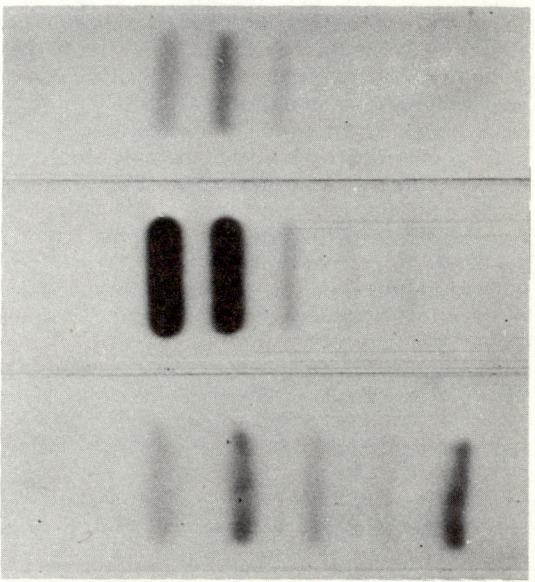

Fig 9.7. **Separation of lactate dehydrogenase isoenzymes on agar-gel electrophoresis of normal serum: (top), serum from a patient who had suffered a myocardial infarction 3 days earlier (middle), and serum from a case of infective hepatitis (lower). Isoenzymes 1 and 2 are at the left and isoenzyme 5 at the right.**

comparing the serum isoenzyme pattern with that of the suspected tissue. The number of possible patterns of isoenzyme distribution is, however, rather limited, and in practice, the method is largely confined to differentiation of the heart and liver patterns. Differentiation between heart and liver as sources of serum lactate dehydrogenase is not often a clinical problem, because other signs and symptoms usually point clearly to the affected organ. However, a few occasions may arise in which differentiation by other criteria is difficult, e.g. when a second rise in enzyme activity follows closely after one due to myocardial infarction. In such a case the electrocardiogram may not be helpful because of the previous infarction, and the possibility of a second infarction must be distinguished from secondary damage to the liver due to impaired circulation as the cause of the renewed rise in serum enzyme activity. The determination of the lactate dehydrogenase isoenzyme pattern in serum is perhaps even more useful in the delayed investigation of suspected myocardial infarction. This is an example of the increased sensitivity of isoenzyme tests referred to earlier, in that the abnormally high percentage of LD_1 in the serum persists after the total lactate dehydrogenase activity has returned to normal. Therefore, it may be possible to demonstrate the abnormal isoenzyme pattern more than a week after a myocardial infarction, when the levels of other enzymes have again become normal.

Most clinical biochemistry laboratories now include in their range of tests at least one method of distinguishing between the several lactate dehydrogenase isoenzymes and of estimating the relative proportions of LD_1 and LD_5. Electrophoresis of the serum sample followed by staining to show the active enzyme zones provides a useful qualitative picture. LD_1 is considerably more stable to heat than LD_5, so that the lactate dehydrogenase activity which remains after heating for 1 h at 60°C is a good indication of the proportion of the total activity due to LD_1. Addition of urea similarly denatures and inactivates LD_5 more readily than LD_1, and various inhibitors (e.g. oxalate) have a differential effect on the two isoenzymes. A further method makes use of the fact that the isoenzymes will act to some extent on substrates other than lactic acid. The next higher homologue of lactic acid is α-hydroxybutyric acid:

$$CH_3 CH(OH)COOH \qquad\qquad CH_3 CH_2 CH(OH)COOH$$

Lactic acid $\qquad\qquad\qquad$ α-Hydroxybutyric acid

All the isoenzymes of lactic dehydrogenase will accept α-hydroxybutyric acid as a substrate, but the activity relative to lactate dehydrogenase activity is rather greater for LD_1 than for LD_5. "α-Hydroxybutyrate dehydrogenase" activity is therefore a measure of the amount of LD_1 in the sample.

While these examples of serum isoenzyme fractionation are those that are most useful clinically at the present time, it is likely that more tests will come into use as research uncovers further organ-specific isoenzyme variants.

10 Inherited Enzyme Variation as a Cause of Disease

That catalytic activity is inseparable from the structure of the enzyme molecule was noted in Chapter 3. The structure of enzymes, like that of all proteins, is in turn determined by the sequence of bases in that section of chromosomal DNA which codes for the protein in question. Therefore, the continuance of a particular enzymic activity in the cells of succeeding generations depends on the faithful transcription and translation of the coded information by each cell into a specific amino acid sequence and on the inheritance of the DNA code in an unchanged form by daughter cells. While no errors or variations in the processes of inheritance and enzyme synthesis can ideally be tolerated, in fact some degree of variation in enzyme structure can take place before the affected molecules cease to be effective catalysts, provided that these modifications do not involve amino acid residues which are critical for structure or activity. This type of variation may arise at the level of the gene, or may be due to modification of enzyme molecules after synthesis. Some examples of genetic and non-genetic isoenzyme formation were discussed in Chapter 3. The isoenzymes do not appear to have different functions in many cases, but in other instances they may represent evolutionary adaptations to the differing metabolic needs of various cells or organs: the isoenzymes of LDH or aldolase may have arisen in the latter way, for example.

The isoenzymic forms of several enzymes are present normally in all individuals of the same species. These forms may perhaps have been so necessary to life, and their appearance during evolution may have conferred such an advantage on the organism, that natural selection ensured their survival and propagation throughout the species. Isoenzymes due to multiple gene loci fall into this category.

On the other hand, a number of enzymes show considerable variations between individuals of the same species. These enzyme variants may be recognized by differences in physical properties of the enzyme molecules (e.g. in electrophoretic mobility or heat-stability) or in catalytic properties such as Michaelis constants, relative activity towards different substrates or response to inhibitors. The patterns of inheritance of these enzyme variants show that they have arisen by allelic mutations of the corresponding structural genes and that these mutations have subsequently been inherited in accordance with Mendelian laws. Often, the inherited enzyme variants provide no more than molecular evidence of individuality and ancestry. Very occasionally, they confer an advantage on their possessor. Sometimes, they may place the individual at a disadvantage when faced with unusual circumstances, such as when challenged by administration of a drug. More important from the medical viewpoint, however, are those instances in which the properties of the enzyme have been so modified by mutation that it can no longer sustain its normal role in metabolism, since in these cases disease or even death are the consequences.

It is possible also that individual differences in enzyme make-up may result from genetically determined alterations in rates of enzyme synthesis or breakdown, as well as from structurally altered enzymes.

The idea that changes in enzymes might be the cause of inherited metabolic disease is a surprisingly old one, being first clearly formulated by Sir Archibald Garrod 70 years ago. Garrod studied the hereditary background of a small group of patients with the rare disorder, alkaptonuria, in which the urine darkens on standing due to the presence of an abnormal metabolite, homogentisic acid. He established that the condition was inherited as a Mendelian recessive character, and concluded that the explanation lay in an inborn alteration in the metabolism of tyrosine.

Further pedigree data, from cases of albinism,

pentosuria and cystinuria, enabled Garrod to suggest that these also were "inborn errors of metabolism", and he postulated that an alteration in a Mendelian genetic factor (the term "gene" had not then been proposed) resulted in lack of a specific enzyme. This deficiency could lead to the accumulation of an abnormal metabolite, which might be toxic, or to failure to produce some essential constituent of the body.

Garrod was not able to identify the specific enzymes, the lack of which caused these classical inborn errors. This was achieved in 1958 for alkaptonuria, when LaDu demonstrated that the enzyme homogentisate oxidase was indeed deficient, as Garrod had predicted. Rather earlier, in 1953, Jervis was able to demonstrate that phenylketonuria is due to absence of phenyl-alanine hydroxylase activity from the livers of affected individuals. This condition, first described in 1934, is characterized by abnormal excretion of large amounts of phenylpyruvic acid and other metabolites of phenylalanine, because of the failure to convert phenylalanine to tyrosine (p-hydroxyphenylalanine). Its most important clinical consequence is severe mental retardation. A specific enzyme lesion has now been con-firmed in more than 100 inborn errors of meta-bolism extending over the whole range of human metabolism.

The concept of molecular disease is thus firmly established. The range of forms which this may take, from benign to serious, and the molecular changes which underlie these forms have been demonstrated most clearly, not for an enzyme, but for haemoglobin. The presence of large quantities of this protein in red blood cells facilitates its purification and studies on its structure. In 1949, Linus Pauling proposed that sickle-cell anaemia results from an alteration in the physical properties of haemoglobin as a result of mutation; it is thus an inherited molecular disease.

The specific molecular change which differen-tiates sickle-cell haemoglobin (HbS) from the normal molecule present in the adult (HbA) was shown by V. M. Ingram in 1958 to consist of the replacement of a single glutamic acid residue by a valine molecule in the β-polypeptide chain of the HbS molecule. The technique used by Ingram was two-dimensional peptide mapping or "fingerprinting" (Chapter 3).

At the present time some 200 variants of human haemoglobin have been described, which are usually designated by the geographical locality of their discovery. Since the amino acid sequence and three-dimensional configuration of the normal haemoglobin molecule are known in detail, the changes in properties which are shown by each newly discovered molecular variant can be related to the position of the altered amino acid residue within the molecule and the function of the amino acid residue which occupies that position in the normal molecule. Certain substitutions have been shown to affect the oxygen-carrying function of the molecule, i.e. its ability to combine with and to release oxygen under appropriate conditions. Some of these substitutions involve amino acids lining the pocket in the molecule into which the oxygen-carrying haem group fits. A hydrophobic environment is essential in this region of the molecule, so that replacement of a hydrophobic by a hydrophilic residue (as in Hb Hammersmith, in which a serine residue replaces phenylalanine in this critical region) has a marked effect on the function of the molecule as an oxygen carrier as well as on its stability. The haemoglobin molecule, like other protein molecules, retains its shape because of numerous hydrophobic interactions which take place in the interior of the molecule (Chapter 3). The amino acid residues between which these interactions occur in the molecule of haemoglobin have been identified. In some haemoglobin variants, e.g. HbE and Hb Shepherd's Bush, the amino acid substitution has been shown to weaken significantly the structure of the molecule, so accounting for its impaired biological effectiveness. In HbE, inter-chain bonding is affected, whereas in Hb Shepherd's Bush it is the stability of a segment of α-helix that is reduced. However, a clear division cannot always be drawn between modifications which affect stability and those which affect function.

Most of the changes which are characteristic of abnormal haemoglobins can be seen to involve only a single base change in the DNA code, i.e. they are the products of point mutations. For example, the RNA code for valine is GUA (or G), whereas for glutamic acid it is GAA (or G), so that a mutation which changes a single base in the appropriate place in DNA (and thus in RNA) is the only alteration necessary to produce HbS instead of HbA. The haemoglobin variant Hb Constant Spring has an extra 30 amino acid residues attached to its α-chain. This may be due to a point mutation affecting the "stop" codon. More extensive changes involving deletion or duplication of sections of the haemoglobin gene may be responsible for some haemoglobin variants: the Lepore haemoglobins, for example, apparently result from non-homologous pairings between two chromosomes, followed by unequal crossing over. One variant haemoglobin molecule, Hb Koellicker, appears to result from environmental factors. When released from the cells by intravascular haemolysis HbA is susceptible to attack by carboxpeptidase A in the plasma with the production of a modified molecule.

Although haemoglobin is not an enzyme, the many well-studied haemoglobinopathies provide models for the types of variation which can be expected to affect the multitude of structural genes and their product enzymes which each organism contains. Changes in amino acid residues which are involved in substrate binding or substrate transformation will produce enzymes of altered catalytic function, analogous to those haemoglobin variants which have impaired oxygen-transport ability.

Changes elsewhere in the enzyme molecule may result in altered properties of stability, or in a molecule that cannot be incorporated into its normal intracellular location, so that increased rates of enzyme degradation result in a deficiency of the active catalyst.

Another inference that the enzymologist can draw from the study of haemoglobin variants is the great range of altered molecules which can in theory and in practice arise by mutation and the correspondingly wide spectrum of properties which they display. Consequently, if alteration of a particular enzyme is the cause of an inherited disease, patients who suffer from that disease need not have identically modified enzymes. Variations in the severity of the disease may be associated with different types of mutant enzyme. For this reason descriptions of a particular disease affecting one family may not match exactly the manifestations of that disease as it affects a different family, whose members may possess another form of the mutant gene.

This phenomenon is shown very clearly by the 50 or so variants of the enzyme glucose-6-phosphate dehydrogenase which have now been reported. Each of these may be attributed to a different mutant allele at the same structural gene locus, each mutant gene producing a structurally different form of the enzyme protein with its own characteristic properties. The clinical effects of the many variants range from none that is apparent, through haemolytic attacks which are experienced only when the patient is exposed to certain drugs (e.g. the anti-malarial Primaquine) or foodstuffs (e.g. the fava bean),

to mild or severe chronic haemolytic anaemia. Among the changes which have been demonstrated in the properties of the enzyme in different patients or families with mutations affecting glucose-6-phosphate dehydrogenase are the following: increased or decreased electrophoretic mobility, with normal activity and no clinical manifestations; increased electrophoretic mobility, slightly lowered enzyme stability, reduced activity (drug-induced haemolysis); slow mobility, increased affinity for glucose-6-phosphate, double pH optimum, lowered stability (drug-induced haemolysis); slow mobility, reduced stability (congenital haemolytic anaemia); slow mobility, reduced affinity for glucose-6-phosphate and stability, abnormal pH dependence (congenital haemolytic anaemia); very slow electrophoretic mobility, increased affinity for glucose-6-phosphate, double pH optimum (no clinical abnormality).

There is thus no obvious correlation between the alteration in the properties of the enzyme and the occurrence of clinical symptoms. In view of the wide range of altered properties of mutant glucose-6-phosphate dehydrogenase, demonstrated even by this incomplete list, it is not surprising that affected patients should respond differently to various drugs or foods. (It is interesting to recall that Garrod forecast that idiosyncratic response to drugs would be found to be attributable to inherited variations in enzymic make-up.)

An example of an abnormal sensitivity to a particular drug which results from inheritance of an altered enzyme is seen in those individuals, about one in 2000 of the population, who show a prolonged apnoea after receiving the muscle relaxing drug suxamethonium (succinyl dicholine). This drug is extremely valuable in anaesthesia because of its brief action, which in turn is determined by its inactivation by cholinesterase. The normal function of the form of this enzyme found in the plasma is not known, but its substrate specificity suggests that its physiological substrate is probably a choline ester (e.g. acetylcholine). The structure of suxamethonium is that of a double choline ester:

$$CH_2-\overset{\overset{\textstyle O}{\|}}{C}-O-CH_2CH_2\overset{+}{N}(CH_3)_3$$
$$CH_2-\underset{\underset{\textstyle O}{\|}}{C}-O-CH_2CH_2\overset{+}{N}(CH_3)_3$$

Cholinesterase removes the two choline groups successively and, in patients with normal levels of the enzyme, respiratory paralysis lasts only a few minutes. However, when the plasma enzyme activity is less than half the lower limit of normal, apnoea may last for several hours, so that respiration has to be maintained artificially.

The low serum cholinesterase activities have been shown to be determined genetically and to be due to the presence of a mutant enzyme. As well as being less active *in vitro* against all substrates, the mutant enzyme is more resistant to inhibitors, e.g. dibucaine. The demonstration of a lower than normal degree of inhibition by this compound helps to differentiate a low activity of the normal enzyme from a low activity due to the mutant enzyme.

As with many other enzyme abnormalities, the possible range of cholinesterase variants is now known to extend beyond a normal and a single abnormal form. Four variants, controlled by four allelic genes at one locus, probably exist: the normal enzyme, the dibucaine-resistant enzyme, a fluoride-resistant enzyme, and an absent or inactive enzyme produced by a "silent" gene. Normal serum cholinesterase is itself heterogeneous, being separable into several zones on starch-gel electrophoresis. However, these may be due to reversible polymerization and not to different genetic forms.

Rapid inactivation of suxamethonium by the more common form of cholinesterase is an advantage in its clinical use. However, in some instances rapid metabolism of a drug might reduce its effectiveness and increase the dose that is required for successful therapy. This possibility arises with the anti-tuberculosis drug, isoniazid. This compound is inactivated by acetylation, and it has been found that the population can be divided into two groups, "fast" and "slow" acetylaters, who make up about 40 per cent and 60 per cent of the total, respectively, in European and negro populations. The difference is due to an inherited variation in acetylase, and the "fast" acetylaters have a greater requirement for the drug which, while negligible when therapy is frequent, becomes significant in attempts to develop preparations which need only be given at weekly intervals or longer.

So far, we have referred to inherited abnormalities which are produced by changes affecting a single enzymic process. However, when it is remembered that metabolic pathways involve the cooperative action of many enzymes, each

of which is potentially subject to the effects of mutation, it is apparent that a single disease which is characterized by failure to transform one metabolite into another may arise from changes which affect different enzymes in the pathway, if this is normally a multi-stage process. In other words, different molecular causes may underlie a single disease. Thus, a metabolic disease manifested as a failure to transform A to E —

$$A \xrightarrow{a} B \xrightarrow{b} C \xrightarrow{c} D \xrightarrow{d} E$$

could be due to inherited abnormalities of enzymes a, b, c or d.

Examples of the way in which specific deficiencies of different enzymes in a metabolic sequence can give rise to clinically similar diseases are found amongst congenital haemolytic anaemias. Erythrocytes rely on the metabolism of glucose for their supply of energy, producing ATP by glycolysis and by way of the hexose monophosphate shunt pathway. These cells are among the few in the body which do not contain the enzymes of the tricarboxylic acid cycle. Therefore, deficiency of almost any one of the enzymes in either pathway might be expected to have the same consequences — depletion of ATP and reduced coenzymes, leading to premature red-cell lysis. Haemolytic anaemia has been shown to result from congenital deficiencies of hexokinase, phosphohexose isomerase, glucose-6-phosphate dehydrogenase, pyruvate kinase, and other enzymes.

These enzymes and metabolic pathways are present in other cells than red blood cells, but the deleterious effects of enzyme mutation are usually only manifested in the red cells. This is probably because other cells have the ability to synthesize proteins, and can thus maintain the level of an altered enzyme, which might be broken down more rapidly than normal, whereas red cells, which do not synthesize proteins, cannot do so. In certain instances (e.g. pyruvate kinase), a different gene locus may be active in tissues other than red cells.

When a defect affecting a widely distributed enzyme or metabolic pathway is not expressed equally in all tissues, a spectrum of diseases may result. For example, the glycogen storage diseases comprise a group of conditions in which glycogen accumulates in one or more tissues, with impaired function of the affected tissue, e.g. the heart or liver, hypoglycaemia, etc. Although these diseases have a similar underlying cause, closer examination of the structure of the stored glycogen and measurement of the activities of glycogen synthesizing and degrading enzymes in the affected tissues reveals that several different enzyme deficiencies can be recognized, with rather different metabolic and clinical consequences depending on the tissue or organ principally affected (Fig. 10.1).

An even more complex group of related congenital metabolic abnormalities is made up of the lipidoses, or lipid storage diseases. All

A spectrum of diseases can also arise from enzyme deficiencies which affect a single metabolic pathway according to whether the defect in a particular case occurs before or after a branchpoint, or whether an essential metabolite has been reached or not before the stage at which the block becomes effective. The metabolism of phenylalanine has provided many examples of congenital abnormalities, beginning with Garrod's own study of alkaptonuria. This is a branched

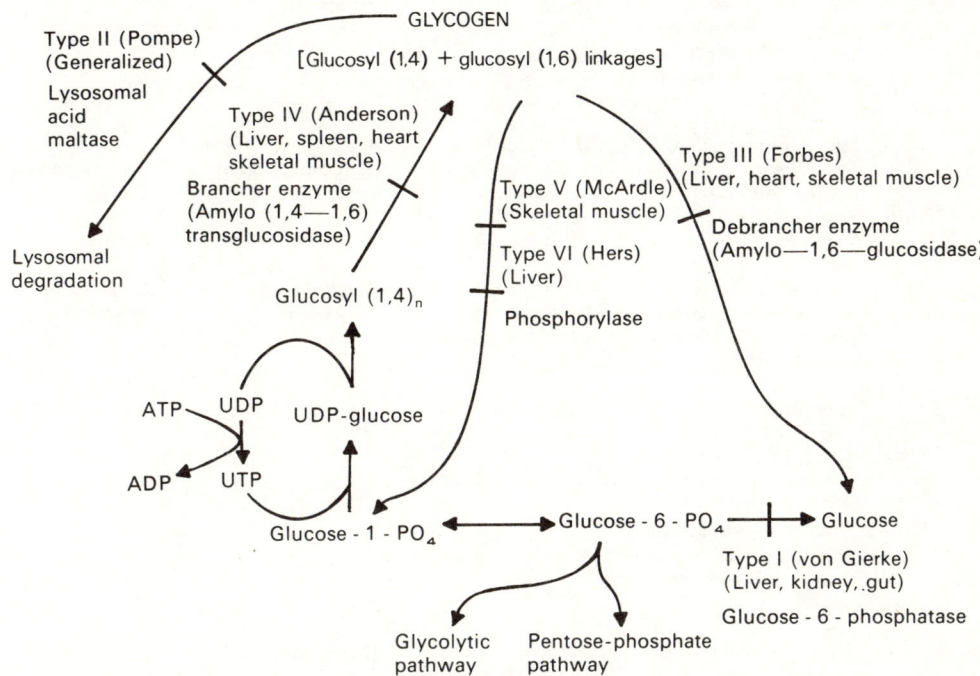

Fig 10.1. Pathways of glycogen metabolism showing positions of enzyme deficiencies in glycogen-storage diseases.

these diseases are due to reduction in the activity in the lysosomes of specific catabolic enzymes required for different stages in the breakdown of complex lipids and glycolipids. The lipids consequently accumulate in the affected tissues, disrupting their structure and impairing their function. The clinical effects vary with the tissues in which the lipids principally accumulate: since complex lipids are components of nervous tissue, the central nervous system is the site of lipid deposition in many of these conditions, with rather more severe consequences than when peripheral tissues are principally involved. The deficient enzymes have now been identified in several lipidoses (Table 10.1).

pathway (Fig. 10.2), and metabolic defects occurring near the ends of the branches are not associated with serious clinical symptoms. However, this generalization does not apply universally. A defect affecting the first enzyme in the metabolism of galactose, galactokinase, is less serious than one affecting a later enzyme in the sequence, galactose-1-phosphate uridylyl transferase.

In albinism, the enzyme tyrosinase, a copper-containing aerobic oxidase which converts tyrosine to dihydroxyphenylalanne (DOPA) and to DOPA-quinone is absent from melanocytes. These compounds are stages on the pathway of melanin formation so that affected individuals show the characteristic lack of pigmentation of

TABLE 10.1 SOME LIPID STORAGE DISEASES

Disease	Clinical features	Sites of lipid accumulation	Major lipid type accumulated	Enzyme defect
Gaucher's	Hepatosplenomegaly, Hypersplenism, bone lesions. May be chronic or acute and rapidly progressive.	Spleen, bone marrow, lymph nodes, liver, brain	Glucocerebroside (glucosyl ceramide)	Glucosyl ceramide β-glucosidase
Niemann-Pick	Hepatosplenomegaly. Several forms with varying degrees of CNS involvement and rate of progression. Typically fatal within 2 years.	Spleen, liver, lymph nodes, bone marrow	Sphingomyelin (phosphoryl choline ceramide)	Sphingomyelinase
Krabbe's	Rapidly progressive, fatal disease of infants. Mental and motor deterioration; usually blindness.	Globoid cells rich in galactocerebroside infiltrate white matter.	Galactocerebroside (galactosyl ceramide)	Galactocerebroside β-galactosidase
Metachromatic leucodystrophy	Progressive paralysis and dementia; usually fatal within 2-10 years.	White matter of CNS and peripheral nerves; also in kidney, liver, etc.	Ceramide galactose 3-sulphate	Cerebroside sulphatase
Fabry's	Paraesthesia of extremities, ectasia of vessels of skin, oedema, albuminuria; death usually from renal failure or cardiac or cerebrovascular disease in 5th or 6th decade.	Most tissues and plasma	Ceramide trihexoside	α-Galactosidase
Tay-Sach's	Progressive retardation of development, paralysis, dementia and blindness; fatal usually by 3-4 years.	Cerebral tissues; retina and ganglion cells of CNS	Ganglioside GM_2	Hexosaminidase A

the skin, eyes and hair (or only of eye pigmentation in a variant form). A different enzyme, tyrosine hydroxylase, catalyses the conversion of tyrosine to DOPA in the course of catecholamine formation in the adrenals and nervous tissue and this pathway is normal in albinos. Impairment of visual acuity and sensitivity to sunshine are the only disadvantages experienced by the affected individual.

Alkaptonuria is without clinical consequences except for deposition of dark pigments in various tissues beginning in the second or third decade of life, with arthritis developing in middle age. This is presumably due to an increased diversion of the metabolism of tyrosine along the melanogenic pathway compared with the normal situation. The deficient enzyme in alkaptonuria is the benzene ring opening enzyme, homogentisate oxidase. A case designated tyrosinosis, without definite signs and symptoms, except for the excretion of abnormal amounts of p-hydroxyphenylpyruvic acid has been described. Excretion

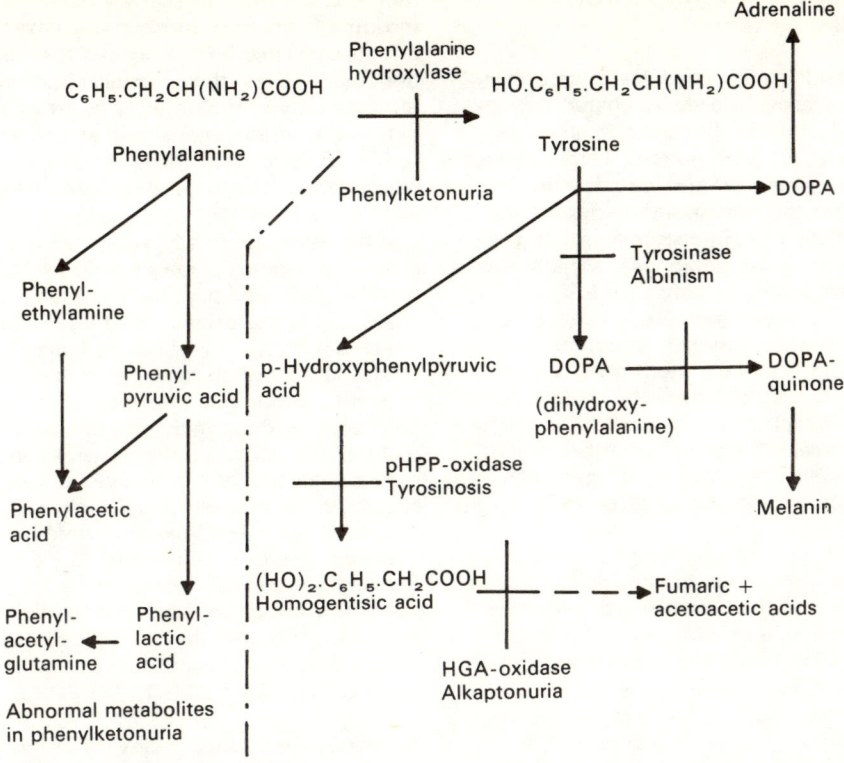

Fig 10.2. Pathways of phenylalanine metabolism showing positions of various enzyme deficiencies.

of p-hydroxyphenyl lactic acid was not observed in tyrosinosis, in which the deficient enzyme was presumed to be a p-hydroxyphenylpyruvate oxidase. However, normal children on a Vitamin C-deficient diet excrete both p-hydroxphenyl-pyruvic acid and p-hydroxyphenyl lactic acid, while normal subjects may also excrete homo-gentisic acid when deficient in this vitamin. Vitamin C thus appears to be a cofactor in the activity of both p-hydroxyphenylpyruvate and homogentisate oxidases.

The most serious defect of phenylalanine metabolism is phenylketonuria, which occurs right at the beginning of the pathway and is due to the absence of the liver enzyme phenylalanine hydroxylase. This is a multi-component enzyme system belonging to the class of mixed function oxidases, but only the specific enzyme protein normally present in the liver is deficient in phenyl-ketonuria. The most important consequence of

phenylketonuria is greatly impaired mental development. The incidence of the disease has been variously estimated in different populations; an average figure for European population is about 1 per 12 000, while 1 per 30 000 has been estimated for the population of New York.

Failure to convert phenylalanine to tyrosine results in accumulation of phenylalanine in the blood. Some is excreted unchanged; some gives rise to abnormal excretory products such as the phenylpyruvic acid from which the condition derives its name. Although the exact means by which the mental defect is developed is not known, it seems to be related to the accumulation of phenylalanine. The accumulated phenylalanine also competitively inhibits enzymes catalysing later stages in the metabolism of tyrosine, which phenylketonurics receive in the diet. Thus, these patients tend to be fair haired and blue-eyed because of interference with melanin formation.

INHERITANCE OF INBORN ERRORS OF METABOLISM

In classical Mendelian genetics, traits are either dominant or recessive. Recessive characteristics are only manifest when the recessive gene has been contributed by both parents. Heterozygotes are indistinguishable from individuals who are homozygous for the corresponding dominant gene. Many of the mutant enzymes which give rise to inborn errors of metabolism are inherited as recessive characters, and are thus fully expressed only in subjects who are homozygous for the mutant gene. However, dominance of the normal gene over the mutant gene is often not complete so that, although the heterozygous individual is clinically normal, careful biochemical investigation reveals the presence of the mutant enzyme, if for example this has different electrophoretic, stability, or kinetic properties from the normal enzyme.

Usually, the amount of enzyme produced by the normal gene is enough to provide for the metabolic needs of the heterozygote. However, when an abnormal load is imposed, the fact that a proportion of the total enzyme present consists of less efficient molecules contributed by the mutant gene, resulting in a lowering of the effective total activity, may become apparent. A transient metabolic stress of this nature can be produced for diagnostic purposes by a tolerance test; administration of the substance, the metabolism of which is potentially impaired may result in higher blood-levels in the heterozygote than in an individual who is homozygous for the normal gene. Although the heterozygotes are generally symptom-free, detection of the heterozygous state is clinically of great importance to enable advice to be given to prospective parents who may be carriers of a potentially lethal or crippling gene.

A number of the enzyme mutations which give rise to disease are sex-linked; that is, they are determined by genes located on the X-chromosome. Since the male has only one X-chromosome compared with the female's two, the chances of an X-linked recessive trait manifesting itself as clinical symptoms are much greater in male offspring than in females.

Half of the male children of a heterozygous mother will receive an X-chromosome which carries the affected gene and, since for this chromosome no normal copy is received from the father, the disease will be expressed fully in these children. On the other hand, those female children who receive a defective X-chromosome from their mother will have its effects cancelled by the normal X-chromosome donated by the father. Female children will be affected only in the unlikely event of a mating between a heterozygous mother and an affected father, in the case of those females who receive an abnormal X-chromosome from the mother as well as the father.

This situation, in which females act as carriers of a disease which is only manifested in males, is seen in haemophilia. A further example is provided by the mutations affecting glucose-6-phosphate dehydrogenase. The incidence of affected offspring, half of all male children of a heterozygous mother and a normal father, may be contrasted with an incidence of one quarter of children of both sexes where *both* parents are heterozygous for the defective gene which is characteristic of diseases determined by an autosomal (non-sex linked) mutant gene. The identification of sex-linked recessive conditions with serious clinical consequences is correspondingly important so that intending parents can be informed of the high risk to male children.

DETECTION OF HETEROZYGOTES

As already mentioned, many inborn errors of metabolism are incompletely recessive in a Mendelian sense. Consequently, individuals who are heterozygous for one of these conditions can be distinguished from persons who are homozygous for the normal gene. Often, the mutant allele produces no enzyme activity at all: perhaps because it has become changed to a "nonsense" code, or perhaps becuase the protein which it produces is altered in a structurally minor, but functionally decisive manner, so that an inactive protein is produced. The presence of an inactive protein can sometimes be recognized by immunochemical means, when the mutant enzyme, although sufficiently altered to be inactive, retains the antigenic determinants which enable it to react with an antiserum raised against the normal enzyme. The normal allele present on the other chromosome of the pair produces its normal enzyme, and this is usually enough to protect the heterozygous individual from developing the disease. However, the total enzyme activity is usually less than that present in persons with two normal alleles, so that the presence of an abnormal allele, i.e. the heterozygous condition, can at least in theory, be detected by the demonstration of a lower than normal enzyme activity in a suitable sample of the blood or tissues of the suspected carrier.

In practice, difficulties may arise. When the enzyme deficiency (or rather, partial deficiency) is confined to only one tissue, it may not be practicable to obtain enough tissue for a reliable determination of enzyme activity to be made. This is the situation in phenylketonuria, in which the deficiency is of a liver enzyme. Although needle-biopsies of the liver can be taken safely, sufficiently precise methods by which a low activity of phenylalanine hydroxylase can be demonstrated reliably on the small amounts of tissue thus obtained have not been developed. In this condition, therefore, changes in the excretion or metabolism of phenylalanine must be sought in suspected heterozygotes. On average, heterozygotes have a higher fasting level of phenylalanine in the blood than normals, but the two populations overlap considerably in this respect so that a reliable distinction cannot be made in an individual case. The separation between normal homozygotes and heterozygotes becomes much clearer in a phenylalanine tolerance test, in which blood phenylalanine levels are measured at intervals after a loading dose of phenylalanine (preferably intravenous). A phenylalanine tolerance test is claimed to distinguish between the two groups in 90 per cent of cases.

When the affected enzyme is not confined to one tissue only the problems of obtaining a suitable sample for enzyme assay are much reduced. Such is the case in the important metabolic disease, galactosaemia. This autosomal, recessively inherited condition occurs in about one in 70 000 births in the United Kingdom and affected infants fail to thrive and have a high blood galactose level, galactosuria and aminoaciduria. Fatty infiltration of the liver and cirrhosis may develop and if the patient survives there is mental retardation and cataract formation. The metabolic failure lies in the conversion of galactose into glucose because of a deficiency of the enzyme galactose-l-phosphate uridylyl transferase (Fig. 10.3). Consequently, the galactose received by the infant as part of the lactose of milk accumulates together with its first metabolite, galactose-1-phosphate. The latter compound is a competitive inhibitor of enzymes of glucose metabolism, particularly of phosphoglucomutase which converts glucose-1-phosphate to glucose-6-phosphate. Galactose-1-phosphate thus interferes with energy-production in all glucose-metabolizing tissues and this probably accounts for the failure to thrive and the mental retardation of galactosaemics. Galactose is converted to the sugar-alcohol, galactitol, by the enzyme aldose reductase and accumulation of this compound may have an influence in the development of cataract. A further, minor, pathway of galactose utilization, by the enzyme UDP-galactose pyrophosphorylase, probably provides the UDP-galactose needed for the synthesis of cerebrosides (e.g. in myelin formation). The activity of this enzyme increases with age in rat liver, but not in human liver.

The presence of alleles other than the normal one and that which determines the suspected pathological condition can confuse the detection

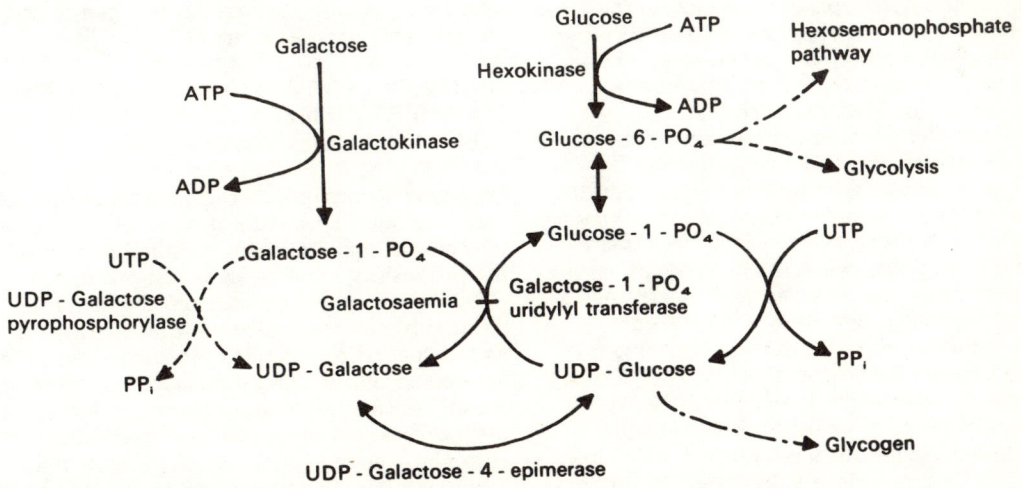

Fig 10.3. Galactose metabolism showing location of enzyme deficiency in galactosaemia.

of heterozygotes, if this is based solely on measurements of total enzyme activity. For example, a variant form of galactose-1-phosphate uridylyl transferase has been observed, called the Duarte variant, which has similar kinetic properties to the normal enzyme but which migrates faster on starch-gel electrophoresis. Individuals with this variant enzyme have a lower than normal total enzyme activity, however, and this can cause difficulty in the identification of heterozygotes for the important galactosaemia gene.

Galactose-1-phosphate uridylyl transferase occurs in the red blood cells as well as in the liver, and these cells form a readily-sampled population of cells in which the greatly reduced enzyme activity of the abnormal homozygotes can be demonstrated, thus confirming the diagnosis. Excessive levels of galactose in the urine often provide the first clue in clinical practice, showing up as a strongly positive test for reducing substances. However, it is important to remember that specific tests for glucose which make use of glucose oxidase have largely replaced the older, non-specific copper-reduction tests such as Fehling's or Benedict's in many hospitals and clinics, and that glucose oxidase does not react with galactose.

Galactose tolerance tests have been used to detect heterozygous carriers of the disease and, although these tests show a lower than average ability to metabolize the sugar in heterozygotes, overlap with normal homozygotes is so great that individual discrimination becomes difficult or impossible. A much clearer distinction can be based on galactose-1-phosphate uridylyl transferase activity in red blood cells. When accurate enzyme-assay methods are used discrimination between heterozygotes and normal homozygotes is about 75 per cent certain.

In both phenylketonuria and galactosaemia the mutant allele produces virtually no enzyme at all, or a product with little activity. When the mutant gene produces a protein which is enzymically active but which differs in some respect from the normal enzyme, the problem of identification of heterozygotes is less difficult since a qualitative change as well as, or instead of, a quantitative one is rather easier to demonstrate than solely a difference in total activity. Unfortunately, conditions in which qualitative changes in the enzyme have occurred are in general the less serious ones. The electrophoretically fast Duarte variant of galactose-1-phosphate uridylyl transferase has already been mentioned.

Individuals who are heterozygous for the normal and variant alleles show three enzyme zones on electrophoresis of haemolysates of their red cells, the normal zone together with two abnormal faster bands. In spite of a reduced total enzyme activity there is no clinical disorder associated with the Duarte gene. The altered resistance to inhibitors shown by mutant forms of serum cholinesterase can also be used to demonstrate the presence of both normal and altered enzymes in the heterozygous individual, but again these enzyme variants are of minor clinical importance.

It has been assumed in this discussion that heterozygotes possess one normal and one modified allele, and that abnormal homozygotes (i.e. those in whom a metabolic abnormality is expressed) possess two abnormal alleles of the same type. While this is probably the most usual genetic make-up in metabolic diseases, it is also possible that two different mutant alleles might combine in a single individual, the sum of their effects giving rise to a metabolic disease. Thus, instead of being homozygous for a single abnormal allele at a given locus, the patient is heterozygous for two different mutant alleles. This possibility is easily understood when the great range of potential gene and enzyme variations is recalled. Heterozygotes with two different abnormal alleles have already been identified in anaemias due to phosphohexose isomerase or pyruvate kinase deficiencies and probably exist also among phenylketonurics. This phenemenon almost certainly contributes to the wide clinical spectrum seen in patients who suffer from what is apparently the same inherited metabolic disease, both in symptoms and in response to treatment.

DIAGNOSIS, PREVENTION AND TREATMENT OF INHERITED METABOLIC DISEASES DUE TO ENZYME DEFICIENCY

Diagnosis of inborn errors of metabolism depends typically on the presence of an excessive amount of a normal metabolite in the blood or urine of the propositus, or of an abnormal metabolite. A definite diagnosis comes from the demonstration of a specific enzyme defect, when suitable tissue samples can be obtained. In searching for excessive or abnormal metabolites chromatographic techniques — paper chromatography, or the more modern thin-layer and gas-liquid chromatography — are especially useful. The indications for suspecting a metabolic disease may include failure to feed or thrive in an infant, slow development; or clinical signs such as enlarged liver.

For those diseases which can be treated, early diagnosis is essential, and screening programmes by which all newborn infants are examined are justified in these instances. This is true for phenylketonuria. Earlier colorimetric tests for the detection of phenylpyruvic acid in the urine have now largely been abandoned because of the large proportion of false negatives recorded. A test now widely used is that devised by R. Guthrie: a spot of blood from a heel-prick is collected on filter-paper, which is placed on an agar plate inoculated with *B. subtilis,* the growth of which is inhibited by the presence of the amino acid analogue, β-thienylalanine. The organism requires phenylalanine for growth and if the child's blood contains an excess of this amino acid the inhibition of growth is overcome and a correspondingly large ring of the micro-organism appears around the blood sample.

Since many metabolic defects are at present untreatable and are totally disabling or incompatible with all but brief life, pre-natal diagnosis, if made sufficiently early, offers the parents of the affected foetus the choice of an abortion. This possibility has now become a reality in many metabolic defects by the discovery that samples of cells removed from the amniotic cavity during pregnancy and cultured *in vitro* can be used to ascertain the presence or absence of key enzymes. The cells present in amniotic fluid are mainly of foetal origin, deriving mainly from foetal skin and amnion. Since the amniotic fluid is both ingested by the foetus and also receives its urine, abnormal metabolites can in some cases be detected by analysis of the amniotic fluid itself, in a manner analogous to the detection of abnormal metabolites in the urine of the new-born infant. Pre-natal diagnosis of the adreno-genital syndrome, an automosal recessive defect in the synthesis of adrenocorticosteroids, has been achieved by the demonstration of increased amounts of 17-oxosteroids and pregnanetriol in amniotic fluid. However, the ability of the placental circulation to remove accumulating metabolites complicates the interpretation of amniotic fluid composition (e.g. with respect to amino acids) so that pre-natal diagnosis will probably concentrate in most cases on the amniotic cells.

For the diagnosis of many diseases the number of cells obtained by a single amniocentesis (amniotic fluid sampling operation) is too few for reliable biochemical or histochemical enzyme studies. However, human fibroblasts such as those from foetal skin can be cultured *in vitro* for up to 50 to 70 generations, retaining their characteristics and thus increasing the amount of material available for analysis. The time required for cell growth does impose some restrictions, however: the amniotic fluid can safely be sampled at about 16 week's gestation, when its volume has reached about 250 ml, leaving 3-4 weeks for cell culture and analysis, if 20 weeks is regarded as the latest date for abortion.

Pre-natal diagnosis is now possible in about forty genetic abnormalities, including galacto-saemia and several lipidoses and glycogen-storage diseases of those mentioned in this chapter. Although the demonstration of specific enzyme defects has been emphasized in this discussion, it is important to remember that the availability of foetal cells allows the morphology of the chromosomes to be examined, thus disclosing gross abnormalities such as Down's syndrome (trisomy of chromosome 21, resulting in mongolism) and many others which together probably occur once in every 200 live births.

Amniocentesis is a safe procedure, but one which carries a small but finite risk. Therefore, it is unlikely at present to develop into a general screening procedure, but rather to be applied in selected cases where a previous abnormal child or an unexplained death in infancy suggests an increased risk of inherited disease, or when the mother is in an age-group with a greater incidence of chromosome abnormalities. The specialized biochemical and especially cytological techniques needed to examine amniotic-cell samples, particularly the maintenance of cell-lines of known enzymic make-up, are likely to favour the establishment of regional or national reference centres for this type of diagnosis.

An even earlier stage of prevention of genetic disease is represented by genetic counselling, i.e. informing prospective parents of the magnitude of the risks of bearing a diseased child. This again is most likely to be indicated when an earlier, unexplained neonatal death or affected child has drawn attention to the possibility of genetic disease, and when investigations on the parents have established their heterozygous status.

When an inherited metabolic abnormality has been diagnosed few therapeutic measures are available at present. However, one or two potentially serious conditions can be treated successfully by careful dietary regulation which is designed to avoid as far as possible ingestion of the toxic metabolite or its precursor. Perhaps the most dramatic advance in this respect which has been recorded in recent years is the dietary control of phenylketonuria. Provided that the diagnosis is made early enough and a low-phenyl-alanine diet instituted (preferably within one week) affected children develop normally and show normal intelligence. Phenylalanine is an essential amino acid needed for protein biosynthesis so that an entirely phenylalanine-free diet cannot be used. Careful monitoring is required to ensure that adequate, but not excess, phenylalanine is being taken and that the rather unpleasant diet is being adhered to.

In contrast with phenylalanine, galactose can be removed completely from the diet without ill effects so that dietary treatment of galactosaemia is somewhat more simple and is also effective in preventing the development of symptoms.

Supplementation of the diet with vitamins is beneficial in some inborn errors of metabolism. In some of these conditions (e.g. cystathioninuria, in which there is a deficiency of the enzyme cystathionase with high urinary excretion of the amino acid cystathionine, with sometimes mental retardation) the mutant enzyme appears to have reduced affinity for its coenzyme or prosthetic group. Since vitamins are essential components of many cofactors, increasing the vitamin intake increases the availability of cofactor and so favours the formation of active holoenzyme from inactive apoenzyme.

Although augmentation of vitamin intake (Vitamin B_6 in the case of cystathioninuria) produces a reduction in amino acid excretion in cystathioninuria, reports of increased activity of the enzyme in patients' livers before and after vitamin treatment have not been altogether consistent.

Attempts to remove a stored product have had some success in Wilson's disease (hepato-lenticular degeneration). In this condition an unknown enzyme defect leads to the accumulation of large amounts of copper in the liver and other tissues with disruption of their structure and function and eventual hepatic cirrhosis and nervous-system dysfunction. Administration of chelating agents such as penicillamine promotes urinary excretion of copper and reduction in stored copper, and alleviates the symptoms of the disease markedly.

Treatment of genetic diseases by direct replacement of missing enzymes is not possible in the present state of knowledge, although a number of approaches have been tried. Stimulation of enzyme production (enzyme induction)

can be effective in those situations when the cells are potentially able to make active enzyme but have not done so because of immaturity or lack of stimulus. For example, jaundice is sometimes apparent in the newborn infant because the ability to conjugate bilirubin to form its water-soluble glucuronide derivatives has not been acquired, i.e. the activity of the specific enzyme needed for this step, bilirubin UDP-glucuronyl transferase, remains low. This enzyme is present in the smooth endoplasmic reticulum of liver cells (Chapter 8), together with enzymes of drug inactivation. Administration of drugs such as barbiturates stimulates proliferation of smooth endoplasmic reticulum and its associated enzymes. Therefore, barbiturates have been given to mothers just before giving birth in order to promote the production of detoxicating enzymes in the foetal liver and so help the baby through a period of risk, e.g. when rhesus incompatibility may increase the load of unconjugated bilirubin. However, this approach is unlikely to be successful in inborn errors of metabolism in which the power to make specific enzymes is irreversibly lost, but if conditions are identified in which an enzyme defect is due not to changes in the structural gene but to a failure to respond to the normal inducer of enzyme production, synthetic inducers may be found which will switch on enzyme production.

Barriers to attempts to raise the level of missing enzymes by direct infusion of enzyme preparations exist in the difficulty of ensuring that the enzymes find their way into tissue cells, since metabolism is almost entirely intracellular, and in maintaining effective levels of activity because of the rapid turnover and short half-lives of enzymes in the body. In addition, the problem of immunological reaction is present if enzymes from non-human sources are used. Chances of success in direct enzyme replacement are probably greatest in those genetic diseases in which the deficient enzymes are normally amongst the degradative enzymes of the lysosomes. Lack of these enzymes (e.g. in some lipidoses and glycogen storage diseases) leads to accumulation of un-degraded large molecules. Lysosomes act phago-cytically, taking in material from outside the cell, so that these organelles might be able to incorporate active, exogenous enzymes if these are presented in a suitable form. This has in fact been attempted by incorporating specific gluco-sidases into microscopic lipid droplets ("lipo-somes") which have been infused into isolated, perfused animal livers. First results indicate that a small, though temporary, increase in tissue enzyme levels is obtained. Further development of the technique may enable conditions to be found under which enzyme activity persists long enough for a significant reduction in stored material to be obtained.

Organ transplantation offers a means of introducing into the body of an enzyme-deficient patient, healthy tissue which is capable of producing the missing enzyme, and thus of protecting itself from invasion by accumulating metabolic products. However, organ transplantation is itself a procedure which carries such high operative and post-operative risks, because of the ever-present danger of immunological rejection, that it can only be contemplated at the moment to alleviate terminal failure of a vital organ, e.g. the liver in Wilson's disease or the kidneys in Fabry's disease. Furthermore it is not absolutely certain in all cases that the amount of enzyme produced by the transplanted organ will be adequate to protect it from the ravages of the disease. Where organ transplantation has been attempted in patients with Fabry's disease and Wilson's disease, however, there have been indications that the transplanted tissue is capable of maintaining its functions.

In the distant, but not unforeseeable future, correction of inborn errors of metabolism may depend on the much-discussed "genetic engineering", that is, on the introduction into the cellular DNA of a normal structural gene which can direct the synthesis of normal enzyme, or on the reversal of the base change that has given rise to the mutant enzyme in the first place. Any such procedure must depend basically on a knowledge of the structure of the gene (its sequence of bases) and the position of the affected structural gene on the chromosome. Progress is being made in both these respects. However, selective editing of the DNA code is likely to prove more difficult.

The ability of chemical agents or ionizing radiation to produce alterations in DNA structure is well established, but such changes are of a random and unselective nature. A great improvement in the selectivity and specificity of these procedures is needed before they can be used to correct particular genetic defects. Another possibility might be the introduction into the diseased cell of normal sections of DNA, in a manner similar to that by which bacteriophage introduces its nucleic acid into and transforms bacterial cells. In mammalian cells, however, the new DNA must be introduced not only into the cell but into the nucleus if a permanent transformation is to be achieved. Cell fusion may have an application in genetic modification.

SUMMARY

A great diversity of protein molecules may be associated with a particular type of enzyme activity. Some of these modified enzymes arise from non-genetic phenomena but most are the result of modifications in the structural genes. Qualitative and quantitative differences in enzyme make-up underlie individual differences in such matters as responses to drugs.

When inherited differences in enzyme molecules are such as to impair seriously their functions as catalysts, errors of metabolism result. These in turn may range from the trivial to the fatal, depending on the role of the normal enzyme in metabolism or on the toxicity or site of accumulation of abnormal metabolic products. Where successful treatment of inherited disease is possible it usually depends on early and accurate diagnosis. The prevention of untreatable diseases depends on genetic counselling or intra-uterine diagnosis which, to be fully useful, is dependent on the recognition of the heterozygous state.

Index